MOLECULAR STRUCTURE

MOLECULAR STRUCTURE

Understanding Steric and Electronic Effects from Molecular Mechanics

Norman L. Allinger

A JOHN WILEY & SONS, INC. PUBLICATION

Copyright © 2010 by John Wiley & Sons, Inc. All rights reserved

Published by John Wiley & Sons, Inc., Hoboken, New Jersey
Published simultaneously in Canada

No part of this publication may be reproduced, stored in a retrieval system, or transmitted in any form or by any means, electronic, mechanical, photocopying, recording, scanning, or otherwise, except as permitted under Section 107 or 108 of the 1976 United States Copyright Act, without either the prior written permission of the Publisher, or authorization through payment of the appropriate per-copy fee to the Copyright Clearance Center, Inc., 222 Rosewood Drive, Danvers, MA 01923, (978) 750-8400, fax (978) 750-4470, or on the web at www.copyright.com. Requests to the Publisher for permission should be addressed to the Permissions Department, John Wiley & Sons, Inc., 111 River Street, Hoboken, NJ 07030, (201) 748-6011, fax (201) 748-6008, or online at http://www.wiley.com/go/permission.

Limit of Liability/Disclaimer of Warranty: While the publisher and author have used their best efforts in preparing this book, they make no representations or warranties with respect to the accuracy or completeness of the contents of this book and specifically disclaim any implied warranties of merchantability or fitness for a particular purpose. No warranty may be created or extended by sales representatives or written sales materials. The advice and strategies contained herein may not be suitable for your situation. You should consult with a professional where appropriate. Neither the publisher nor author shall be liable for any loss of profit or any other commercial damages, including but not limited to special, incidental, consequential, or other damages.

For general information on our other products and services or for technical support, please contact our Customer Care Department within the United States at (800) 762-2974, outside the United States at (317) 572-3993 or fax (317) 572-4002.

Wiley also publishes its books in a variety of electronic formats. Some content that appears in print may not be available in electronic formats. For more information about Wiley products, visit our web site at www.wiley.com.

Library of Congress Cataloging-in-Publication Data:

Allinger, Norman L.
  Molecular structure : understanding steric and electronic effects from molecular mechanics / Norman L. Allinger.
       p. cm.
    Includes index.
    ISBN 978-0-470-19557-4 (cloth)
  1. Molecular structure. 2. Physical organic chemistry. I. Title.
    QD476.A648 2010
    547′.13–dc22

                                                                                          2009049263

Printed in Singapore

10 9 8 7 6 5 4 3 2 1

CONTENTS

Foreword	xi
Preface	xv
Acknowledgments	xix

1 INTRODUCTION — 1
What, Exactly, is a Molecular Structure? — 3
References — 6

2 EXPERIMENTAL MOLECULAR STRUCTURES — 7
Electron Diffraction — 7
Microwave Spectroscopy — 10
X-Ray Crystallography — 11
    The Phase Problem — 12
    Rigid-Body Motion — 14
    Molecular Mechanics in Crystallography — 16
Neutron Diffraction — 17
Nuclear Magnetic Resonance Spectra — 19
Bond Lengths Depend on Method Used to Determine Them — 20
References — 25

3 MOLECULAR STRUCTURES BY COMPUTATIONAL METHODS — 28
A Brief History of Computers — 28
Computational Methods — 33
Semiempirical Quantum Mechanical Methods — 34
    Self-Consistent Field Method — 35
Ab Initio Methods — 37
Density Functional Theory — 41
Molecular Mechanics — 42
References — 47

4 MOLECULAR MECHANICS OF ALKANES — 51

Potential Energy Surface — 51
Force Constant Matrix — 57
    Diagonal Part — 57
    Vibrational Spectra — 58
    Off-Diagonal Part — 63
Stretch–Bend Effect — 65
Urey–Bradley Force Field — 66
van der Waals Forces — 66
    Dr. Miller's Nuclear Explosion — 73
    van der Waals Interactions between Nonidentical Atoms — 74
Congested Molecules — 75
    Tetracyclododecane — 76
    Vibrational Motions of Compressed Hydrogens — 80
    Other Very Short H···H Distances — 81
Alkanes Summary — 84
Extension of The Alkane Force Field — 85
Alkenes — 88
Functional Groups in Molecular Mechanics — 89
References — 90

5 CONJUGATED SYSTEMS — 92

Introduction — 92
Structures of Conjugated Hydrocarbons — 98
    in-$[3^{4,10}][7]$Metacyclophane — 102
Aromatic Compounds — 104
    Simple Benzenoid Compounds — 105
    Corannulene — 108
    C_{60}-Fullerene — 109
Aromaticity — 112
    Cyclooctatetraene — 115
    [10]–[16]Annulenes — 116
    [18]Annulene — 116
    Triquinacene (Homoaromaticity) — 118
Electronic Spectra — 122
Structures of Conjugated Heterocycles — 124
    Porphyrins — 127
References — 129

6 "EFFECTS" IN ORGANIC CHEMISTRY — 134

- Electronegativity Effect — 137
  - Electronegativity Effect on Bond Lengths — 138
  - Electronegativity Effect on Bond Angles — 143
- C–H Bond Length versus Vibrational Frequency — 144
- Hyperconjugation — 147
  - Baker–Nathan Effect — 151
- References — 160

7 MORE "EFFECTS"—NEGATIVE HYPERCONJUGATION — 162

- Bohlmann Effect — 163
- Anomeric Effect — 167
- Dimethoxymethane — 169
  - Energetic Effects — 169
  - Structural Effects — 170
  - Angle Effects — 173
- 2-Methoxytetrahydropyran — 173
- α-Halo Ketone Effect — 177
  - Molecular Mechanics Model — 182
  - Energetics — 185
  - Summary — 186
- References — 186

8 ADDITIONAL STEREOCHEMICAL EFFECTS IN CARBOHYDRATES — 189

- Glucose — 189
- Gauche Effect — 193
  - Polyoxyethylene (POE) — 195
- Delta-Two Effect — 197
- Glucose Diastereomers — 200
- Cellobiose Analog — 201
- External Anomeric Torsional Effect — 204
- References — 210

9 LEWIS BONDS — 212

- Hydrogen Bonds — 212
- Quantum Mechanical Description of a Hydrogen Bond — 215
- Hydrogen Bonding Models in Molecular Mechanics — 218

Hydrogen Fluoride Dimer	221
Water Dimer	222
Methanol Dimer	223
Ethylene Glycol	223
Other Lewis Bonds	227
Amine–Carbonyl Interactions	228
References	237

10 CRYSTAL STRUCTURE CALCULATIONS — 239

Crystalline Phase	240
Anticipation of Unit Cell	241
A Priori Calculations of Crystal Structures	241
Molecular Mechanics Applications to Crystals	242
Comparison of X-Ray Crystal Structure with Calculated Structures	242
Benzene Crystal	243
Biphenyl	246
Ditrityl Ether	247
More of the [18]Annulene Story	249
References	254

11 HEATS OF FORMATION — 257

Benson's Method	258
Statistical Mechanics	261
Heats of Formation of Alkanes from Molecular Mechanics	265
Tim Clark Story	269
Thermodynamic Properties of Alkanes	273
Heats of Formation from Quantum Mechanics: Alkanes	274
Strain Energy	282
Ring Strain Energy	285
Dodecahedrane	287
Heats of Formation of Unsaturated Hydrocarbons	291
[18]Annulene, Aromaticity	294
Fullerene	295
Heats of Formation of Functionalized Molecules	296
References	297

Concluding Remarks — 301

Appendix   **302**

Introduction	302
Jargon	302
Basis Set Superposition Error	303
Carbohydrate Conformational Nomenclature	304
Conformational Search Routine	305
Driver Routine	308
Molecular Mechanics Programs	311
Nuclear Explosion Preventer	313
Quantum Chemistry Progam Exchange	313
Ring Counting	315
Stereographic Projections	317
References	319

Index   **321**

FOREWORD

Some time ago, knowing nothing about molecular mechanics (MM), I visited the University of Georgia for a year with the intent of doing research on something else. I soon heard that there was something very interesting going on in the chemistry department under the direction of N. L. Allinger. It seemed that members of this research group were applying the principles of classical mechanics to, not only vibrations and rotations of molecules, but to the intimacies of the molecule itself, particularly its chemical behavior as a reflection of its structure. By optimizing the coordinates of an initial structure (essentially a Dreiding model) in a specified force field, they were able to arrive at a molecular structure that accurately reflects experimental structures. Moreover, they found that the force field is transferable to other molecules with a reasonable similarity to the base set, offering endless computational possibilities. Atomic positions within the optimized structure are near but not exactly at their individual potential energy minima (stretching, van der Waals etc.), being slightly displaced to compromise positions that represent the energy minimum for the entire molecule. This engenders some strain enthalpy relative to a standard, resulting in the enthalpy of formation of the molecule, a property I was very interested in. I soon met Lou Allinger and came to respect him for having the qualities of a first rate scientist along with those qualities summed up in the regrettably unfashionable word "gentleman".

Like many others, I have waited with anticipation for the revised edition of Allinger's 1982 ACS Monograph on *Molecular Mechanics*. Well, here it is.

But not exactly. The present work is more than a rewrite of the former work. Rather, it is a survey of the impact of computational chemistry on organic chemistry, a development in which Allinger has participated since his student days with D. J. Cram (Nobel Prize, 1987).

After a general and sometimes amusing history of molecular mechanics along with comments on computation and computers in the short initial chapters, we have a grasp of the MM problem to be solved and the strategy to be pursued in solving it. In Chapter 4 we come to a beginning tactical approach to MM calculations of the properties of the alkanes with emphasis on the force constant matrix, diagonal or with off diagonal cross terms. The subtle and difficult problems of van der Waals force constants, electrostatic interactions, and crystal packing forces are treated. I was astonished to find how little crystal packing forces invalidate MM calculations. Bond lengths are hardly changed at all, angles a little more, though torsional angles are substantially changed. The end result is that the geometry comes through quite well in most cases.

Chapter 5 treats extension of the method to alkenes, alkynes and aromatic compounds by generalizing the force constant matrix. The relationship between bond orders and molecular geometry in conjugated systems is discussed along with VESCF inclusion in MM4. The valence bond description of molecular structure is discussed with special regard to MM4 results for phenanthrene, corranulene, and the C 60 fullerenes along with a general review of aromaticity and electronic spectra.

The strength of the book is clearly delineated in chapters 6 and 7 which treat inclusion into MM4 of various "effects" known to organic chemists for a long time, the electronegativity effect, the anomeric effect, the hyperconjugative effect, and the Bohlmann or negative hyperconjugative effect. Chapter 8 is a continued examination of ever smaller and more subtle effects, anomeric, gauche, delta-two and external anomeric effects as operative in carbohydrates and smaller model compounds. The point here is that the parameters derived from known chemical behavior of the model compounds are transferable to the larger ones where chemical behavior is unknown or little known. The chemical behavior of ever larger molecules influenced by ever smaller internal molecular influences supports the idea that various effects found in the laboratory are grounded in *geometric determinants of chemical behavior*.

As a nonspecialist, I gradually became aware that much of what we call chemistry is really geometry or, more properly, manifestations of the classical geometry of molecules. Never losing sight of the essentially quantum mechanical nature of the chemical *bond*, Allinger has stuck to his credo that molecules are assemblages of *atoms* which are large enough to be treated using simple classical physics. This has resulted in a series of programs which reproduce the chemical nature of molecules much more rapidly and efficiently than the numerous quantum mechanical programs in present use. The essence of the book lies not in that it describes a quick and easy way of approximating the chemical nature of the world around us, but in its fundamental exposition of organic chemistry as a *manifestation of molecular geometry*.

In Chapter 7 the author makes a critical point: Molecular orbital programs provide very accurate answers in favorable cases but various effects known to organic chemists are bundled in the answer. In parameterizing the MM force constant matrix one *must* separate effects, which may cooperate or oppose one another, in order to know which on- or off-diagonal elements should be modified and how they should be changed. There is no lumping of geometric or energetic properties when one must account for the influence of torsional flexing on adjacent bond lengths and angles. To the reader not engaged in building a matrix, the secrets of a molecule seem to be revealed layer by layer, like peeling an onion.

Chapter 9 is a detailed and exhaustive examination of hydrogen bonds and, more generally, Lewis bonds. Reparameterization to give a quasi van der Waals interaction is sufficient to handle these energetically small but by no means negligible bonds. Crystal packing is the subject of Chapter 10 with the introduction of a packing program called CRSTL. A thorough examination of the on-going [18]annulene debate is given with a proposed resolution of conflicting D_3 and D_6 experimental symmetries.

The final Chapter treats the important topic of enthalpies of formation, comparison with ab initio and experimental work, and the important distinctions among the various results, including special energies due to conjugation and strain. Subsequent to this, a section of about Chapter length entitled *Concluding Remarks* is followed by an *Appendix* which summarizes present force fields, basis set superposition error, conformational search methods and pitfalls, the driver routine, carbohydrate nomenclature, and some of the unavoidable Jargon of the trade (which is mercifully rare in the rest of the book). All of this is concluded with the happy comment "That's all there is to it." and indeed it is.

In summary, I recall a time when there were two distinct cultures in the chemical world. One culture was called "test tube chemists" with the implication that there wasn't much thinking going on, and the other was that of the theoreticians who were seen as spending years to obtain increasingly accurate answers to questions no one had asked. Fortunately, we no longer see these unfair distinctions. Theory and "real" chemistry have merged. Nowhere is that confluence of interests and talents more clearly expressed than in this fine book. Throughout, Prof. Allinger's wide and thoughtful reading, along with his appreciation of the chronology of the advance of chemical knowledge, is evident. The writing style is relaxed but informed, giving the reader the impression of a casual discussion with a friendly, knowledgeable mentor.

PREFACE

If we look back some 40–50 years, say to about 1965, at the way chemistry was, we note that we could have divided chemists into two different groups: experimental chemists and theoretical chemists. The experimental chemists generally stood in awe of the theoreticians, and there were rather few of the latter. The solution of the hydrogen atom problem via the Schrödinger equation was overwhelmingly impressive to an experimental chemist. However, the experimentalists really saw no connection (or at most a minimal connection) between theoretical chemistry and the real world in which they lived. At that time, computers were largely inaccessible to chemists, although their potential abilities for solving numerically complicated problems were evident to many. A few organic chemists, in particular physical organic chemists, had begun to more seriously consider approaching the structures and properties of molecules from a theoretical and computational point of view. As far back as the 1930s, various theoretical concepts such as "resonance theory" were utilized to try to better understand such things as the reactivity of benzene as compared with linear polyenes. But once pencils and paper and tables of logarithms gave way to the desk calculator, the physical organic chemists began to note with more interest some of the work done by theoretical chemists. But still, few physical organic chemists were willing to spend the laborious effort that was needed to obtain any kind of a new theoretical result by hand calculation. Already in the 1930s, but especially through the 1950s, many theoreticians began to publish what were then very useful tables containing numerical values for things such as overlap and two-centered Coulomb integrals. With the aid of these tables, theoretical calculations were brought down to the point where the average physical organic chemist could realistically carry out useful approximate calculations on rather simple systems. But few were willing to do even that because the labor involved still seemed disproportionately large relative to the information gained. All of that changed as computers became available. Computers in which all of the information was stored made things like Hückel calculations, and a little later molecular mechanics calculations, doable at the practical level. Thus arose the subject that is now referred to as *computational chemistry*. What had previously been within the realm of theoretical chemistry not only became practically useful, it became available and usable to those who wanted to more intimately understand more of why chemistry was as it was. Throughout the 1980s and 1990s, the number of computational chemists increased substantially, and it was found that in many cases they could point the way in which experimental chemists could profitably move forward.

As chemistry has advanced throughout the years, the amount of information available has grown enormously. But the amount of material that an individual chemist can really master has not grown proportionately. As a result, there has been much more specialization. Today, there continues to be a substantial gap between what the physical organic chemists know, think about, and work with compared to the corresponding related information with which the modern theoretical chemists deal. The gap between theory and experiment has in some ways gotten larger.

This book is in part directed at reducing that gap. The author has attempted here to bring together many of the theoretical or computational methods and the experimental methods commonly used for studies involving molecular structure. The primary theoretical methods here discussed are quantum mechanics, namely *ab initio* and semiempirical molecular orbital theory (including density functional theory), and molecular mechanics. The primary experimental methods are electron diffraction, X-ray (and neutron) crystallography, and microwave spectroscopy. The level here is not to try to teach the reader to do all of these things, but rather to outline and discuss enough so that the general usefulness and limitations of each of these things can be understood by the reader.

There will be considerable emphasis on molecular mechanics in what follows. There are at least two really good reasons why we should study molecular mechanics. One is obvious and most readers will recognize it. Molecular mechanics is a computational method that allows us to calculate structures of molecules, not from first principles exactly but from a mixture of first principles, second principles, and a knowledge of chemistry. It has the advantage that it requires relatively little computing time, so it can be applied to much larger systems than can be conveniently studied by quantum mechanics. Molecular mechanics also offer the advantage of reasonably high accuracy in cases where experimental problems limit the accuracy that can be obtained in that way. So it is a practically useful and powerful method for studying structures and subsequently for studying some of the physical and chemical properties of those structures.

The second major reason for developing molecular mechanics, which is less widely appreciated, is because it is a *model* of a large part of chemistry. Even if one can solve a particular problem using quantum mechanics, or if one can do the experiment, there is still an advantage to having a relatively intuitive and straightforward conceptual model that is widely applicable, which interrelates large amounts of chemical information and which can be easily applied to a given problem.

In quantum mechanics one has a few constants of nature that must be known if one is to solve a problem at the practical level. Generally speaking, their values are well known and can be looked up in handbooks. There are similarly constants that must be known in molecular mechanics, but here they are force field dependent, and they are not separately known at the outset. In other words, they must be treated as parameters. So the development of molecular mechanics has proceeded by first developing a conceptual model, together with the necessary equations that quantify that model, to describe organic molecular systems in a very general way. Then it was necessary to determine the values required for the parameters so that the model describes real systems in an accurate way. This parameter development was originally (before about

1980) done from experimental data, but quantum mechanical data have for the most part superceded experimental data for this purpose now, and experimental data are used more for confirmation to see that nothing falls through the cracks.

One might note that an important difference between the structures determined by quantum mechanics and by molecular mechanics depends upon the fact that in quantum mechanics one works with atomic nuclei and electrons. In molecular mechanics, the basic units are atoms. Molecular mechanics can handle many structural problems very well without going into the detail and complexity required when one deals with subatomic particles.

The above is an outline, at least to the author, of where it is that we wish to go, and why it is that we would want to go there.

The book *Molecular Mechanics* by U. Burkert and N. L. Allinger was published in 1982. Regrettably, Dr. Burkert died just before the book actually appeared in print. The present author has subsequently considered over the years whether it might be advantageous to do a new edition of this book.

The early part of the 1982 edition of *Molecular Mechanics* was devoted to a relatively full outline of the method, to the extent that the important details of it were known at that time. The latter part of that book consisted of many examples of applications illustrating various kinds of calculations that were possible and a comparison with many agreements, and some disagreements, that were obtained from experimental methods. Since many of the examples in the book became dated after some years, and the scope of the subject had advanced, there was a consideration as to whether a second edition would be desirable.

The 1982 edition was limited to what biological chemists nowadays would refer to as "small molecules," which means up to say 50 atoms or so. (In 1982, a small molecule meant a diatomic molecule, or thereabouts.) We specifically avoided a discussion of macromolecules, particularly proteins, at that time for two reasons. The first was that because of computer limitations, force fields being developed for proteins were very "stripped down," and much of the effort at that time was being spent trying to decide what could be left out, or what simplifying approximations could be used, so as to make it possible to actually carry out the calculation. There was, of course, also the overwhelming problem of solvation. What could actually be calculated was at a pretty low level. Accordingly, we decided not to try to include this topic at all in the 1982 book.

Subsequently, a great deal of progress was made in the protein field of computational chemistry, which led us to reconsider the possible inclusion of macromolecules in a second edition of *Molecular Mechanics*. The present author discussed this matter in some detail with Peter Kollman, who agreed quite a few years ago to write that part of the book. The planning got only as far as the outline stage when regrettably and unexpectedly Dr. Kollman died. Accordingly, the present author has decided that it is not feasible to include macromolecules in the present work, and that will have to be left for another time and another author.

While this present volume could have been a second edition of *Molecular Mechanics*, it was decided to change the emphasis somewhat for two reasons. The first reason is that the first part of the earlier book still covers in an adequate manner the general principles of the subject. Such principles tend to change very little and very slowly with time. The second reason is that molecular mechanics has simply worked itself into the mainstream of chemistry, so that it is one of several methods, computational and experimental, that can be applied to the solution of various types of chemical problems. It is often important to know which method is the best to use, with respect to the solution of a particular problem. Most chemists today have considerable expertise in a particular field, but relatively few have as much expertise as they might like, or could probably use, in related fields concerning molecular structure. Molecular mechanics gives us a model that attempts to tie together in a convenient way the various computational and experimental methods that are commonly used to study molecular structure. Hence the scope of the present volume.

Finally, there are two points to make. These constants that are used in molecular mechanics are in fact much more constant than one might have thought, and they are transferable from small model systems to larger compounds, pretty much to within chemical accuracy, as far as we are aware. In some cases it turns out, however, that they are not really constants, but they are at least simple, well-defined functions.

The second point is that with this high degree of transferability of constants in molecular mechanics, there is very little with respect to the structure of organic molecules that needs to be put in terms of quantum mechanics. It can pretty much all be put on a classical mechanical basis, using the pictures of valence bond structures that have been well known to organic chemists since the 1930s. This furnishes us, at least in the author's view, a quantitative description of structure that correlates well with the descriptions that physical organic chemists have always used for these phenomena. There are, to be sure, exceptions. And these exceptions will probably continue to make chemistry interesting for a long time.

ACKNOWLEDGMENTS

The author is indebted to a number of people for specific suggestions regarding the subject matter of this book and for reading parts of the book. These include Professors Wesley D. Allen, M. Gary Newton, Henry F. Schaefer, III, Paul von R. Schleyer, and Peter R. Schreiner. Another larger group of people contributed indirectly to the book through wide-ranging discussions with the author. These include, particularly, Janet Allinger, Lawrence S. Bartell, Paul D. Bartlett, Donald B. Boyd, Richard H. Boyd, Ulrich Burkert, James Cason, Charles A. Coulson, Richard W. Counts, Donald J. Cram, Michael J. S. Dewar, Alfred D. French, Robert B. Hermann, Tommy Liljefors, Kenny B. Lipkowitz, Max M. Marsh, Mary Ann Miller, John A. Pople, Harold A. Scheraga, Joseph A. Sprague, Julia C. Tai, David H. Wertz, Frank H. Westheimer, Kenneth B. Wiberg, and Young H. Yuh.

I would particularly like to acknowledge the help of Dr. Jenn-Huei Lii and Dr. Kuo-Hsiang Chen, who actually carried out most of the numerical calculations discussed, to Dr. Lii for the drawings in the book, and especially for his extensive computer programing work for our research group over the last twenty years. And finally to Professor Donald W. Rogers for his careful reading of and detailed comments on the entire manuscript. I also would like to thank Ms. Shawn Stephens for her diligence in converting a large amount of dictation and pen scratchings into the actual book manuscript and to my wife Irene for her patience and constant encouragement.

1

INTRODUCTION

An *atom*, for present purposes, consists of a positively charged nucleus plus negatively charged electrons that are distributed about the nucleus in a more or less spherical cloud. Important properties of atoms as far as chemistry is concerned are largely determined by the charges and masses of these particles. One might, therefore, conclude that problems in chemistry are really problems in electrostatics (or electrodynamics). While in a sense that is true, at the practical level it is convenient to approach the problem in a different way.

To a first approximation, a *molecule* may be regarded as a simple assemblage of atoms. The details of that assemblage may be largely understood from the familiar laws of classical mechanics and electrostatics, and such a model is now usually referred to as a *molecular mechanics model*. A more proper solution to molecular problems uses quantum mechanics (which automatically includes electrostatics as well). But this more sophisticated (and harder to understand) method is not needed for the most part for understanding practical chemical problems. Much of the time it is found that molecular mechanics is adequate for solving the problem at hand.

Within its area of applicability, molecular mechanics has two definite advantages over quantum mechanics. These are related and depend upon how hard we look at a particular problem. The first advantage is what we might call "intuitive." With a

---

Molecular Structure: Understanding Steric and Electronic Effects from Molecular Mechanics,
By Norman L. Allinger
Copyright © 2010 John Wiley & Sons, Inc.

molecular mechanics formulation of a molecule at hand, the chemist can often look at a problem and decide that a certain change will lead to a certain result in a qualitative (or semiquantitative) sense. It is normally much more difficult to look at a quantum mechanical problem and come to a similar conclusion in a simple and direct way.

Part of the difficulty with quantum mechanics is that its quantitative application to the solution of a problem is usually much more demanding mathematically and computationally than the corresponding molecular mechanics application. If one wants "chemical accuracy" in solving a problem for a small molecule, the quantum mechanical calculation may require something like 10^4 times as much computation time as the equivalent molecular mechanics calculation. For problems involving rather small systems, this may not pose much of a difficulty, but this ratio increases rapidly with molecular size. For problems involving larger systems, the quantum mechanical solution may require shortcuts or approximations at the practical level, which often interfere with the attainment of the desired accuracy.

The application of quantum mechanics to chemical problems involves solving equations that describe laws of nature. At our current level of understanding, that is just the way it comes out. If one wants to implement exactly those laws of nature, one solves those equations. There are a lot of choices as to how one might go about that, but overall it is clear what has to be done.

On the other hand, molecular mechanics gives us a *model* that describes the molecular system. Our model attempts to reproduce the results of the laws of nature, but there is really unlimited flexibility in just how one may go about that. The most important things about a model are: The model must not violate any laws of nature or fundamental principles, and the model should give results to the desired accuracy in a relatively simple and easy way while being as intuitive as possible.

The invention of actual physical models of molecules has been ascribed to A. W. Hofmann. He is said to have built these models by taking a croquet set, drilling holes in the croquet balls, and using pieces of the croquet handles to connect those balls into large ball-and-stick models that he used in a lecture demonstration given at a meeting of the Royal Institution of Great Britain, Friday, April 7, 1865, entitled "On the Combining Power of Atoms."[1] In this lecture Hoffman discussed the laws governing chemical combinations from the atomic point of view and introduced the concept of valency. However, at that time (1865) chemistry was considered to be two dimensional, and Hofmann's models were also two dimensional. These ball-and-stick models subsequently evolved into smaller (and three-dimensional) desktop models that we still use today.

The earliest three-dimensional molecular models known to the present author were those constructed by van't Hoff in connection with his stereochemical work[2] (1874). These models (which the author has had the privilege of holding in his hands) were constructed of orange-colored cardboard triangles approximately one inch along an edge, which were glued together to form tetrahedra. With assemblages of these tetrahedra, van't Hoff described the geometries of simple molecules, as is now so well known. In the early 1890s, Sachse illustrated the boat-and-chair structures of cyclohexane as three-dimensional models of essentially the Hofmann type.[3]

While the ball-and-stick models are quite useful for understanding in a three-dimensional sense, how various parts of a molecule interact with one another, or with other molecules, in trying to refine this model, a few other things were noted. The most important thing was that atoms have a physical volume. That is, they occupy space and hence bump into one another. The *space-filling* model was thus developed and was quite useful for understanding steric effects. But this space-filling model still has deficiencies. Perhaps the most conspicuous deficiency is that an atom is not very well represented by a hard sphere. Atoms are, on the contrary, better represented by soft spheres that exert weak attractive forces on one another at longer distances, which then become strongly repulsive at short distances (van der Waals description).

We also know from classical mechanics that if we have two weights held together by a spring, the system can vibrate. And there are extensive consequences from that vibrational motion, which can be well described by classical mechanics. Thus we have Hooke's law, and vibrational frequencies, and the model of a molecule that is considered as an analog of a classical system, where the atoms and bonds are similar to weights and springs. And as the system becomes more complex (many weights connected by many springs), classical mechanics deals with it very well. Such a description of a molecule allows one to calculate the vibrational spectra (Raman and infrared) of that molecule. With that information one can proceed to calculate thermodynamic properties of molecules. Thus, with these refinements, the classical mechanical model of a molecule moves up from the rigid ball-and-stick model to the soft sphere vibrating model that is normally used in molecular mechanics today.

WHAT, EXACTLY, IS A MOLECULAR STRUCTURE?

In order to understand a structure, we have to understand a lot of component pieces. Listed in order of importance, it would seem that what we have to know would be as follows:

1. The numbers and kinds of atoms present
2. The connectivity of the atoms
3. The stereochemistry
4. The internal coordinates of the molecule*
5. The **F** matrix of the molecule

---

*It takes three coordinates to describe the location of an atom, say in a Cartesian coordinate system. If a molecule contains N atoms, it will take $3N$ coordinates to describe their locations. The *internal coordinates* (of which there are $3N - 6$ for a molecule containing N atoms, except for linear molecule, where there are only $3N - 5$), namely the bond lengths, bond angles, and torsion angles, define the structure of the molecule completely. The *external coordinates* (of which there are 6, or 5 if the molecule is linear) define the location and orientation of the molecule in space. The *Cartesian coordinates* of a molecule specify the location of each atom in the molecule in a Cartesian coordinate system. These Cartesian coordinates include both the internal and the external coordinates of a molecule. The Cartesian coordinates of the molecule can be converted into internal coordinates in a straightforward and unique way, and the internal coordinates can be converted into Cartesians by deciding where to locate the molecule. Hence, knowing either set of the coordinates is equivalent to knowing both sets, as far as the structure is concerned.

An understanding of these items tells us with increasing accuracy the structure of the molecule. It is essentially in historical order. The earliest chemists who thought about molecular structure worried only about the numbers and kinds of atoms that were present in the molecule. Next, with the discovery of isomerization, they worried about the connectivity, and they thought that took care of the problem. However, as the three-dimensional nature of molecules became clear, it was realized that stereochemistry was an essential extension of the connectivity, and it had to be considered as well. And in classical organic chemistry, having that much of the information about a molecule meant that the structure was known and understood.

However, once it was realized that structures could be determined quantitatively using the tools of X-ray and electron diffraction and microwave spectroscopy, it was recognized that there was in fact more to the problem. Conformational analysis showed us that there were three-dimensional aspects of the structure that had to be considered beyond classical stereochemistry, if one were fully to interpret experimental results, mostly spectroscopic and thermodynamic results, that were structure dependent. Hence, it became necessary to know the internal coordinates of the molecule, including the torsional coordinates that define the conformations. Finally, since the molecules do not have a rigid motionless structure such as our ball-and-stick models, the structures and, in particular, interconversion rates between conformations are temperature dependent. So this means that one must understand the potential energy surface of the molecule, not just the minimum energy points, if one is to fully understand the structure. Hence, the little list above gives both the historical development of the theory of structure and also the increasingly quantitative progression of just what we mean by structure.

If we have an atom in a Cartesian coordinate system, it will take three coordinates to describe its location. If we have N atoms in a coordinate system, it will take $3N$ coordinates to describe their locations. It doesn't matter if the atoms are bound together or not. Hence the exact geometry of any molecule including its spatial location and orientation can be described by $3N$ coordinates.

The molecule can translate, rotate, and vibrate. To discuss the translation and rotation of the molecule, we need only describe the location of the center of mass of the molecule in the coordinate system and the orientation of the molecule with respect to the coordinates. We can do this with the aid of three coordinates for translation and three for rotation. (The special case of a linear molecule will not be discussed here.) These six coordinates simply define where we are positioning the molecule and how we are orienting it, and they have nothing to do with the structure of the molecule.

Thus, there are $3N - 6$ coordinates that represent vibrational degrees of freedom of the molecule, and they are necessary to define its structure. If we want to know its structure, what we have to determine are the $3N - 6$ coordinates that define the positions of all of the atoms when the molecule is in its ground state. If the molecule is a little bit complicated, there may be several different stable structures (conformations). In that case we will wish to know the structures of all of the conformations. We will also need to know their relative energies so that we can calculate a Boltzmann distribution among them. Also, we frequently wish to know the torsional barriers on the potential surface that separate these structures from one another.

Calculations that are now to be described are usually done in Cartesian coordinates, but they can be done in internal coordinates (which has the advantage that there are six fewer coordinates needed). However, since we can translate from one set of coordinates to the other in a unique way, in principle it makes no difference which set we use. Conceptually, it's much easier to understand what is happening if one thinks in terms of internal coordinates, and that's what we will do here.

To determine a molecular structure, the problem to be solved then is to find that point on the potential energy surface where the energy is at a minimum. And if there are several such points, we want to find them all. And if we are interested in transition states, we want to find also saddle points, which are in effect minima in all directions except one, and in that direction they are maxima.

The mathematics here is pretty straightforward, although it was impossibly tedious except for very simple cases before computers were available. With computers of current availability, the problem can be solved in typical cases for molecules containing up to 100 atoms or so, in a short time, typically of the order of seconds or minutes. The problem is always solved by successive approximations, so the time required is dependent on how good a structure one uses to start. And the problem can be solved for systems containing many thousands of atoms, although it may take quite a while.

How does one solve the problem? First, it should be said that there is an extensive literature available that answers this question. I will give a somewhat superficial overview here. If one wishes the answer to this question from the viewpoint of physics, the book *Molecular Mechanics* by Machida[4a] is recommended. Other descriptions can be found in *Molecular Mechanics* by Burkert and Allinger[4b] and elsewhere.[4c] In general, what one wishes to do is to find that structure (the atomic positions) where the energy of the system is at a minimum. That is, one wants to find the point on the potential surface where the derivatives of the energy with respect to each of the coordinates is equal to zero. This will, for a system of N atoms, consist of solving $3N - 6$ simultaneous equations. These equations can ordinarily only be solved not by using a "closed" method but by successive approximations. The usual way to carry out these calculations is by utilizing matrix algebra. There are many details here that are of interest for those who develop methods for actually solving this kind of problem. However, here we will only say that such methods exist, are well known, and one can read about them elsewhere if interested.[4] For our purposes here, it will be sufficient to describe the force constant matrix, commonly referred to as the **F** matrix, which is a general description of the simultaneous equations to be solved, and to understand in a chemical sense what kinds of things make up the elements of this matrix. Our classical model system contains weights and springs, and at the energy minimum the net force acting on each atom must be zero. That is, the system is stationary at the energy minimum. There is one difference between that model system and an actual molecular system in that the atoms are never actually at rest at the energy minimum position, but they vibrate about that position in a prescribed way. The problem can best be dealt with by first finding the energy minimum and then adding the vibrational information to the minimum energy solution.

In the force constant matrix, there appear terms that are associated with each of the degrees of freedom in the molecule. If these degrees of freedom are the internal coordinates, then the force constants associated with each of those degrees of freedom are what we have to know to put into the force constant matrix.

The **F** matrix of a molecule is part of the description of the molecular structure. If we want to know the structure as a function of temperature, we have to know the thermodynamics, and for that we have to know the **F** matrix. So here we consider that what is described by the **F** matrix is an essential part of the structure.

Finally, we certainly do want to know the heat of formation of the molecule. The heat of formation of a molecule is probably better thought of as a property, rather than as part of the structure. This is a fundamental quantity, arguably the most important property of a molecule, and if we are interested in actual chemistry (i.e., interconversions between molecules), then we need to understand heats of formation. To be able to calculate and understand those, we want to understand bond energies as the component parts of those heats of formation.

So these are some of the basic things that we will want to learn to help us to understand molecular structure. And there are fundamentally two different ways that we can study these things: by experiment (Chapter 2) and by theory using computational methods (Chapter 3). We will use both of these tools to study organic molecules in the remainder of this book.

REFERENCES

1. A. W. Hoffman, *Proc. R. Inst. Great Britain*, **40**, 187 (1865).
2. J. H. van't Hoff, *Bull. Soc. Chim. France*, **[2] 23**, 295 (1875); the original (Dutch) version appeared in 1874.
3. H. Sachse, *Chem. Ber.*, **23**, 1363 (1890); *Z. Physik. Chem.*, **10**, 203 (1892).
4. (a) K. Machida, *Principles of Molecular Mechanics*, Kodansha and Wiley, co-publication, Tokyo and New York, 1999. (b) U. Burkert and N. L. Allinger, *Molecular Mechanics*, American Chemical Society, Washington, D.C., 1982. (c) A. R. Leach, *Molecular Modeling* Principles and Applications Pearson Education, Essex, England, 1996. (d) A. F. Carley and P. H. Morgan, *Computational Methods in the Chemical Sciences*, Horwood, Chichester, 1989. (e) D. W. Rogers, *Computational Chemistry Using the PC*, 3rd ed., Wiley, Hoboken, NJ, 2003. (f) D. Frenkel and B. Smit, *Understanding Molecular Simulation—From Algorithms to Applications*, Academic, San Diego, 1996. (g) U. Dinur and A. T. Hagler, *New Approaches to Empirical Force Fields, Reviews in Computational Chemistry*, Vol. 2, K. B. Lipkowitz and D. B. Boyd, Eds., Wiley, New York, 1991, p. 99.

2

EXPERIMENTAL MOLECULAR STRUCTURES

There are numerous physical methods that can be used to determine partial molecular structures. But here we will limit our discussion to the four methods that are in wide current use and that may be utilized to accurately determine a total molecular structure. These are three diffraction methods, namely electron diffraction, neutron diffraction, and X-ray diffraction, and one spectroscopic method, microwave spectroscopy. (There are, of course, many other methods for studying molecular structure, some of which are extremely powerful, although usually in limited areas. But, not all of chemistry can be discussed here, so we will limit the topics to those mentioned.) Each of these methods measures something that is a little different, sometimes quite a bit different, from what the other methods measure. The calculational methods, described in the next chapter, generally calculate something that is still different from any of the above. The structures of the same molecule determined by these different methods are not, in general, identical. Hence, we need to understand how the structures obtained from each of these methods are interrelated.

ELECTRON DIFFRACTION[1-3]

The procedure for determining a structure by electron diffraction consists of allowing the molecules that are to be studied in the gas phase to flow into an evacuated chamber

---

Molecular Structure: Understanding Steric and Electronic Effects from Molecular Mechanics,
By Norman L. Allinger
Copyright © 2010 John Wiley & Sons, Inc.

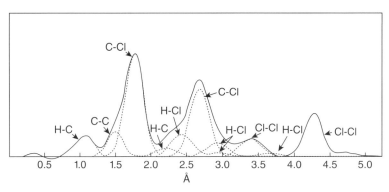

Figure 2.1. Radial distribution curve for 1,2 dichloroethane. (From Braude and Nachod.[3] Copyright 1955. Reprinted by permission of the authors.)

and an electron beam is focused on the stream of molecules. The molecules diffract the electron beam and generate a diffraction pattern that can be recorded photographically or otherwise. Processing this pattern yields a radial distribution function that gives the distances between the atoms in the molecule. A typical radial distribution function, that of 1,2-dichloroethane, is plotted as the solid line in Figure 2.1.

This molecule is a mixture of gauche and anti conformations. The strongest part of the electron scattering occurs from single atoms, and this does not give information useful for present purposes and is removed in the processing. The next strongest part comes from pairs of atoms, and the diffraction pattern can be converted into the distances between the pairs, as represented by a radial distribution function. The amount of scattering by a particular pair of atoms falls off exponentially with the distance between them and is proportional to the product of the atomic numbers of the two atoms involved. Thus, it is relatively easy to see the distance between two carbons (weight 36) and less so to see the distance between a carbon and a hydrogen (weight 6), and it is quite difficult to see the distance between two hydrogens (weight 1). Second-row atoms scatter considerably more than do first-row atoms and tend to dominate the diffraction pattern if present. Third-row atoms, and higher, scatter the electron beam so much that it is often difficult to determine much about the rest of the structure of the molecule accurately if they are present. Accordingly, the technique is most useful for molecules containing first-row atoms or perhaps one or two second-row atoms in addition.

If the atoms in a molecule were rigidly held at certain distances, then the radial distribution function would consist of a series of lines, corresponding to those distances, with intensities that would depend on the atomic numbers of the pair of atoms and the distance between them. But, of course, the atoms vibrate. Thus, instead of a line, one obtains a Gaussian function where the area under the curve depends upon the variables mentioned.

In Figure 2.1, if we start at the left side and go toward the right, the first large peak that we come to is seen at a distance of 1.1 Å. This corresponds to the shortest distance between two atoms in the molecule, the C–H bond length. The C–H bonds will be similar in length, but in principle not identical unless the atoms involved occupy identical (or enantiotopic) positions, in a stereochemical sense [as in nuclear magnetic resonance (NMR) spectra at very low temperatures]. In the case at hand, the four hydrogens in the trans conformer are equivalent and will show a single line. But in the gauche case, there are two different types of hydrogen, both of which are different from those in the trans case. Accordingly, there are two equally intense lines from that conformer, which together have a total area equal to that of the single line from four hydrogens in the trans conformer. These intensities then have to be multiplied by the mole fraction of the appropriate conformer. But, because of the vibrations of the atoms, instead of getting three closely spaced lines, all we see is one Gaussian peak, under which the three lines all lie and are unresolveable. One can measure the average C–H bond length quite accurately, perhaps to about 0.002 Å, but one cannot measure the individual C–H bond lengths separately.

The next distance occurs at 1.50 Å and is the C–C distance. It is not resolved from the C–Cl peak that follows it in the experimental curve. However, since we know that the curves are Gaussian, and we know their relative areas, we can resolve them as shown by the dashed lines in Figure 2.1. Then at about 1.75 Å, we see the C–Cl distances. The two chlorines in the gauche form are equivalent to each other, but different from the two chlorines in the trans form, which are also equivalent to each other. But again, we cannot resolve these small differences, and we see both of the C–Cl bond distances superimposed. As we continue on, we see the increasing distances, the hydrogen to the carbon to which it is not attached, and so on. An interesting feature here are the Cl···Cl distances, which are quite different in the gauche and anti forms, and quite conspicuous. These are found as two peaks at 2.7 and 4.3 Å corresponding to the gauche and anti chlorines, respectively. From the areas under those two bands, it is possible to measure rather accurately the ratio of the two conformations in the gaseous mixture, and hence their free energy difference from the Boltzmann distribution. There is some uncertainty here because the temperature of the measurement is not exactly known. The temperature of the nozzle through which the compound is injected into the measuring chamber can be measured and controlled, but as the gas expands, it cools. This introduces some uncertainty into the temperature and hence in the free energy difference.

Another problem occurs with electron diffraction (to take an example), which is referred to as *shrinkage* (Fig. 2.2).[4] This quantity can be easily understood by looking at a simple vibrating linear triatomic molecule such as CO_2. The bond lengths have the same value, call it d. The bond angle OCO is 180°. What is the distance between the oxygens? One might assume that it is simply $2d$, but it is not. It is noticeably smaller. In other words, the molecule "shrinks" relative to what simple geometry appears to require. In this case, the angle-bending vibrational motion of the molecule causes the oxygen to move along the arcs that are indicated. The distance $2d$ corresponds to the

Figure 2.2. Cause of shrinkage.

maximum separation of the oxygens. When we look at the vibrational motion as shown, we can see that the mean separation is clearly less. This is a general phenomenon, and older electron diffraction studies that did not properly take shrinkage into account yielded inaccurate results.

This experimental technique of electron diffraction is quite difficult to apply in practice, and the equipment for making the measurements is not commercially available. Accordingly, there are only a few groups worldwide that do this kind of work. The molecules studied have to be rather small or alternatively highly symmetrical, because otherwise the radial distribution function shows so many overlapping bands that it cannot be unambiguously interpreted with much accuracy. However, for small molecules, the method can be quite accurate, and bond lengths to an accuracy of the order of $0.002\,\text{Å}$ can be obtained under favorable circumstances.

MICROWAVE SPECTROSCOPY[1]

This is quite a different technique from electron diffraction, but similar in usefulness. It contains a number of limitations, but for the most part these limitations are somewhat different from those with electron diffraction. Accordingly, it can be very advantageous to have both an electron diffraction structure and a microwave structure for a given molecule.

In the case of microwave spectroscopy, what one actually determines are microwave frequencies from which one can determine the energy differences between the rotational states of the molecule. These give rise to the rotational constants, which are inversely related to the moments of inertia of the molecule. As with electron diffraction, microwave spectra are measured in the gas phase. A major disadvantage here is that a molecule has only three moments of inertia, so one can obtain only three pieces of information. An advantage is that these moments can be measured with very high accuracy to five or six significant figures. But the problem then is how to utilize the three numbers that are measured experimentally in order to get as much information as possible about the structure of the molecule under study. There are no experimental techniques available that will permit us to determine structures for molecules larger than triatomic, except in very special cases, with the kind of accuracy with which moments of inertia can be obtained from microwave spectra. It is, however, possible to calculate by quantum mechanical methods structures that have bond lengths accurate to four or five significant figures for molecules that contain up to three heavy atoms plus hydrogens. The general method used in these calculations is in principle applicable to larger molecules, and such calculations will presumably be made as computer speed and availability continue to increase. This approach will be discussed later under ab initio quantum mechanical calculations (Chapter 3).

A significant limitation in the microwave case is that the intensity of the spectra obtained are proportional to the dipole moments of the molecules under examination. If the molecule does not have a dipole moment, then it does not show a microwave spectrum, and hence this technique is inapplicable. Unfortunately, many simple molecules (methane, ethane, benzene, ethylene, acetylene, etc.) do not have dipole moments for reasons of their symmetry. Such molecules cannot be studied in a simple direct way by microwave spectroscopy. (Substitution of one of the hydrogens with a deuterium is sufficient to generate a dipole moment, so such substituted molecules may often be studied by this technique. However, such a substitution also introduces asymmetry and another type of atom into the molecule, which complicates the interpretation of the data.)

One needs to have far more than three moments of inertia to determine the structure of most molecules. There are two ways around the fact that for most molecules one can obtain only three pieces of data. One way to proceed is called the *isotopic substitution method*. One replaces an atom in the molecule with a different isotope of the same atom, for example, deuterium for hydrogen, and then one can get three more moments of inertia for a molecule with the same structure. If one replaces a number of different atoms in the molecule one at a time, with different isotopes, one can generate a large number of moments of inertia for the same "structure." Eventually, one has as many pieces of data as there are atomic coordinates, and the structure (coordinates) can be solved for using simultaneous equations (referred to as *Kraitchman's equations*[5]). This would be straightforward if the vibrations were harmonic, but, of course, they are not. So, unfortunately, this procedure introduces still another uncertainty. Since the vibrations are anharmonic, the structure of the molecule changes slightly with isotopic substitution because the heavier isotopes lie further down in the potential well and vibrate over smaller amplitudes. These differences change the average location of the vibrating atoms. The more anharmonic the vibration, the more the change in structure. Furthermore, in practice, the problem turns out to be large enough to spoil the otherwise very high accuracy that can be obtained by microwave methods. This method is still useful; it is just much less accurate than one would like.

The other alternative that can be used to obtain a structure from just three moments of inertia is that if the structure contains relatively few atoms, and hence relatively few coordinates to be determined, and if some parts of the geometry of the molecule are already known, or can be assumed from some other available information involving similar molecules, then the part of the structure that is not known can be determined (as long as only three pieces of data are required). This method usually contains considerable uncertainty because one is never quite sure just how accurately part of the structure can be transferred from another molecule. So the result is that the microwave method gives moments of inertia very accurately. However, one must make assumptions and approximations if one is going to determine a real molecular structure by this method. For this reason, the simultaneous use of microwave and electron diffraction data is most useful because if one can fit both of these kinds of data at the same time, one can normally increase very much the reliability and accuracy of the determination.

X-RAY CRYSTALLOGRAPHY[1a,2,6-8]

X-ray crystallography is by far the most widely used experimental method for determining molecular structure. A few hundred or so structures have been determined by electron diffraction, and another few hundred or so by microwave spectroscopy, but about 300,000 have been determined by X-ray crystallography and are available in the

Cambridge Structural Database.[8] The number of entries in the database increases by approximately 10% per year. For reference, the total number of compounds in the Chemical Abstracts database exceeded 50 million in 2009.

The Phase Problem

Put into simplest terms, an X-ray crystal structure is determined by directing a beam of X rays onto the crystal. These are scattered and yield a diffraction pattern, which is recorded. The common analogy is when a wave moving in a lake hits a series of piles (posts) holding up a dock, one sees a diffraction pattern generated by such waves interacting first with the piles and then with each other. A problem that faced crystallographers from the beginning was that they could tell the difference in phases (the degree to which two waves are out of phase), but they could not tell the sign of that difference (which wave was in front and which was behind). If one has two waves, there are only two possibilities, and one can easily "solve" the problem. But in a typical X-ray diffraction pattern there are very large numbers of waves that are out of phase by varying degrees, and the problem is to determine the signs of these phases. This problem must have been given a lot of attention by crystallographers in earlier times, but it proved to be intractable. And in those times, solving a crystal structure was usually a major operation. There were various special methods used for solving structures, but in general, the problem was formidable.

All of this was subject to change after work by Karle and Hauptman,[9] who worked out the general solution for the phase problem (they received the Nobel Prize for this work). They indicated in their publication (in 1950), however, that the mathematical complexity of their proposed method appeared to be overwhelming, and they did not foresee immediate applications. As it turned out, computers became available, and increasingly powerful, just at this time (as will be discussed at the beginning of Chapter 3). The implementation of their "direct methods" for solving crystal structures thus occurred only slowly over a period of many years, but became the standard method for solving X-ray structures after about 1980. So now many crystal structures can be solved in a routine way, although, of course, there are still cases that cause special problems. But the generation of the enormous library of crystal structures in the Cambridge Database[8] has occurred after these direct methods for solving crystal structures were developed.[10]

Additional discussion and examples of the usefulness of the information that can be obtained from X-ray structures will be furnished in Chapter 10.

One problem, which is easy to understand, concerns a difference between X-ray work, in which one determines the distance between the mean positions of atoms, and electron diffraction, where one determines the average distance between atoms. These quantities may sound like they are the same thing from what is stated, but they are not quite the same, as the following simple considerations will show. Suppose that we have three atoms, A, B, and C, in a line. For simplicity, imagine that the two end atoms, A and C, are fixed, and B is vibrating between them along a vertical axis as shown in Figure 2.3. The mean position of B is the center as labeled, and it moves to extreme positions at the ends of the arrows as indicated. The distance between the mean

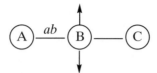

Figure 2.3. Bond length.

positions of A and B is indicated by the line segment *ab* (which is called r_α, an X-ray quantity). But this is not the average distance between the atoms (r_a or r_g, as determined by electron diffraction). Rather, *ab* is actually the *minimum* distance between the atoms. The average distance between them is significantly longer.

This X-ray method has major advantages and disadvantages relative to the other methods. One disadvantage is that the structure is determined in a crystal, rather than in the gas phase. Accordingly, one is not looking at an isolated molecule, but a molecule that is being subjected to various distortions by neighboring molecules. By way of an orientation, while we might in favorable cases obtain bond lengths with an accuracy of say 0.002 Å in electron diffraction, and perhaps with similar accuracy from the rotational constants in many cases, the accuracy is normally very much less in crystallography. There are several different problems that affect the accuracy of crystallographic structures. One obvious one is the effect of neighboring molecules in the crystal. Since the force constants for stretching bonds are quite high, bond length distortions from neighboring molecules are normally small, 0.001–0.002 Å or so. But force constants for bending angles are much smaller, and torsional force constants are often smaller still, so molecules tend to distort significantly in crystals with respect to bond angles (up to a few degrees), and often very much more in the torsion angles, depending mainly on the heights of the torsional barriers that are holding the molecule. Some of these distortions are much larger than the accuracy of the measurement. But there are further problems that are even worse. The structures determined by the other methods previously discussed find the nuclear positions of the atoms, but X rays are scattered by the electrons. Hence, to take a simple example like an amine, one does not measure the position of the nitrogen nucleus by X-ray methods, rather one measures the position defined by the center of the electron density of the nitrogen atom. Because of the lone pair, the center of electron density will be offset from the nucleus about 0.01–0.02 Å. Hence, one locates the nitrogen in a position that is different from the nuclear position found by the other methods.

The situation is even worse when it comes to locating hydrogens by the X-ray method. A hydrogen atom has only one electron. When it forms a bond, that electron is largely pulled away from the simple spherical 1s orbital and into the bond. Hence when one measures a C–H bond length using X rays, the C–H bond is about 0.12 Å shorter than that defined by the nuclear positions. In the case of an O–H group, the bond is shorter by approximately 0.18 Å! Other differences in structure that result when X-ray structures are compared with nuclear position structures are further discussed under neutron diffraction, in the next section.

Unfortunately, many chemists do not realize the fact that "bond lengths" determined by X-ray crystallography are significantly different from "bond lengths" that correspond to internuclear positions and that are determined by other physical methods. One accordingly frequently sees tables of bond lengths in which X-ray values are compared with other values. There are very large discrepancies in many cases, and these are often attributed by the unknowledgeable to experimental error. It is not really an experimental error. The error is made when one uses centers of electron density to define atomic positions, and then compares them with nuclear positions. Those kinds of numbers should not normally appear together in the same table.

Rigid-Body Motion

Finally, there is the thermal motion problem in X-ray crystallography.[11] Molecules vibrate internally, and molecules in a crystal lattice also vibrate with what is referred to as the *rigid-body motion*. That is simply the whole molecule moving back and forth and undergoing rotational types of oscillation in the crystal lattice. (These are motions that correspond to translations and rotations in the isolated molecule that are converted into vibrations by the constraints imposed by the crystal lattice.) The molecule as a whole is quite heavy, so these vibrational frequencies tend to be quite low, and the vibrational amplitudes are large. The most serious part of this problem comes not from the translational motions but rather from the rotational oscillations of the molecule in the crystal lattice. The effect of these is to shorten the apparent bond lengths and compact the apparent size of the molecule.

Of course, where there are problems, there are usually solutions to problems. In modern crystallography, when one determines the coordinates of the "atoms," one also obtains *thermal ellipsoids*. The atoms (usually other than hydrogen) are not represented by only spheres of electron density, but by ellipsoids that also include the rigid-body motions of the molecule. Having the relationships between the shapes and the orientations of these ellipsoids relative to one another, it is possible to calculate bond length corrections from the data.[11] These corrections are not highly accurate, but certainly the bond lengths can be determined much more accurately if the corrections are applied than if they are ignored. An accuracy of 0.002 Å is desired by the other structural methods, and an accuracy of 0.003–0.004 Å is considered pretty good. In X-ray crystallography, for the bulk of the structures to be found in the Cambridge Database (determined at room temperature), estimated standard deviations in the bond lengths in favorable cases (other than for hydrogen) are about 0.003–0.004 Å. But the average deviations, when systematic deviations including thermal motions are included, are more like 0.006–0.012 Å. If the thermal corrections are made, they will improve the accuracies in the bond lengths in these structures by perhaps 0.002 Å on the average.

But we can do better. The thermal motions cause a problem because the amplitudes are so large, which is in part due in turn to the close spacing of the vibrational levels. The problem can be most easily understood with reference to the two ellipses shown at the top of Figure 2.4. These two ellipses* represent the electron densities of two

---

*Usually about 30–40% of the electron density associated with an atom is represented by an ellipsoid, as this gives a convenient size. Ellipsoids are often used rather than spheres, as the usual thermal motion can be more accurately represented using them.

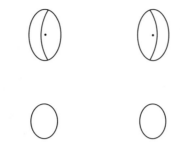

Figure 2.4. Thermal motion (schematic).

atoms that are bonded together, and they are undergoing a quasi-rotational kind of vibrational motion, where the rotational axis is in between them and perpendicular to the plane of a page. (Sometimes this type of motion is referred to as a *libration*, or a *rotational oscillation*.) The two nuclei are moving back and forth along the curve lines shown in the ellipses during the oscillations. Often the data will refine to yield two thermal ellipsoids, more or less parallel to each other, as shown in Figure 2.4.

The centers of the electron densities are then considered to be at the "centers" of the ellipsoids, as shown by the two dots in Figure 2.4, so the distance between those two dots is taken as the bond length. But because of the thermal motion, the nuclei are actually vibrating in arcs, as indicated in the figure. If the two ellipsoids were simply translating in an up–down direction on the page, those lines would be straight, and go through the dots. However, if there is a rotational type of oscillation about an axis in the center between the two dots, then the paths are in fact as indicated by those curved lines. The actual atomic positions (the arcs) are consequently further apart on the average than the centers of electron density as defined by ellipsoids (the dots). Hence, the bond lengths are always determined to be systematically shorter than they actually are, as a result of this thermal motion.

While the results of the thermal motion effect are somewhat exaggerated in the top two ellipsoids in Figure 2.4, this thermal motion leads to errors, often in the range of 0.005–0.010 Å, with the experimental bond length always being found to be too short. When one reduces the temperature, the thermal ellipsoids become smaller, as shown with the lower set of ellipsoids. Even so, the two atoms are still following curved paths, and the bond length obtained will be too short, but the effect will obviously be much reduced at lower temperatures.

Thus, we find that if the X-ray measurements are made at liquid nitrogen temperatures, the total errors are reduced by about half, say to about 0.003–0.005 Å on average. Unfortunately, although X-ray crystallography is largely done at liquid nitrogen temperatures currently, most of the structural information in the Cambridge Database was determined in earlier times and at room temperature.

In 1965 there was described in the chemical literature[12] a computer program called ORTEP that could be used to draw chemical structures obtained from X-ray crystallography. It would draw a structure as a pair of stereographic projections, which made it easy to see structures in three dimensions (see Fig. 2.5). (If the reader is not able to see Figure 2.5 in three dimensions, please look at the instructions given in the Appendix

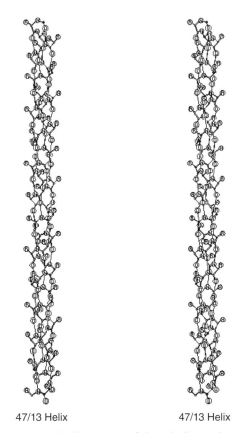

Figure 2.5. Stereographic illustration of the α-helix produced with ORTEP.

under Stereographic Projections). Note how clearly the three-dimensional nature of the structure can be seen compared to what can be seen from just one of the structures. The program can represent the individual atoms as a ball-and-stick model or by their thermal ellipsoids.

The thermal ellipsoid option is shown by the stereo plot of the cubane molecule in Figure 2.6. The ellipsoidal nature of the carbons is clearly evident, especially so because of the way that they are projected with the segment cut out. This was certainly one of the most widely used programs in chemistry—and still is. (The hydrogens are usually represented only as spheres as their scattering power is not sufficient to give enough data to make it useful to try to construct ellipses for them.)

Molecular Mechanics in Crystallography

Molecular mechanics can be used to study crystal structures (or liquids or solutions) as well as isolated molecules. The lattice forces of a crystal distort the molecules in the crystal, sometimes rather severely. Accordingly, if one wants to relate crystal structures

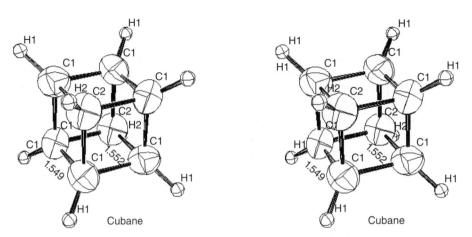

Figure 2.6. Cubane (stereographic projection).

to the structures of isolated molecules, this can be done conveniently with molecular mechanics. Ab initio calculations can also be used similarly. These topics will be further discussed in Chapter 10.

NEUTRON DIFFRACTION[1,2]

One way to avoid two of the major problems that result in X-ray crystallographic work is to not use X rays but rather to use a neutron beam at low temperatures to study diffraction in crystals. A major disadvantage is that to do this, one must have a source of neutrons, and there are not many of these worldwide. Such measurements do offer some real advantages over ordinary X-ray crystallographic data. First, neutrons are scattered by atomic nuclei, not by the electrons. Hence, they determine where the nuclei are, rather than where the centers of electron density are. And neutron diffraction crystallographic studies have been and are almost always carried out at liquid nitrogen, or liquid helium, temperatures, so the thermal problems are much reduced compared with those that affect most available X-ray data. (There is current interest in He cooling for X-ray work, but it is still under development.) What kind of difference does this actually make? An interesting set of experiments was carried out and discussed by Coppens[13a] to try to answer that question.

In much of theoretical chemistry, it is convenient to consider the valence electrons and the core electrons of atoms separately. Similar division of the electrons into these two groups has proven to be very instructive in X-ray crystallography, as well. Crystallographers routinely refer to "high-angle data" and "low-angle data" with reference to the scattering angle of the X rays (the numerical values of $\sin(\theta/\lambda)$. What has been found is that when one looks at high-angle data, the center of electron density is usually located at a somewhat different point than when one looks at low-angle data. The data actually come from a continuous range of angles, so the division is somewhat arbitrary, but nonetheless quite useful. What is found is that the high-angle X rays tend to go right through the valence shell and are scattered mainly by the core electrons, whereas the low-angle X rays are scattered more strongly by the valence electrons. This observation led to a "two-atom model" of an atom, where the valence electrons are represented as being at one position and the core electrons at a different position, where each of these positions is optimized independently. Obviously, one would expect that the center of the core electron density should be spherically distributed around the nucleus, whereas the distribution of the valence electron density would be much more dependent on just what kind of bonding was present in the part of the molecule concerned. And it is generally found that one's expectations are borne out. The high-angle data give core electron distributions that correspond more closely to the nuclear positions than when all of the data are used in the structure refinement. The low-angle data show the electron distribution of the valence electrons in a generally well-anticipated way.

If one is interested in studying the structure (in a geometric sense) of a molecule, of course, the high-angle data are more useful because they define the atomic positions much more accurately. On the other hand, if one is interested in studying bonding, the low-angle data are more useful because that is exactly what they show (as the title of Coppens' book indicates). Ordinary X-ray data (a mixture of low- and high-angle data) show overall electron density and, hence, require some interpretation, depending on the interests of the user. Just how big are these effects? We can summarize by saying that they are pretty big.

We might consider as an example the molecule tetracyanoethylene and also the epoxide formed from that molecule, both of which are discussed by Coppens[13a] and are shown in Figure 2.7.

In looking at the discrepancies between the X-ray atomic positions, and those determined by neutron diffraction in the case of the oxide, the X-ray position of the oxygen is shifted relative to the neutron position as shown by the arrow in Figure 2.7 by 0.013(4) Å. In the parent nitrile, the nitrile carbon positions are shifted (X ray relative to neutron) in the direction shown by the arrow by 0.0085(15) Å.

The epoxide information is straightforward. Electron density from the lone pairs is simply located on oxygen as expected, somewhat out away from the nucleus so that the center of electron density for the oxygen atom is shifted by 0.013 Å. The nitrile

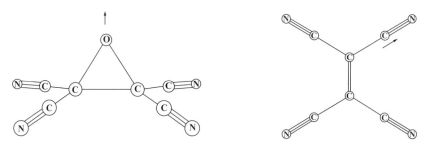

Figure 2.7. X-ray vs. neutron structures for tetracyanoethylene and its oxide.

case is a little more complicated. Certainly, one expects that the lone pair on nitrogen in the sp orbital will be away from the attached carbon, and along the C–N line, and that should lead to a shift of the total nitrogen electron density in that direction. But it's not that simple. Indeed, electron density from the lone pair does tend to shift the apparent nitrogen position in that direction. However, the carbon to which the nitrogen is attached is joined by a triple bond to nitrogen. The triple bond means that three electrons, one each from the σ orbital and the two π orbitals, are pulled from the nitrogen into the C≡N bonding area by the bonding interaction, and this pulls the center of electron density on nitrogen back toward the carbon. It seems that the shift in electron density from these three bonds away from nitrogen just about cancels out the density from the lone pair of electrons on the other side, so that the nitrogen essentially does not move, when one compares the neutron and X-ray positions. But the nitrile carbon is attached to the nitrogen by a triple bond. On the other side, that carbon is joined by only a single bond to another carbon. Thus, the electron density of the carbon attached to the nitrogen moves in the direction indicated. And it moves significantly, 0.0085 Å, because the triple bond to nitrogen involves three electrons, versus only one on the single-bond side.

It should be clear from these examples that X-ray data can be utilized to study bonding in molecules, and it can be very informative. However, if one is concerned with the specific question of the structures of molecules, as defined by their nuclear positions, it should also be evident that X-ray data require much more interpretation than do electron diffraction or microwave data, and one can easily be misled by a superficial examination of information obtained by X-ray crystallography. Neutron diffraction, on the other hand, avoids much of the problem here, although it still suffers from the fact that the structure is determined in a crystal lattice and that it will show thermal motion. If neutron data are lacking, the separation of high- and low-angle X-ray data can give similar information, although less cleanly.

It might be added that the scattering of different atoms by X rays depends on the atomic numbers of the atom (actually upon the electron densities). But with neutron scattering, the amount of scattering is dependent on something referred to as the "neutron cross section." This is not related to atomic number in any simple way. In particular, it is found that hydrogen, which scatters X rays very weakly because of its low atomic number, is quite a strong scatterer of neutrons. Hence, the positions of hydrogens can be located from neutron diffraction with reasonable accuracy, whereas from X rays, the scattering is too weak to yield more than an approximate position for the hydrogen atom.

Nuclear Magnetic Resonance Spectra

Although we will not discuss this topic in detail in this book, we want to mention that it is possible to determine accurate structures of molecules (in liquid crystals) utilizing NMR spectra.[14a,b] The more general NMR methods are, of course, well known to organic chemists and widely used for solving all sorts of structural problems.[14c] Extending these methods to liquid crystals and using distance geometry,[14d] it is possible to determine complete accurate molecular structures. The accuracy of the method is quite high, comparable to microwave methods, although the structure obtained is that of the molecule in the liquid crystal, and not exactly that of the isolated molecule. The structural distortions from the isolated molecule are generally small and can be corrected for.[14b]

BOND LENGTHS DEPEND ON METHOD USED TO DETERMINE THEM

The ancient Greeks were very good at philosophy, which is the forerunner of "theoretical science." But this got them only so far. In the 1700s, scientists in general, and chemists in particular, began increasingly to use experimental methods to decide how nature really worked. Chemistry subsequently became known as an *experimental science*. And it has been like that, and still is, in the eyes of most chemists. Theoretical chemists have long studied the subject, but until relatively recently they were more concerned with using theory to understand the results of experiments, rather than to try to predict new things. In the 1960s many chemists, including the present author, began using chemical theory and computational chemistry for large-scale predictive purposes.

Chemistry is defined as a study of the elements and their properties. Most of the time chemists are interested not so much in the elements themselves but rather in molecules, which are combinations of elements into discreet units. The organic chemist normally thinks of a molecule by reference to a rather simple model, usually of the ball-and-stick type. Such models are extremely useful for aiding in the visualization of three-dimensional interactions between molecules and between parts of molecules. But they are also a little bit misleading. A real molecule is not static, the way the usual mechanical model is. Rather, it is undergoing rapid motions, both internally and externally. Molecular structure can be defined at different levels, which are useful for different kinds of purposes. As we go past the ball-and-stick model, to the hard sphere model (space-filling model), and then to the soft sphere model (computational chemistry), we obtain an increasingly accurate description of the molecule and its properties.

Most organic chemists are more interested in chemical reactions than they are in molecular structures. But the two are intimately related. The molecular structure of a molecule is more simple than the chemical reaction of two molecules, so let us start with the more simple case. What, exactly, is a molecular structure and how do we determine it?

A molecular structure, at a minimum, consists of the molecular formula of the molecule, together with a list of the coordinates of each atom in the molecule. The structure for a given molecule can be determined by a variety of ways, some of them experimental and some of them computational. But it is important to note at the outset that the structure that is determined for a particular molecule depends to a considerable extent on just what method is used to determine it. The reason for this is that the molecules undergo an internal vibrational motion. What that means is that the atoms do not occupy fixed points, as in the mechanical model, but they are moving about some average point. But alas, the location of that average point depends on just what kind of method is used to determine it. The way that the average motion is expressed is a function of just what kind of experiment or calculation was used to determine the molecular structure. Many organic chemists will think that the vibrational motion is quite small and can be ignored. For many purposes that is true, but sometimes it is not. Hence, it is important to know something about the different methods used to determine structures, and how they relate to one another, in order to understand what is relevant for the particular case at hand.

There are four different methods that are in common use for determining accurate molecular structures experimentally, as discussed earlier in this chapter. There are also four different kinds of computational methods that are commonly used to determine molecular structures, as will be discussed in Chapter 3. Each of these methods will give a certain structure for a particular molecule. The computational methods usually give some approximation to the same structure (the equilibrium structure). In practice, these structures differ because of different computational errors in the different methods. But the experimental methods each give somewhat different structures from one another, apart from experimental errors, and all of the experimental structures differ from the computational structure. Thus, we might say that a typical C–C single-bond length is 1.54 Å, which has an accuracy of 0.02 Å or so. If that is sufficient accuracy, then all of the methods mentioned will give that number, give or take a hundredeth or two of an angstrom. But most of these methods can actually measure or calculate that bond length in a given molecule to within a few thousandths of an angstrom. Hence, numerous studies have been carried out to try to understand (accurately) just why a particular bond length is the way that it is because there is often a considerable amount of information in that last decimal place. Hence, if we want to find out as much as we can about a given system, we need to be cognizant of these different ways of determining bond lengths (or of atomic positions in general), so that we can more fully utilize this available information for chemical purposes.

A few examples can be given here to outline some of the complexities that come into play when one wants to know an accurate bond length. As an overview, recall that if the two atoms that are bonded together to define the bond length have their positions calculated quantum mechanically and accurately, one will be at the minimum energy position on a Morse curve (Fig. 2.8), and that distance is referred to as the *equilibrium bond length* (r_e). But for any experimental number, the vibrational motion of the atoms comes into play. If the motion were a simple harmonic motion, then the vibrational amplitudes would not change the average atomic positions. But since the motion is always anharmonic, the further up the potential well one goes, the longer the bond length becomes.

This can be illustrated by using the C–H bond length as an example. Because of the light mass of the hydrogen, the lowest vibrational level is relatively high above the equilibrium energy level. The vibrational amplitude is therefore large (almost 0.1 Å), and the anharmonicity is quite significant. If one compares the r_e bond length in methane, for example, with the r_g bond length obtained by electron diffraction, the values are 1.089 and 1.107 Å respectively, a difference of 0.018 Å. Because the levels

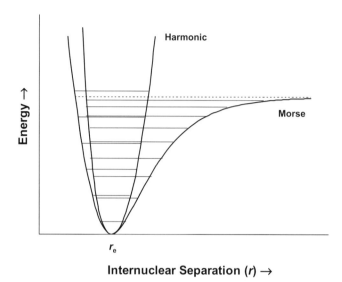

Figure 2.8. Morse and harmonic potentials for bond stretching.

are far apart in this case, one does not have to worry about the contributions from the excited vibrational levels. But suppose one looks at the C–D bond length? The potential function is the same. The only difference is that the deuterium nucleus is approximately twice as heavy as that of hydrogen. This means that the lowest vibrational level lies much lower down in the potential well, and the C–D bond length (r_g) is 1.102 Å, some 0.005 Å shorter than that of C–H. The r_e lengths are, of course, the same. The C–H bond also has a much higher vibrational frequency than the C–D bond (2888 versus 2178 cm^{-1}, respectively).

There are some additional points of interest here. Notice that the relatively small change in the bond length from C–H to C–D (of 0.005 Å) is accompanied by quite a large change in the vibrational frequency (710 cm^{-1}). This means the C–D stretching frequency is well removed from the C–H band envelope that occurs in most organic molecules about 3000 cm^{-1} in the infrared spectrum. Hence, if one wishes to study a particular hydrogen in a complex molecule, if it can be replaced by deuterium, the infrared spectrum will probably be an excellent diagnostic tool for such studies. This variation of stretching frequency with bond length has been studied in some detail for C–H bonds[15]. Changes in bond lengths that are quite small lead to large changes in vibrational frequencies. Hence, an accurate way to measure C–H bond lengths can be to simply measure their vibrational frequencies, which is usually very much easier to do. This relationship between bond length and stretching frequency is ubiquitous and can be quite useful. It will be discussed further in Chapter 4.

It is also perhaps of interest that the small bond length difference obtained by going from C–H to C–D is sufficient to induce other changes in physical properties. For example, molecules having the type of structure shown in **A** and **B** are nonplanar because of the puckered cyclohexadiene ring in the center:

[Structure A: 4,5-dimethylphenanthrene with CH₃ CH₃ groups] [Structure B: same with CD₃ CD₃ groups]

A   **B**

The methyl groups are accordingly above and below the general plane of the molecule, leading to the existence of enantiomers for such molecules. The rates of racemization of **A** and **B** have been measured,[16] and, because of the shorter C–D bond lengths, the two methyls interfere less seriously with one another in **B** than in **A**, and the racemization rate is, therefore, significantly faster for **B**, as is well reproduced by molecular mechanics.[17]

In the 1960s, it was noted by various workers that when one determined structures by the microwave method, and by the electron diffraction method, on simple molecules where the accuracy was well known for both measurements, there were significant disagreements between the same bond length measured by the two methods. The disagreement was systematic. That is, the microwave bond length was always shorter than the electron diffraction bond length for the same bond in the same molecule and sometimes by more than the sum of the experimental errors. Obviously, experimentalists in both fields were concerned about this and began asking, "what is the *true* bond length and how does it differ from what we are measuring?" In an effort to answer this question, a symposium series was founded, mainly under the leadership of James E. Boggs, at the University of Texas at Austin. Various electron diffraction and microwave people (two different groups, generally speaking) met to discuss the situation and try to and resolve the question. As a historical note, there were more than 20 of these symposia, held biannually in Austin, Texas, and they not only solved the problem mentioned but they additionally contributed enormously to our understanding of molecular structure.

The basic problem stemmed from the fact that molecules vibrate, and that vibrational amplitudes for most bonds are fairly large (approaching 0.1 Å), and they are anharmonic. If one is going to measure bond lengths to an accuracy of something like 0.002 Å, one needs to somehow average the atomic positions over the vibrational motion. But, as it turns out, some of the averaging that is carried out in the experiments differs, depending on what type of experiment it is that one is doing.

There are currently in wide use at least five different kinds of bond lengths that we will discuss here (Table 2.1) and later in this book. One of these corresponds to the point on the potential surface at the bottom of the well and is called r_e (the equilibrium bond length). This is the quantity that is usually calculated quantum mechanically. It is not observable directly because the atoms never remain at rest. The other four are experimentally determined bond lengths. The vibrational motions are averaged in different ways by different experimental methods, and hence the same bond length has

TABLE 2.1. Bond Lengths of Some Typical Bonds (Å)

Bond	Compound	r_e	r_g	r_z	r_s	r_α
Butane	C_1–C_2	1.530	1.537	1.535	1.532	1.534
Propanol	C–O	1.423	1.429	1.425	1.424	1.424
2-Butanone	C=O	1.208	1.213	1.210	1.209	1.209
2-Butene	C=C	1.337	1.342	1.339	1.338	1.339
Acetonitrile	C≡N	1.153	1.157	1.152	1.153	1.151
2-Butyne	C≡C	1.209	1.213	1.209	1.210	1.208
Chloromethane	C–Cl	1.774	1.781	1.780	1.777	1.779

different values, depending on the method used to measure it. They are typically determined by the experimental methods indicated: r_g from electron diffraction, r_z from microwave spectroscopy, r_s from microwave spectrocopy when isotopic substitutions are used, and r_α from X-ray (or neutron) crystallography. There are less commonly used bond lengths as well, and for information on these, or additional information on the above, one may consult the book by Cyvin.[18] For historical accuracy and completeness, Kozo Kuchitsu was the major contributor to the solution of the interrelationship between the different types of bond length problems. This is documented in the book by Cyvin and also by numerous later studies.

Let us first make a cursory examination of Table 2.1. Take the simple molecule butane and consider the bond length of the terminal carbon–carbon bond, as in the first entry. The MM4 molecular mechanics program has been developed to fit insofar as possible, and as accurately as possible, all of the experimental and computational data on many small molecules, including those in Table 2.1. The bond lengths shown in the table were all calculated by MM4 for the sake of consistency. And these numbers all agree well with the various available experiments and theoretical calculations. Note that for the C_1–C_2 bond for butane (the first entry), r_e has a value of 1.530 Å. The other butane r_x values are all slightly larger.

The r_g value is ordinarily the longest bond length because of the averaging that is used in electron diffraction measurements. The value of r_z is normally smaller than that for r_g, although in butane it is by only 0.002 Å. And, as shown in Table 2.1, the total bond length range for this C_1–C_2 bond over the various types of r values spans 0.007 Å. While this is not very much, it is well beyond the accuracy of good experiments and good calculations, and so one may or may not want to worry about it, depending on the problem at hand.

In Figure 2.8 is shown a Morse curve, which is well known to be a rather accurate description of the variation of the energy of a bond with the bond length. The equilibrium bond length (r_e) is the bond length at the minimum of the curve. It is the length the bond would have if we could freeze out the vibrational motion. This is the value one would obtain from a quantum mechanical calculation using an infinite basis set and an infinite amount of electron correlation (the Schrödinger limit). But because the Morse curve is anharmonic, the vibrational levels tend to stretch out more toward the right-hand side of the curve the higher we go. The harmonic approximation to the Morse curve is also shown, and it can be seen that even the zero-point vibrational level has its average position shifted very slightly in the latter. For a carbon–carbon bond, at room temperature, the population of butane bonds is almost completely confined to this first level. At higher temperatures, or with bonds involving heavier atoms, or weaker bonds, there may be greater or lesser population of the higher levels. But in any case, the higher up in the well we go, the longer the bond becomes, and the poorer the harmonic approximation.

Except for r_e, the other r values in Table 2.1 are all averaged in some way over the vibrational motion of the bond. The symbol r_g has been used to represent bond lengths in electron diffraction since about 1970. A slightly different definition was used prior to that time (r_a) and is found in the older literature. It is found that r_g is always approximately 0.002 Å larger than r_a.

The r_z is the usual way of defining the bond length in microwave work, although r_s is obtained if the isotopic substitution method is used. Finally, the r_α value is from X-ray diffraction (or other work in crystals, such as neutron diffraction).

Note that the next entry in the table, the C–O bond in propanol, starts somewhat shorter than the C–C bond in butane, and has its smallest value for r_e and its largest for r_g, with the others in between. The last entry, chloromethane, again is found to show r_g as the longest bond and r_e as the shortest, with a difference of 0.007 Å. So this is a rather common pattern for ordinary single bonds in organic molecules.

If we look at multiple bonds, since they are shorter and stronger than single bonds, we would expect that their potential wells are deeper and narrower. Their vibrational amplitudes would, therefore, be smaller, and hence the difference between r_e and r_g would be expected to be somewhat smaller. If we look at the C–O bond in 2-butanone (the third entry), for example, the r_e and r_g values are 1.208 and 1.213 Å, respectively, with a difference of 0.005 Å. For acetonitrile, this difference is only 0.004 Å, so the difference between the double bond and the triple bond is only marginally discernable.

We can see the general trends of the variations in bond lengths with the experimental techniques from the simple cases described in Table 2.1. Since bond lengths are rather strongly dependent on the molecular environment in which the bond is located, these bond length variations are similarly environment dependent and have to be calculated on a one-at-a-time basis. The methods described by Kuchitsu and Cyvin[18] were programmed by Hedberg and Mills[19] in 1993. All of this, plus some additional extensions,[20] are included in the MM4 program.[21] The work by Kuchitsu was done in the days of hand calculations, and to make the calculations practically useful, a number of simplifying approximations were made. These approximations were unnecessary for computer calculations, and they were omitted in the MM4 version.

REFERENCES

1. (a) A. Domenicano and I. Hargittai, Eds., *Accurate Molecular Structures*, Their Determination and Importance Oxford University Press, Oxford, 1992. (b) J. Laane, M. Dakkouri, B. van der Venken, and H. Oberhammer, Eds., *Structures and Conformations of Non-Rigid*

Molecules, Kluwer Academic, Boston, 1993. (c) I. Hargittai, *Struct. Chem.*, **16**, 1 (2005). (d) The catalog of structures from diffraction/microwave spectroscopy is the Landholdt Bornstein Series, which contains volumes under the heading of *Structure Data of Free Polyatomic Molecules*. This is an ongoing series of books (tables) in which the detailed structures determined by the methods mentioned can be found, together with a small number of these structures determined by less commonly used methods. The latest volume published as of this writing is the New Series II, Vol. 28C, Springer, Berlin, 2007. Volume 28D is about to appear.

2. G. A. Sim and L. E. Sutton, Senior Reporters, *Molecular Structure by Diffraction Methods*, Vols. 1–3, Chemical Society, London, 1973, 1974, 1975.
3. J. Karle and I. L. Karle, in *Determination of Organic Structures by Physical Methods*, E. A. Braude and F. C. Nachod, Eds., Academic, New York, 1955, p. 444.
4. Y. Morino, S. J. Cyvin, K. Kuchitsu, and T. Iijima, *J. Chem. Phys.*, **36**, 1109 (1962).
5. J. Kraitchman, *Am. J. Phys.*, **21**, 17 (1953).
6. J. P. Glusker and K. N. Trueblood, *Crystal Structure Analysis A Primer*, 2nd ed., International Union of Crystallography, Oxford University Press, Oxford, 1985.
7. W. Clegg, A. J. Blake, R. O. Gould, and P. Main, *Crystal Structure Analysis—Principles and Practice*, International Union of Crystallography, Oxford Science Publications, Oxford, 2001.
8. Cambridge Structural Database, maintained by the Cambridge Crystallographic Data Center, Cambridge, England.
9. J. Karle and H. Hauptman, *Acta Cryst.*, **3**, 181 (1950).
10. J. P. Glusker and A. Domenicano, in *Accurate Molecular Structures*, A. Domenicano and I. Hargittai, Eds., Oxford University Press, Oxford, 1992, p. 126.
11. (a) D. W. J. Cruickshank, *Acta Crystallogr.*, **9**, 747 (1956), and following papers. (b) K. N. Trueblood, in *Accurate Molecular Structures*, A. Domenicano and I. Hargittai, Eds., Oxford University Press, Oxford, 1992, p. 199. (c) W. R. Busing and H. A. Levy, *J. Chem. Phys.*, **26**, 563 (1957).
12. C. K. Johnson, *ORTEP—A FORTRAN Thermal-Ellipsoid Plot Program for Crystal Structure Illustrations*, Oak Ridge Natl. Lab., Oak Ridge, TN, Avail. CFSTI (1965) (AEC Accession No. 33516, Rept. No. ORNL-3794.). From *Nucl. Sci. Abstr.* **19**(17), 4153 (1965). Report written in English. CAN 64:98554 AN 1966:98554 CAPLUS [Copyright © 2007 ACS on SciFinder (R)].
13. (a) P. Coppens, *X-Ray Charge Densities and Chemical Bonding*, International Union of Crystallography, Oxford University Press, Oxford, 1997, p. 51, Table 3.1. (b) P. Becker, P. Coppens, and F. K. Ross, *J. Am. Chem. Soc.*, **95**, 7604 (1973).
14. (a) A. E. Torda and W. F. van Gunsteren, *Molecular Modeling Using Nuclear Magnetic Resonance Data, Reviews in Computational Chemistry*, Vol. 3, K. B. Lipkowitz and D. B. Boyd, Eds., Wiley, New York, 1992, p. 143. (b) P. Diehl, *Nuclear Magnetic Resonance Spectroscopy and Accurate Molecular Geometry*, in *Accurate Molecular Structures*, Eds., A. Domenicano and I. Hargittai, Eds., Oxford University Press, Oxford, 1992, p. 299. (c) J. B. Lambert and E. P. Mazzola, *Nuclear Magnetic Resonance Spectroscopy, An Introduction to Principles, Applications, and Experimental Methods*, Pearson Education, Upper Saddle River, NJ, 2004. (d) J. M. Blaney and J. S. Dixon, *Distance Geometry in Molecular Modeling, Reviews in Computational Chemistry*, Vol. 5, K. B. Lipkowitz and D. B. Boyd, Eds., Wiley, New York, 1994, p. 299.

15. (a) D. C. McKean, J. L. Duncan, and L. Batt, *Spectrochim. Acta*, **29A**, 1037 (1973); D. C. McKean, *Spectrochim. Acta*, **31A**, 861 (1975); D. C. McKean, J. E. Boggs, and L. Schäfer, *J. Mol. Struct.*, **116**, 313 (1984). (b) H. D. Thomas, K. Chen, and N. L. Allinger, *J. Am. Chem. Soc.*, **116**, 5887 (1994).
16. K. Mislow, R. Graeve, A. J. Gordon, and G. H. Wahl, Jr., *J. Am. Chem. Soc.*, **86**, 1733 (1964).
17. N. L. Allinger and H. L. Flanagan, *J. Comput. Chem.*, **4**, 399 (1983).
18. K. Kuchitsu and S. J. Cyvin in *Molecular Structures and Vibrations*, S. J. Cyvin, Ed., Elsevier, New York, 1972.
19. (a) L. Hedberg and I. M. Mills, *QCPE Bull.*, Program QCMP 128, **13**, 37 (1993). (b) L. Hedberg and I. M. Mills, *J. Mol. Spectroscopy*, **160**, 117 (1993).
20. B. Ma, J.-H. Lii, K. Chen, and N. L. Allinger, *J. Am. Chem. Soc.*, **119**, 2570 (1997).
21. The MM4 program is available to all users from Dr. J.-H. Lii, Department of Chemistry, National Changhua University of Education, No. 1, Jin-De Road, Changhua City 50058, Taiwan, jhrobert.lii@gmail.com

3

MOLECULAR STRUCTURES BY COMPUTATIONAL METHODS*

A BRIEF HISTORY OF COMPUTERS

We might begin this discussion with the following question: Why do you want to calculate something? The answer to this question, as given by Lord Kelvin, president

---

*Two general references on the subject of computational chemistry will be cited here, as they will be useful to just about everyone who is interested in this kind of work. The first of these is *Encyclopedia of Computational Chemistry*, P. v. R. Schleyer, N. L. Allinger, T. Clark, J. Gasteiger, P. A. Kollman, H. F. Schaefer III, and P. R. Schreiner, Eds., Wiley, Chichester, UK, 1998. This set of five volumes contains a discussion of just about every imaginable topic on the subject. The topics are listed in alphabetical order. However, a topic may not be listed under itself but may be considered a subtopic of something else. If what is desired is not found, a perusal of the extensive index at the end of Volume 5 is recommended and may yield the location of the desired information. For example, much of what one might look for under Molecular Mechanics is not listed there but will be found under Force Fields.

The second general reference is D. C. Young, *Computational Chemistry. A Practical Guide for Applying Techniques to Real World Problems*, John Wiley and Sons, New York, NY, 2001. The last part of the title defines the scope of the book. Enough theory is given for most chemists to understand what it is that they are going to do (if they didn't already know), but that is not the point. The point is to solve real problems by available methods. The discussion concentrates on what methods are available, how they are used, how well they work, and their strengths and weaknesses.

Molecular Structure: Understanding Steric and Electronic Effects from Molecular Mechanics,
By Norman L. Allinger
Copyright © 2010 John Wiley & Sons, Inc.

of the Royal Society, in 1895 is thought provoking. He stated: "If you cannot calculate something accurately, you probably don't understand it very well."

We might look at the definition of the word *computer*. In 1950, the dictionary definition read as follows: "A person (usually a woman) who computes (does calculations)." Earlier, the *computer* was usually armed with a table of logarithms. But in the 1940s, desk calculators became available. These were substantial mechanical devices, approximately 1 foot square, and 3 inches tall, and they weighed perhaps 10–15 pounds. They typically sat on a desk and so were referred to as *desk calculators*. These machines carried out the four principle arithmetic functions: add, subtract, multiply, and divide. They were electrically powered, although hand-cranked "adding machines" were in use prior to that. They were also very noisy, and to multiply two large numbers together might have required 5 seconds or so. They were the forerunners of the *hand calculator* of today. The hand calculator, especially one that could do logarithms or take square roots would have been regarded as a miraculous invention in the 1940s. The meaning of the word *computer* evolved from the pre-1950 period from the person into the name of the electronic device that we know today.

From one point of view, computers are hundreds of years old. Modern electronic computers were, from the 1950s into the 1980s, largely driven by cards, with holes punched in them. These cards contained instructions, and the computer carried out those instructions. It seems that there were long ago in the Mideast, particularly in Persia, machines designed to weave carpets that similarly took instructions from punched cards. From the instructions, the machine put the various colored threads in the appropriate places, and the threads went into the loom and generated the desired pattern in the product. Such machines are said to have existed at least since the 1700s, and they certainly bore a significant resemblance to computers of the 1950 period. And there were mechanical computing machines in the latter half of the nineteenth century. They would certainly have been fast and accurate compared to hand calculations, but for whatever reasons, they were never widely used.

Apart from the above, the first computer in the modern sense of the word (electronic computer) appears to have been developed at Bletchley Park in Buckinghamshire, England, specifically for solving the problem of decoding the complicated codes generated by the German military during World War II.[1a] It was given the name Colossus, and a group of British intelligence code breakers utilized it to crack the famous Enigma code of the German military. Their success with this was not known to Germany until after the war had ended, and so for quite a long time, the code breakers were able to read many of the messages being sent in the supposedly secret code from one German military group to another. The British group utilized in part what you might call a *brute force technique* to solve this problem. That is, one simply does trial-and-error testing of various combinations of letters until one gets something that makes sense. The Germans assumed that this would be impossible with the Enigma code, and by hand work it more or less was impossible. What made the Enigma code so difficult to solve was the fact that it depended on a machine, and both the sender and receiver had to have one of those machines. And there was a key number that had to be set by the receiver, to match the key number set by the sender. And there were millions of choices. So what this meant was that solving the code in any one message was of no help at all for the next message because the key used would be different. So while the code could be broken, and a given message solved (if it were sufficiently long, and given enough time), every message would be a new problem. And if it took a month, say, to solve each message, this would not be militarily useful. But the primitive computer that was invented for the purpose was enormously faster than a human for this kind of work. The result was that intercepted messages were being deciphered in a matter of hours.

Part of the potential importance of computers was recognized very early, although the only use foreseen originally was for calculations that were intensely numerical. For example, when asked about a potential usefulness of computers, Thomas Watson the chairman of IBM, stated in 1943, "I think there is a world market for maybe five computers."

In the late 1940s it became fashionable for universities and others to build their own computers. These were typically enormous machines that utilized vacuum tubes and miles of wire, and they generated a lot of heat. For example, in the magazine *Popular Mechanics*, in forecasting the relentless march of science, the statement was made in 1949 that "computers in the future may weigh no more than 1.5 tons." Such computers were very primitive by modern standards, but they were lighting fast when compared with the desk calculator (or a table of logarithms). The first commercially available computer was built by the Remington-Rand Corporation and was delivered in 1951 to the Census Bureau, which utilized it for processing the massive amounts of data just collected in the 1950 census.

Surely the invention and development of the electronic computer should be regarded as one of the most important developments in human history. This recalls the statement made by Commissioner Charles H. Duell of the U.S. Office of Patents in 1899. He recommended that the patent office be closed because "everything that can be invented has been invented." Fortunately, Congress reacted by doing what they often do—they did nothing. For further information on the history of the giant computers, the book by Waldrop may be examined.[1b,c]

In Table 3.1 are listed some specific computers and some of their characteristics. These are machines that happen to be familiar to this author. There were several computer manufacturers in the 1950s and 1960s. Most of these companies no longer exist, at least not under the same name, with International Business Machines (IBM) being a notable exception. Other companies of the time included Westinghouse, Burroughs, Sperry-Rand, Control Data Corporation, and Digital Equipment. Later arrivals on the scene were Apple, Cray and Amdahl, Gateway, Hewlett-Packard, Dell, and Compaq. Of these, IBM was probably the most important. Its machines were the most widely used over the longest time period, and more scientific progress overall seems to have been contributed by IBM. The several early machines listed in Table 3.1 are each more or less representative of a milestone in computer development.

From the point of view of the computational chemist, Digital Equipment deserves special mention. The machines listed prior to the mid-1980s were so-called *mainframes*, very expensive machines housed in *computer centers*.* Typically, a university or large

---

*Computer center is a term used in the southern part of the United States beginning in the 1950s. The northern equivalent was computing center.

TABLE 3.1. Machine Performance Test

Year	Machine	OS	Memory	Disk Storage	CPU Time[a] (sec)
1952	Giant	—	—	—	6×10^8
1956	IBM 650	—	—	2 k bytes	3×10^7
1960	IBM 7040	—	10 k bytes	—	3×10^5
1967	IBM 360 total/single user	JCL/OS	512 k bytes/100 k bytes	29 M bytes/1 M bytes	3×10^4
1985	DEC MicroVAX II	VMS	8 M bytes	110 M bytes	2581
1990	IBM RISC6000 340	AIX	64 M bytes	1 G bytes	313
1998	DEC Alpha 21164/500 MHz	UNIX	512 M bytes	10 G bytes	9
2007	PC Core 2 Duo/1.66 GHz (Intel)	DOS	2 G bytes	200 G bytes	2

[a]Central processing unit (3×10^7 sec = 1 year).

company had one such machine, and it was shared by dozens, later perhaps hundreds, of users. There were available, to be sure, small (less powerful) computers that were used for various applications, for example, data processing in analytical chemistry. However, the VAX machines of Digital Equipment were the first less-than-mainframe machines that were powerful enough for routine work in computational chemistry. The VAX machines were also priced at about one-tenth the price of the mainframe (roughly $100,000 versus $1,000,000) and, hence, began to spread into university chemistry departments and other similar sized organizations.*

The rapid development of ever smaller, faster, and cheaper computers, especially after the introduction of the IBM Personal Computer (PC) (1980s), really created a major change in our society, as is, of course, now well known to all.

The computers in Table 3.1, except for the first, were all used by this author's research group. Each machine is identified by number, and the first year of use by the author is given. The next column gives the name of the operating system (OS). The amount of memory and/or disk space is then given in kilobytes, megabytes, or gigabytes. The relative times that these machines required for a specific job are then given.

The machine we called Giant was a homemade computer at Wayne State University when this author went there in 1956. It filled a large room that had been a gymnasium and was a vacuum tube machine, where the tubes were the size of one's fist. Its usage was being phased out at that time, and it was not actually used by this author.

The first computer used by this author was the IBM 650, beginning in 1957. The machine was relatively new, at least for Wayne State University, at that point. The

*Perhaps it should be added that most university chemistry departments did not have $100,000 lying about that could be used to purchase one of these machines. Universities typically receive large discounts from manufacturers on various kinds of equipment because there is an enormous sales advantage for a manufacturer to having large numbers of students graduating and coming into the labor market that had been trained on its machine rather than on a competitor's machine.

machines listed after 1961 were all binary machines. That is, the arithmetic they used internally involved binary numbers (only 0's and 1's). The IBM 650 used duodecimal arithmetic. It was also a vacuum tube machine, but here, the vacuum tubes were reduced to the size of one's thumb, and here is where this author learned about *preventive maintenance*. The tubes (probably a few thousand of them) were mounted and wired into trays, probably about 100 tubes in a tray, and the tray was perhaps about 2 feet square. Unfortunately, vacuum tubes tended to burn out with use. When one had thousands of vacuum tubes in the computer, such burnouts and the required replacements shut down the computer and were very wasteful of time. Accordingly, vacuum tube manufacturers had good quality control over their product and knew pretty well how long one could expect the vacuum tube to last. (The main reason they burned out was because as they emitted electrons from a hot wire, the wire slowly vaporized.) The entire tray of vacuum tubes was accordingly replaced at some fixed time before that point was likely to be reached, so that shutdowns because of tube burnouts were not frequent. Nonetheless, electronic machines of that era, which contained wires with many soldered joints, were subject to rather frequent failures. This machine was lightening fast, compared to calculations using a desk calculator, and required perhaps a few minutes (largely for reading and punching cards) to do what otherwise took all day. (The desk calculator was also very fast compared to tables of lagorithms, which this author had used as a student not many years before.) Thus, *Popular Mechanics'* estimate that "computers in the future would weigh no more than 1.5 tons" was accurate. The IBM 650 consisted of two units, each of which was about 3 feet wide, 6 feet high, and 7 feet long, plus another desk-sized unit for reading and punching cards. These units were joined by cables. I do not know how much they weighed, but I have the impression that it would have been the equivalent of two upright pianos, plus a desk, and somewhat less than 1.5 tons.

These earlier machines did not have random access memory, which had not been invented yet. Data were stored on a rotating drum, so they could only be accessed sequentially. The first memory became available to us for the IBM 650 in about 1959. The memory device was the size of a desk and had a capacity of 12 bytes, but it sped up the calculations by perhaps a factor of 2 or 3. (Yes, that is bytes, not kilobytes or megabytes.)

The usefulness of memory became evident by 1960. Just how much memory would be useful was, however, not immediately clear (see Table 3.1), but for sure, more was better. Bill Gates, the chairman of Microsoft, made the statement in 1981, "640 K ought to be enough for anybody."

We had a standard compound [actually it was d,l-3,4-di(1-adamantyl)-2,2,5,5-tetramethylhexane(s,s) for those interested] that we ran a geometry optimization on as our test case on our computers after 1984, using the MM3 program to determine relative machine speed for that specific application. The MicroVAX II required 2581 seconds (about 20 minutes) to optimize that particular structure from a given starting geometry. The speeds of the earlier machines were extrapolated back from our general knowledge of other optimizations since MM3 had not been available earlier. Even if it had, we could not have afforded the cost of running it for this test case. For example, the cost to us of central processing unit (CPU) time on the IBM 360 was about

$400/hour in 1980. (And $400 was a lot of money in 1980.) The cost was, fortunately, reduced considerably for nights and weekends, which led to computational chemists working a lot of nights and weekends.

There has long been something called *Moore's law* in the computer field.[1d,e] Gordon E. Moore, who was a co-founder of the Intel Company, and one of the leading scientists involved in the early development and use of computers, stated in 1965 (paraphrased by the present author) that the speed of computers had increased by a factor of 2 every 2 years for a long time, and would likely continue to do so. It turned out that the speed of computers did continue to increase at approximately that rate for many years thereafter, and that rate still continues today. As long as that increase continues, problems that we cannot solve at present because of computational limitations will gradually become solvable as time goes on. It is a little surprising to most chemists that this "law" has continued in effect for such a long time. This reminds this author that Niels Bohr once said "predictions are difficult, especially when they are about the future."

The technological change in computers over time has been immense. The first computers became commercially available in 1951 (they existed already in the 1940s, but if you wanted one then, you had to build it yourself). Computers originally used vacuum tubes and miles of wire in circuits to transfer electrons, and information, from place to place. Transistors were then invented, and the computers became much more compact. This sped up the calculations considerably because eventually one is dependent upon how fast electrons can flow from point A to point B. Obviously, they flow faster when other things being equal, point A can be moved closer to point B. Then integrated circuits were developed in which the electrons no longer traveled for miles and miles, but they traveled across a small silicon plate, through an etched line. Transistors made from graphene, which is a one-atom-thick sheet of carbon that has a structure similar to chicken wire (see Chapter 5 under C_{60} Fullerene), have the potential to eventually replace the silicon devices in current use. These are expected[2] to yield additional increases in speeds of a factor of 100.

But as computers get smaller, various quantum phenomena began to come into play. So in some cases computers are being hindered by the very laws that we are using them to study. However, it seems that as one way of doing things reaches a physical limit, another way is found to go beyond or around that limit. At least up until now, the process has been unending. An interesting example concerns computer memory, as commonly used today. A "box" either is or is not filled with electrons to give the required 0 or 1 for binary arithmetic. About the year 2000 Bell Labs announced that it had reduced the size of the box to where only a single electron was required to fill it. It seems unlikely that that number will be reduced.

COMPUTATIONAL METHODS

For present purposes, it will be convenient to divide the computational methods used to determine molecular structure into three categories: (1) semiempirical calculations, (2) ab initio calculations, and (3) molecular mechanics calculations.

Before the importance of quantum mechanics in atomic structure had been fully recognized, a classical mechanical description of the hydrogen atom was developed by Bohr[3] in 1913. While this model explained adequately a number of things regarding the hydrogen atom, its extension to more complicated systems (the Bohr–Sommerfeld model[4]) quickly became unmanageable. The quantum mechanical treatment for the hydrogen atom by Schrödinger (1926)[5] involved the solution of Eq. (3.1), which subsequently became known as the Schrödinger equation:

$$H\Psi = E\Psi \qquad (3.1)$$

where Ψ represents the wave function (Ψ^2 gives the electron density as a function of the spatial coordinates, E is the energy of the system, and H is called the Hamiltonian operator).

The solution of this equation for the hydrogen atom alleviated the problems with the classical mechanical treatment, and suggested a much more general solution, that in principle would be extendable to larger atoms and then eventually to molecules. The hydrogen molecule proved to be manageable,[6] but the calculations quickly bogged down with the apparently insuperable difficulty created by 3- and 4-centered electronic integrals. The situation prompted the following remark by Paul Dirac: "The underlying physical laws necessary for the mathematical theory of a large part of physics and the whole of chemistry are thus completely known, and the difficulty is only that the exact application of these laws leads to equations much too complicated to be soluble."[7] These 3- and 4-centered integrals constituted a problem that was still not adequately dealt with into the 1960s. Various methods were used to get around the problem that these functions simply could not be integrated, but they were painfully slow and/or inaccurate.

SEMIEMPIRICAL QUANTUM MECHANICAL METHODS

Then people looked harder at semiempirical methods. The thinking was, if we cannot solve the Schrödinger equation generally as we would like to, then let us try to solve it any way that we can, by simplifying problems so as to go around the intractable parts of the calculations, and then make empirical adjustments if we can in some way so as to minimize the errors introduced by the approximations. Thus, one might learn more about the ways that one can solve this equation, and also obtain useful information. Historically, quantum mechanical calculations on organic molecules were first applied in considerable detail to conjugated π systems[8,9] as typified by aromatic hydrocarbons long before they were applied to saturated molecules such as the alkanes. The underlying reason for this was purely computational in nature. The interesting chemistry and physical properties of benzene (and aromatic systems in general) stem from the π electrons. Thus, one might try to consider the π system of benzene separately from the rest of the molecule (σ system). Such an approach was published by Hückel in 1931.[8] When one considers only a *planar* π system, the mathematical complexity of the problem can be very much reduced. In such a planar system the σ part and the π parts are orthogonal, which means that they did not overlap. If they did overlap, then there would be complicated "cross terms," which involve simultaneously the σ and the π parts. (There are now known different ways around this problem, as will be discussed later in Chapter 5.) But most aromatic systems are planar anyway; so let us start with the more simple case.

If one considers only the conjugated π system of benzene, it contains 6 electrons. If one considers the entire benzene molecule, it contains 42 electrons. The 6-electron problem could be dealt with by hand calculation to some reasonable extent, whereas the 42-electron problem was hopelessly complicated. In the Hückel method the interactions of each of the electrons with each of the nuclei are accounted for, but the interactions of the electrons with one another are simply ignored. The results are then empirically adjusted in a systematic way so as to approximately allow for this. Although the method is based on a really severe approximation, it nonetheless yielded a great many interesting and suggestive results regarding molecular structure.[9]

Self-Consistent Field Method

Of course, if one extends the Hückel method to include the interactions between the electrons somehow, one will expect to improve the physical model and hence the results obtained. This latter model as developed is referred to as the *Hartree–Fock* or *self-consistent field* (SCF) model. We will begin by applying this model to planar π systems. The basic idea here is that we do not consider the individual electron interactions as a function of the time, but rather we consider the last electron moving in the average field generated by all of the rest of the electrons. And if we think, in turn, of each electron being the last, then we would have an orbital description of all of the electrons that we want. But how, exactly, would we do this? We want to solve the Schrödinger equation in order to find the wave function for the entire π system. To do this, we need to know the corresponding Hamiltonian [the H in Eq. (3.1)]. But the formulation of that Hamiltonian depends upon knowing the wave function. And neither the Hamiltonian nor the wave function is known at the outset, and each depends upon the other. So the usual procedure is to make an informed guess as to the formulation of the Hamiltonian, based on some standard approximations (that we already know and that have been carried over from related simple molecules), and then solve the equation to obtain an approximate wave function. Then we use that wave function to derive a new Hamiltonian. Then we use that Hamiltonian and again solve for the wave function. And we repeat this process in an iterative manner until the solution "converges." That is, when one gets the same wave function by solving the Schrödinger equation that one assumed and started with in constructing the Hamiltonian. The exact procedure actually used is sometimes referred to as the *Roothaan procedure*, which is a specific formulation of the above ideas.[10]

While these ideas are reasonably straightforward, the SCF procedure was agonizingly difficult to actually carry out in practice by hand calculation. It was, however, ideally suited for computer calculations. Of course, as we know now, the things that computers do best are iterative calculations.

At this point one was faced with the problem of the two-electron integrals. They have the general form as shown:

$$\int_{-\infty}^{-\infty}\int_{-\infty}^{-\infty}\int_{-\infty}^{-\infty} \Psi_a(1)\Psi_b(1)\frac{1}{r^2}\Psi_c(2)\Psi_d(2)\,dx\,dy\,dz$$

where the electrons are (1) and (2), the atomic wave functions are Ψ, and the distance between electrons is r. Thus, there are four atomic orbitals (a, b, c, d) that have to be taken into account, and these were usually written as Slater-type orbitals (STO) [Eq. (3.2)]:

$$\Psi = ae^{-br} \qquad (3.2)$$

where a and b are constants and r is the distance from the nucleus.*

To solve the Schrödinger equation for a molecule, integrals of the type shown must be evaluated. And in a typical calculation there are millions of them. If those four orbitals are all centered on the same atom, or if they are centered on two different atoms, the integration problems are much less complicated than if they are centered on three or four different atoms. There was no good way to calculate the values of these 3- and 4-centered integrals for many years, and this held up the usefulness of such calculations. It was known from early studies on π systems that the approximation of simply omitting the 3- and 4-centered integrals from the calculations, followed by some empirical adjustments (the Pariser, Parr, Pople method), introduced relatively small errors into the calculational results.[11,12] (This kind of calculation is referred to as *semiempirical*.) That was mainly because the 1- and 2-centered integrals generally have much larger values than do the 3- and 4-centered integrals. Some of these details will be considered further in Chapter 5 on conjugated systems. For now, we will simply say that there were precedents from this earlier work that suggested that leaving out these 3- and 4-centered integrals followed by some empirical adjustments might be a manageable way in which to carry out practical quantum mechanical calculations on molecules, as this procedure worked well for conjugated systems. It turned out, however, that omission of these 3- and 4-centered integrals led to poorer results in saturated molecules than it did with π-system calculations.

---

*Historically, the Greek letter zeta (ζ) was used to represent the orbital exponent [b in Eq. (3.2)]. For atoms larger than hydrogen, it turns out that an accurate representation of the orbitals requires a somewhat more complicated expression than a simple Slater orbital. Thus, what were called "double-ζ" orbitals:

$$\Psi = a_1 e^{-\zeta_1 r} + a_2 e^{-\zeta_2 r}$$

were developed and are a significantly better approximation for most purposes. After about the early 1970s, Slater orbitals were rarely used for calculations of chemical systems but were instead approximated by a set of Gaussian orbitals. Generally, it takes several Gaussian orbitals to approximate a single Slater orbital. One still commonly sees in the literature expressions such as "double-ζ quality orbitals were used for the calculation." The meaning of triple-zeta orbitals will be self-evident.

While some attempts were made to parameterize such semiempirical methods to reproduce the results of ab initio methods, Dewar insisted that we should not fit to the results that way, rather we should fit directly to experimental results. His view was that this would be the only approach actually useful to chemists. While ab initio methods might fit to experiment eventually, they did not fit at all well at the lower levels of calculation that were possible at the time. If semiempirical methods could be adequately developed and fit to experiment, perhaps one could really reach chemical accuracy with orders of magnitude with less work than would be required by strictly ab initio methods.

At that time it appeared that while ab initio methods could perhaps eventually be applied to the solution of practical problems, it would apparently be a long time before that would be possible. For those who wanted more immediate action, the semiempirical approach seemed to be the only pathway available. The idea here is that one starts with a correct and proper theoretical method, the ab initio method, but one avoids certain of the difficult problems (particularly the 3- and 4-centered integrals), by simply leaving them out. Of course, one then gets poor answers. But if with the aid of suitable parameterization one can adjust those answers back into the correct or at least reasonable range, then perhaps one has a workable and useful kind of a quantum mechanical calculation. In the most simple case, the method used was referred to as CNDO, which stood for complete neglect of differential overlap.[12] The simplest approximation was, of course, the quickest and fastest available approximation, but it was also substantially less accurate than desired for many problems. Accordingly, over the years the CNDO method was modified and replaced by, among others, the MNDO, MINDO/3, and AM1 methods,[13] and later by the PM3[14] and PM6[15] methods. The general theme was always to obtain as good a calculation as one could, consistent with keeping the computing time relatively small (compared with ab initio methods).

AB INITIO METHODS

While a sizable number of chemists pursued the semiempirical pathway during the 1960s, another sizable group was diligently pursuing ab initio calculations for the calculation of molecular structures. This was clearly a slow and difficult general problem, which was crucially dependent upon available computing power. Fortunately, this power became available just as needed.

In 1959 the helium atom was finally "solved" in that its energy was calculated more accurately (to eight significant figures) than it was known experimentally.[16a] Then in 1970, Bender and Schaefer reported[16b] ab initio calculations on the low-lying states of the CH_2 radical (methylene) and concluded that the ground state had a bond angle of 135.1°. The generally accepted structure determined earlier by experiment was linear.[16c] This appears to have been the first molecular structure determined by ab initio methods with sufficient accuracy that it superseded the experimental structure and that one could say was indicative of what was to come.

The real breakthrough in the development of these methods came when John Pople developed his program Gaussian(70) and subsequent extensions.[17] Single-term Gaussian orbitals were known to represent the electron distribution in atoms quite poorly, and

Slater orbitals were used almost completely in precomputer quantum mechanical calculations in chemistry.* But Gaussian functions had been studied over the years[18] as they had the very big advantage that they could be integrated. Pople decided to use a series of three Gaussian functions instead of just one, as had been used by most previous workers. This expanded the work required immensely, but still, it could be done for small systems, and it became increasingly feasible as available computer power increased. Pople's STO-3G (meaning a single Slater-type orbital approximated by a series of three Gaussian orbitals) set of orbitals (a single ζ minimum basis set) proved to be adequate for obtaining at least a fair approximation to molecular geometries for small molecules containing only a few atoms. These calculations were originally applied as what are now referred to as "single-point calculations," using the known experimental geometries and simply carrying out the calculations at those points. But by that time, people working in molecular mechanics had devised methods whereby they could start with a very crude geometry for a molecule, and optimize it, by relocating the atoms to the points on the potential surface where the energy of the system was a minimum. And as one would now expect, these methods could give approximate accuracy very rapidly, or high accuracy with relatively longer computing times. In principle, it was only necessary to extend these geometry optimization procedures to quantum mechanical calculations. That required quite a few years.[19–21] In the meantime, semiempirical methods were further developed, especially by Michael Dewar and his students.[11,13–15] The problem with ab initio methods was still largely the 3- and 4-centered two-electron integrals. While the integrations could now be carried out accurately using Pople's method, they were very time consuming. Much earlier, hand calculations had been carried out omitting these integrals, first for π systems (Pariser–Parr–Pople method, see Chapter 5), and then later this general type procedure was used for ordinary molecules. Since the 3- and 4-centered integrals had relatively small numerical values, compared to the 1- and 2-centered integrals, there was every expectation that this semiempirical procedure could be made to work, at least to a reasonable approximation. And that proved to be true. Early semiempirical calculations gave structural information that was useful at the time, but it was for the most part not competitive in accuracy with the experimental information. One way out was to try to add into the calculation explicitly those few 3- and 4-centered integrals that seemed to be larger and more important, while still omitting the multitude of integrals that had very small values. And indeed, the accuracy of the calculational results for given computer running times was significantly improved by this procedure. Unfortunately, there were not just a few integrals that caused all the problems. There were quite a few different types of integrals, declining in importance, which caused the problems. So one could include a small set of these troublesome integrals with some improvement in the results, or increasingly larger sets, with increasingly large computational demands, and further, but diminishing, improvements in the results. A number of these improvements were developed[10] and were standard calculational procedures for a number of years. But in the long run,

---

*The Gaussian function differs from the exponential function in that the distance relationship depends upon $-r^2$ rather than on $-r$. That is, in the simplest case the Gaussian function is $\Psi = be^{-cr^2}$. The problem is that this function is kind of the right shape, but not really.

as computer power increased, ab initio methods, developed by many, but especially by Pople,[17] came increasingly to the forefront as being the way to carry out the calculations, except for quite large molecules.

However, this was not all as straightforward as one might have thought at the outset. Of course, larger basis sets should give better results, but the problem was, just how large? It was soon found that STO-3G calculations were not very accurate in even simple cases. For example, cyclobutane came out to be planar rather than puckered, and if one were interested in energy differences of 1 or 2 kcal/mol, such calculations frequently gave even qualitatively incorrect answers. One solution to the problem was to increase the number of Gaussians used to mimic a single Slater orbital, and a second possibility was to increase the number of Slater orbitals beyond the number required for a minimum basis set. Such increases helped, but to obtain chemical accuracy it gradually became clear that one needed to increase the total number of orbitals in the calculation by very much more than had been recognized at the outset.

A separate problem was *electron correlation*. Even very large basis sets still often fail to give chemical accuracy in physical properties at the Hartree–Fock level because of the lack of inclusion of electron correlation in such a calculation. (Correlation of electron motions means that two electrons can be in the same place, but at different times. Thus, their motions are correlated.) Take the helium atom as a simple example. At the Hartree–Fock level, each electron is distributed about the nucleus with a spherical symmetry. One electron spends half of its time, say, on the right and half on the left of the nucleus, and similarly for the other electron. When correlation is included (added to the Hartree–Fock method), each electron still spends half of its time on the left and half on the right, but when one electron is on the left, the other is on the right. Thus, the electron distribution (over time) is not changed, but the energy of the system is greatly lowered because the electrons mostly stay relatively far apart. Hartree–Fock calculations deal with average electron densities, and they lead to Hartree–Fock energies. When the electron correlation is included in the calculation for systems that contain more than one electron, the energy of the system always goes down. Importantly, from the point of view of a chemist, is that for different molecules, different structures, and particularly for different conformations, this inclusion of correlation causes the energies to decrease by different amounts. That means that if one wants energies of conformations in general to chemical accuracy, one has to include electron correlation in the calculation.

There were numerous methods available for including electron correlation that had been developed earlier from π-system calculations, and still others were developed later. These were studied and applied to ordinary molecules, and some were better than others.[22] The Møller–Plesset perturbation method[17a,22c] in particular was found to be quite useful. But again, if chemical accuracy is desired, the lowest level of this calculation (now called MP2 for second-order Møller–Plesset) was often insufficient. MP3 and MP4 calculations were sometimes carried out and gave better results, but they were exceedingly time consuming. And for some systems, it was shown that high accuracy would require going to extremely high orders of perturbation, and this just wasn't feasible. Other methods were therefore developed, and are still being developed, to try to attain higher accuracy with shorter computing times. The coupled-cluster methods[22b] are quite effective and currently are widely used for problems that are not too large (few first-row atoms). As with basis sets, it is also necessary to extrapolate the electron correlation to the Schrödinger limit when high accuracy is desired. There are various ways that this can be done, all of which are complicated and time consuming, but possible for small systems (see, e.g., the carefully worked out and thoroughly discussed case of the rotational barrier in butane, and the references therein[23]).

The basic problem in determining structures by quantum mechanical methods, once fast computers were available, came down to the fact that so much of the calculational work depends on carrying out mathematical operations upon infinite series, where the series are not very rapidly converging, and the calculations have to be extended further out in the series than had been recognized earlier. In a sense, the problem is really solved. However, in another sense, we are still sometimes back where Paul Dirac said that we have the solution, but we still often can't carry it out with sufficient accuracy because of practical problems. We can, however, calculate molecular structures very accurately indeed, for molecules containing two or three first-row atoms, plus hydrogens. For larger molecules we can now calculate much of what we want, but perhaps not everything, depending on just how large the molecule is. But the solution to the problem for larger molecules is only a matter of time, as long as Moore's law continues to remain in effect for hardware, and similarly important advances in software also continue. For a review of the history and situation concerning ab initio calculations on large molecules the article by Cioslowski is available.[24]

To carry out the quantum mechanical calculations so that we can obtain four or five significant figures accuracy in the bond lengths of a structure, one carries out the calculation with basis sets of various sizes, where they are increased in a specific and very systematic way using well-balanced basis sets. The process can be continued until the calculational results either converge to within the desired limit of accuracy or can be extrapolated to that limit. The electron correlation must be similarly treated. This procedure can be carried out using the method referred to as *focal-point analysis*.[25] As was mentioned in Chapter 2 on microwave spectroscopy, the moments of inertia of small molecules are typically measured to five or six significant figures, and a very large number of them are known. The quantum mechanical (equilibrium) geometries can be converted to the comparable microwave geometries with vibrational corrections, and the calculated and experimental moments of inertia can be compared. This has been done for a dozen or so molecules containing up to three heavy atoms,[26] and such work will doubtlessly continue and expand to larger systems. One can argue that any one molecule might have errors in the structure, but the moments of inertia still agree with the quantum mechanical calculations just by chance. But when the agreement applies to all of the molecules studied, one is forced to conclude that the calculations and the experimental structures are in fact in agreement to that level of accuracy. One wants to know that this can be done, but for "chemical purposes" these calculations are at least an order of magnitude more accurate than is needed. We can conclude that we are OK here, so far. It seems, however, that there is no limit to how complicated nature can get. For example, the accurate calculations mentioned are for molecules that not only contain few atoms, but they are small atoms. As we go down the periodic table, the atoms become heavier, and they have larger nuclear charges. This means that the inner-shell electrons are pulled in closer to the nucleus and hence must increase their velocities to keep from falling into the nucleus. Since their velocities are then a significant fraction of the speed of light, one then has to be concerned with relativistic effects and include those in the calculations. And so on.

However, as our computational demands increase, so does available computer power. Computers all contain a CPU, which carries out the basic computational work. Current computers also normally contain a graphical processing unit (GPU). This is a massively parallel processing unit (many processors working together simultaneously) that was originally designed to process graphics in video games. By harnessing the GPU's parallel structure for the computational work, it is possible to greatly accelerate certain kinds of calculations, including electronic structure calculations. It was reported at an American Chemical Society symposium in 2008[27] that utilizing this type of processing unit made it possible to speed up molecular structure calculations by approximately a factor of 100. (This would be many years worth of what we are assured by Moore's law!)

DENSITY FUNCTIONAL THEORY[28,29]

It has been known since the 1930s that, in principle, it is possible to determine molecular structures by solving the Schrödinger equation and minimizing the energy of the molecule. For actual calculations to be carried out, the numerical intensity of the problems were such that not much happened until the 1950s with the availability of electronic computers. But it took until the 1970s until the necessary mathematics, and computers, had been sufficiently developed so that such problems could be solved for real systems of interest in a practical way. The general problem proved to be exceedingly difficult because one had to work with a $3N$-dimensional wave function, Ψ, that is used to describe the behavior of each electron in the N-electron system. One is constantly faced with infinite functions that have to be truncated, so that accurate solutions are exceedingly time consuming.

Density functional theory (DFT) is based on the Hohenberg–Kohn theorem, which states that the total energy of a system in its ground state is a functional (a function of a function) of that system's electronic density, $\rho(\mathbf{r})$, and that any density, $\rho'(\mathbf{r})$, other than the true density will necessarily lead to a higher energy. Thus, the variational principle applies here. This means that an alternative methodology to solving Schrödinger's equation is using DFT, which requires only that we minimize the energy functional, $E[\rho(\mathbf{r})]$. The conceptual simplification here is immense. Instead of having to work with a complex $3N$-dimensional wave function Ψ, describing the behavior of each electron in an N-electron system, DFT allows us to work with a simple three-dimensional function, the total electronic density. The only problem is that the exact nature of the energy functional is not known, and it remains unknown after very much effort by many people. However, while it is still not possible to proceed from this point in an ab initio fashion with DFT, we can do what was originally done with the Schrödinger equation and attack the problem with semiempirical methods. Considerable effort has been spent developing the energy functional in an approximate way.

Correlation effects, which are not included in the Hartree–Fock approximation, are also built into the approximate energy functionals that are used in modern DFT applications. DFT methods are thus able, in principle, to treat the entire periodic table.

As with other semiempirical methods, much of the progress has to proceed with trial-and-error approaches. The B3LYP functional[10] is quite commonly used now and often referred to in this book. There are both advantages and disadvantages to the use of B3LYP calculations versus MP2, which are currently the usual choices when dealing with relatively large systems.

We know that with the wave function approach, if we use a large enough basis set and correlation treatment, we will reach the desired accuracy. The problem is usually where can we truncate the calculation, and will this be accurate enough for what we wish to calculate? The limitation is generally computer time. On the other hand, the DFT method is much faster than MP2, let alone more complete correlation methods. A certain level of results is given by the B3LYP functional and procedure at present, and it is usually accepted that this level is not easily or systematically improved. As far as structure, the results are usually significantly less good from B3LYP than they are from MP2 calculations. With the energies, it is less clear. Small energy differences, such as between conformations, are not very well calculated, and MP2 results are usually better, but sometimes not. But over wider ranges, like heats of formation, neither is really of chemical accuracy. We do know how to improve the Schrödinger results in a straightforward way, however, by increasing the basis set size and correlation completeness. DFT can best be advanced with the development of better functionals, and such work is in progress in various laboratories[29] (also see Chapter 11).

MOLECULAR MECHANICS

What is molecular mechanics? Why is it of interest to us? Let us briefly answer these questions here at the outset. *Molecular mechanics* is a computational scheme whereby one can use the methods of classical or Newtonian mechanics to describe the structures and properties of molecules. As will be known to most readers, one can consider a classical system of weights held together by springs as a model for a molecule. Having a classical mechanical model for a molecule gives us the opportunity of understanding molecules in the same way that we understand the properties of physical objects of everyday life. Hence, if we can represent a molecule in terms of a classical system that is simple to understand, and on which we can carry out quantitative calculations regarding properties, then we have gained some kind of an understanding of that system. This model was originally used in a more specialized form to understand vibrational spectra. But with the advent of computers, it was found that we could learn an immense amount about molecules by using this model. One can determine, as early spectroscopists did, properties that are assignable to various atoms and bonds. To make this a really useful computational scheme, we have to do more than describe a given molecule with a computational system. We have to be able to make predictions regarding molecules that we have not studied and about which we have no knowledge. Molecular mechanics does this by making the assumption that various properties of atoms and bonds are transferable from one molecule to another. That is, one can determine for simple molecules sufficient properties of the molecule (bond angles and lengths, conformational energy relationships, etc.) in the form of a set of constants, which can be then transferred to other molecules. At the outset, there was considerable uncertainty as to whether this procedure was in fact workable. Is there reason to suppose that the C–C bond length in ethane, say, can tell us something about the C–C bond lengths in a steroid molecule? In other terms, is this a good enough model to be useful? It has turned out that the model was pretty good at the outset. Moreover, as studies progressed over a period of many years, it was determined that the original model could be improved in a variety of ways. Models are rarely perfect. And the imperfections tell us what it is that we do not know, both about our model and about the object we are trying to model. That's equally true here. Modern molecular mechanics is often competitive with the best theories and experiments for determining various things about molecular structures and properties. But when it does not work correctly, it means that nature knows something that we do not know, and it opens a potential window for understanding.

Some of the important aspects of the behavior of the matter in the universe (at least what physicists now call baryonic matter) is described by a series of natural laws, or rules, which are collectively called *mechanics*. It is convenient to divide the subject into two parts. If the particles constituting the matter under examination have a very small mass, the branch of study is referred to as *quantum mechanics*. If the particles have a large mass, then the subject is referred to as *classical*, or *Newtonian mechanics*. Electrons have small masses, and studies that involve them almost always have to be carried out by quantum mechanics. The ideas behind classical mechanics alone just are not sufficient to accurately describe the behavior of such small particles. If we wish to study the motions of stars and galaxies, for example, these have large masses, and generally speaking, are well described by classical mechanics. (Relativity may also be a problem here in some cases.) Atomic nuclei are then just one step up from electrons in terms of their mass, but it is a pretty big step. And if we put electrons and nuclei together and make atoms, then we find that we have arrived at systems (atoms) where classical mechanics is, or may be, an acceptable way to study most of their properties. The expression "molecular mechanics" was invented[30] to describe the now widely used procedure wherein molecules, or certain kinds of collections of atoms, are treated by essentially classical mechanical procedures.[31] These procedures are not necessarily the best way to treat molecular problems, but they are often the most convenient way. And chemists usually tend to take the pragmatic view here. If the method works (and does not violate any known physical principles), it can be used. The authors' philosophy is to use molecular mechanics when you can, and use quantum mechanics when you must, and to make sure that you know the difference. For a concise discussion of what molecular mechanics is and what it is used for, see the review by Boyd and Lipkowitz,[32] and for a listing of many computer programs available and details of their construction, see the review by Pettersson and Liljefors,[33] the *Encyclopedia* article,[31e] and also the section on Molecular Mechanics Programs in the Appendix of this volume.

How does one determine a molecular structure using molecular mechanics? One first identifies the molecule in question. What are the numbers and kinds of atoms present, and what are their connectivities? Using this information, one could build the structure as a mechanical model, such as a Dreiding model. This would give us a crude approximate structure of the molecule. To obtain the real structure from the Dreiding model, one would have to "optimize" the atomic positions. That is, adjust all of the bond lengths, bond angles, and torsion angles so that they match those of the actual molecule. And how does one do this? What one needs to do, in principle, is to arrange the atoms on the potential surface so that they have the lowest possible energy. That is, each of the bond lengths and angles must be adjusted in such a way that the total energy of the system is a minimum. That will occur where all of the derivatives of the energy with respect to the atomic coordinates are equal to zero. For molecules that have more than one stable conformation (conformer), there will be an energy minimum for each conformer, and these conformers will be separated by saddle points on the potential surface. But the general mathematical problem to be solved is straightforward in principle. One wants to take the approximate structure, such as built with a mechanical model, and minimize the energy of the system by adjusting the atomic positions.

In principle, all we have to do is move from wherever we are on the potential surface with our approximate structure, down to the energy minimum. If we know all of the energies for stretching, bending, and the like, the various bonds and angles, plus the van der Waals and electrostatic forces, we can calculate the total energy of the system at the point where we start. It is then necessary to determine the derivatives of the energy with respect to each of the atomic coordinates. While this is impossibly complicated to do by hand calculations for ordinary molecules, it is straightforward, in principle, and modern computers can do it quickly and accurately. We then follow the slopes (the derivatives) downhill until we can go no further. This type of problem was very well studied by mathematicians over the years, and there are many ways to do this, some of which have been known since the time of Isaac Newton. So here we know what we have to do, we know how to do it, and, with computers, we can now do it.

A very similar scheme, in terms of the mathematics involved, can be applied to determining an exact structure of a molecule from an approximate structure using quantum mechanics. But let's consider only the molecular mechanics case here. What we need to be able to calculate are the forces between the atoms in the molecule. These are the familiar stretching and bending forces, among others. With molecular mechanics, a problem arises because in each of the equations there are numerous constants, and we do not know the values for most of these constants when we start. So what we have to do at the outset is to decide on the form of the set of equations that we will use as our *force field*, that is, to describe our potential surface. Then we have to evaluate the constants that appear in that set of equations, treating them as parameters, to fill out the force field. To do that, we calculate the structures of some relatively simple molecules that are already accurately known (from experiment, from quantum mechanics, or preferably from both), and work the problem backward. That is, instead of using the force field to find the structure, we guess the constants involved in the force field and calculate the structure. And then we improve our guesses in an iterative fashion, and repeat the process until the force field calculates accurately the structures of interest. If we have sufficient accurate structural information, and if we have chosen a proper set of equations for the force field, the force field will then give the structures correctly. Thus, we have devised a usable force field, which can in turn be used to calculate the structures of other molecules that contain those same component atoms and groups. It is important to note here that it is *assumed* that the parameters (all of these constants for stretching, bending, etc.) are transferable from one molecule to another. There is ample evidence that indicates this is a very good approximation, but it is expected to have definite limits, as will be discussed further in the following chapters.

Historically, classical mechanics was well developed and utilized some centuries before the existence of quantum mechanics was even suspected. Hence, in the early decades of the twentieth century, during early studies involving atomic particles, classical mechanics was often invoked. The hydrogen atom was originally viewed as having an electron in an orbit around a proton, quite similar to the Earth revolving about the sun, and heavier atoms were thought of as analogous to the solar system, with many electrons in orbits about the nucleus. But this analogy could be pushed only so far because this is, in fact, the realm of quantum mechanics not of classical mechanics. It was found early in the twentieth century, however, that a general description of a molecule as being analogous to weights connected by springs was a good place to start, and it enabled the calculation and prediction of vibrational spectra and related thermodynamic properties. Note that in this latter case we are considering only atoms as our basic units. We are not trying to go down to the level of electrons and nuclei. At the atomic level, classical mechanics can solve many problems of interest much more quickly and easily than can quantum mechanics. Just how all of this could be done was recognized by 1930 and was summarized in the review by Andrews.[34] He basically described the calculational scheme that is now called molecular mechanics. But to actually determine structures by this method in 1930 was as impossible as it would have been to determine structures using quantum mechanics. The calculations, although orders of magnitude more simple than those required by quantum mechanics, were still prohibitively difficult, except in a particularly ideal case.

Only one actual problem in chemistry was solved using molecular mechanics as we use it now, by hand calculations. In a series of studies, Frank Westheimer described the following situation.[35] Ortho-substituted biphenyl molecules are generally not coplanar because of the steric interference of the ortho substituents. A typical example is 2,2′-dibromo-4,4′-dicarboxybiphenyl. Consider the two structures of this molecule shown in Reaction (1):

HOOC—⟨⟩—⟨⟩—COOH ⇌ HOOC—⟨⟩—⟨⟩—COOH

Racemization of 2,2′-dibromo-4,4′-dicarboxybiphenyl

(1)

Since these structures are nonplanar, they are nonsuperimposable mirror images, that is, enantiomers. One can be converted to the other by a partial rotation about the central bond, and the transition state separating them is the planar form. (The carboxyl groups play no part in what follows, but were "handles" that enabled the enantiomers to be separated.) Experimental chemists had known all about this long before because they were able to separate the enantiomers if the substituents were large enough to make this rotation sufficiently hindered. One enantiomer could be isolated in optically active form, put into a polarimeter tube, and heated in a thermostat, and the rate at which it racimized could be measured. From this information, the barrier to rotation could be calculated. And the rotational barrier was higher as the substituent became larger as was expected.

Westheimer[35] was able to calculate these rotational barriers using known information about the sizes and van der Waals properties of the atoms involved, together with the force constants for bending and stretching certain of the bond angles and lengths in such molecules. And the barrier heights that he calculated were in acceptable agreement with those measured experimentally for several different molecules. This was quite a triumph for molecular mechanics. But it worked only because the number of degrees of freedom in the molecules that had to be used in the calculation could be reduced to quite a small number. Thus, he assumed that the benzene rings were symmetrical and rigid hexagons, for example, removing a huge number of degrees of freedom from the molecule, thereby reducing the calculation to a manageable size.

This was the only real problem that was ever solved by molecular mechanics using hand calculations. Later, Ingold[36] showed that using the ideas of molecular mechanics, and assigning some reasonable parameters, one could explain observed facts about the rates of reaction that were affected by steric hinderance in simple alkyl halides. But much of the calculation could not actually be carried out. One could only make qualitative estimates as to which group was larger and would slow the reaction more. And it seemed likely that this was as far as one could go with molecular mechanics and hand calculations, although following Ingold's lead, some qualitative calculations were carried out to assist in rationalizing results obtained experimentally in conformational analysis.[37]

The next step occurred in 1961,[38] when J. B. Hendrickson published a study of the conformations of cycloalkane rings containing from 7 to 10 members. By that time, conformational analysis had developed very far, and the subject had become a useful tool for predicting many things regarding physical properties and chemical reactivities of molecules.[39] The cyclohexane ring was a well-studied system in conformational analysis because of its rigidity, high symmetry, and the relatively few conformations available to substituted cyclohexanes. But the larger rings were not so simple, and not much progress had been made in studying their conformations until Hendrickson's work. Hendrickson utilized a computer to do the extensive computational work required. And he showed that molecular mechanics could be a viable and highly useful tool, now that it had become possible to effectively utilize computers to do the heavy calculational work.

Westheimer and Hendrickson were pioneers. They utilized a method that they developed specifically to solve the problem at hand. And they were successful. But what remained to be done to make the method widely useful was to develop a *general* method that would be applicable to determining structures for all (or at least most) molecules. The earlier workers had looked at each problem as a special case. This next step was taken by Wiberg,[40] who first developed a general method, and the corresponding computer program, for determining molecular structures.

Wiberg's procedure begins with the number and kinds of atoms involved and their connectivities. This information can be gotten in various ways, for example, by examination of Dreiding models. The coordinates for each atom in the starting geometry had to be determined. Originally, this was often done by actually building a physical model.[37] (Dreiding models were used since they are built to scale and the atomic positions are easily accessible because of the open framework type of structure.) The model was set on a piece of graph paper, and the X/Y coordinates were determined by projecting the positions onto the paper. The Z coordinates were measured with a ruler upward from the plane of the paper. This was a simple and straightforward way to obtain approximate starting geometries and was the method generally used in the early 1960s. But it was soon decided that if a set of standard bond lengths (as are built into Dreiding models) were to be used, the molecular structure could be more quickly calculated by the computer, and hence the familiar input schemes used today were developed.[40,41] Having the approximate starting coordinates, the reference bond lengths and angles (l_0 and θ_0), and the force constants for bending and stretching, plus the van der Waals and electrostatics of the system, one could simply move the atoms around in some systematic way until one located the energy minimum. Mathematicians had already developed sophisticated ways for doing this, so no new mathematics had to be developed. Generally speaking, only that which was already known had to be utilized. In practice, of course, the practical problem originally was to refine existing techniques so as to allow reasonably short running times on existing computers. After about 1990, computer time was no longer a significant problem in optimizing structures of ordinary organic molecules (up to say 100 atoms). We found the *Newton–Raphson method* to be most accurate in such cases.[31b] Bond lengths typically could be optimized to about 0.0005 Å (using computers with a 32-bit binary word in single precision). This accuracy is desirable for the calculation of vibrational spectra, and we have never found need for an accuracy greater than this. For very large scale work using molecular dynamics on proteins, for example, the *conjugate gradient method* is commonly used because it is faster. Increasingly rapid methods are expected to continue to be developed.[21] (Faster methods are not needed for ordinary molecules but are important for proteins.) Further discussion of molecular mechanics will be continued in the next chapter. Additional information is given under Molecular Mechanics Programs in the Appendix.

REFERENCES

1. (a) S. Wilson, Ed., *Handbook of Molecular Physics and Quantum Chemistry*, Vol. 1, Wiley, West Sussex, England, 2003. (b) M. M. Waldrop, *The Dream Machine*, Viking Press, New York, 2001. (c) Also see, S. M. Ornstein, *Computing in the Middle Ages. A View from the Trenches 1955–1983*, First Books Library, Bloomington, IN, 2002. (d) G. Moore, *Electronics Mag.*, 19 April, 1965. (e) For a review, see *Moore's Law*, Wikipedia, The Free Encyclopedia.
2. Y.-M. Lin, K. A. Jenkins, A. Valdes-Garcia, J. P. Small, D. B. Farmer, and P. Avouris, *Nano Lett.*, **9**, 422 (2009).

3. N. Bohr, *Philosophical Mag.*, **26**, 1, 476, 857 (1913).
4. (a) A. Sommerfeld, *Zeit. Elektro. Angewandte Phys. Chem.*, **26**, 258 (1920). (b) A. Sommerfeld, *J. Opt. Soc. Am.*, **7**, 509 (1923); and later papers.
5. E. Schrödinger, *Ann. Phys.*, **79**, 361 (1926).
6. (a) W. Heitler and F. London, *Z. Phys.*, **44**, 455 (1927). (b) H. M. James and A. S. Coolidge, *J. Chem. Phys.*, **1**, 825 (1933).
7. P. A. M. Dirac, *Proc. Roy. Soc. (London)*, **123**, 714 (1929).
8. E. Hückel, *Z. Physik*, **70**, 204 (1931).
9. A. Streitwieser, Jr., *Molecular Orbital Theory for Organic Chemists*, Wiley, New York, 1961.
10. I. N. Levine, *Quantum Chemistry*, 5th ed., Prentice Hall, Upper Saddle River, NJ, 2000.
11. M. J. S. Dewar, *The Molecular Orbital Theory of Organic Chemistry*, McGraw-Hill, New York, 1969, pp. 444–468.
12. J. A. Pople and D. L. Beveridge, *Approximate Molecular Orbital Theory*, McGraw-Hill, New York, 1970.
13. (a) M. J. S. Dewar and W. Thiel, Ground States of Molecules, 38. The MNDO Method. Approximations and Parameters, *J. Am. Chem. Soc.* **99**, 4899–4907 (1977). (b) J. J. P. Stewart, MOPAC, A Semi-Empirical Molecular Orbital Program, *QCPE*, 455 (1983).
14. J. J. P. Stewart, *J. Comput. Chem.*, **10**, 209, 221 (1989).
15. J. J. P. Stewart, *Mol. Model*, **13**, 1173 (2007).
16. (a) C. L. Pekeris, *Phys. Rev.*, **115**, 1216 (1959). (b) C. F. Bender and H. F. Schaefer, III, *J. Am. Chem. Soc.*, **92**, 4984 (1970). (c) F. G. Herzberg, *Proc. Roy. Soc., Ser. A*, **262**, 291 (1961).
17. (a) W. J. Hehre, L. Radom, P. v. R. Schleyer, and J. Pople, *Ab Initio Molecular Orbital Theory*, Wiley, New York, 1986. (b) W. J. Hehre, W. A. Lathan, M. D. Newton, R. Ditchfield, and J. A. Pople, Gaussian(70), Program 236, Quantum Chemistry Program Exchange, Indiana University, Bloomington, Indiana, 1972. (c) M. J. Frisch, G. W. Trucks, H. B. Schlegel, G. E. Scuseria, M. A. Robb, J. R. Cheeseman, J. A. Montgomery, Jr., T. Vreven, K. N. Kudin, J. C. Burant, J. M. Millam, S. S. Iyengar, J. Tomasi, V. Barone, B. Mennucci, M. Cossi, G. Scalmani, N. Rega, G. A. Petersson, H. Nakatsuji, M. Hada, M. Ehara, K. Toyota, R. Fukuda, J. Hasegawa, M. Ishida, T. Nakajima, Y. Honda, O. Kitao, H. Nakai, M. Klene, X. Li, J. E. Knox, H. P. Hratchian, J. B. Cross, C. Adamo, J. Jaramillo, R. Gomperts, R. E. Stratmann, O. Yazyev, A. J. Austin, R. Cammi, C. Pomelli, J. W. Ochterski, P. Y. Ayala, K. Morokuma, G. A. Voth, P. Salvador, J. J. Dannenberg, V. G. Zakrzewski, S. Dapprich, A. D. Daniels, M. C. Strain, O. Farkas, D. K. Malick, A. D. Rabuck, K. Raghavachari, J. B. Foresman, J. V. Ortiz, Q. Cui, A. G. Baboul, S. Clifford, J. Cioslowski, B. B. Stefanov, G. Liu, A. Liashenko, P. Piskorz, I. Komaromi, R. L. Martin, D. J. Fox, T. Keith, M. A. Al-Laham, C. Y. Peng, A. Nanayakkara, M. Challacombe, P. M. W. Gill, B. Johnson, W. Chen, M. W. Wong, C. Gonzalez, and J. A. Pople, Gaussian (03), Revision C.02, Gaussian, Inc., Wallingford, CT, 2004.
18. I. Shavitt, *Isr. J. Chem.*, **33**, 357 (1993).
19. P. Pulay, *Mol. Phys.*, **17**, 197 (1969).
20. P. Pulay, G. Fogarasi, F. Pang, and J. E. Boggs, *J. Am. Chem. Soc.*, **101**, 2550 (1979).
21. For reviews of geometry optimization methods see: (a) G. Fogarasi and P. Pulay, in *Structures and Conformations of Non-Rigid Molecules*, J. Laane, M. Dakkouri, B. van der Veken, and H. Oberhammer, Eds., Kluwer, Dordrecht, The Netherlands, 1993, p. 377; (b) T. Schlick,

Optimization Methods in Computational Chemistry, Reviews in Computational Chemistry, Vol. 3, K. B. Lipkowitz and D. B. Boyd, Eds., Wiley, New York, 1992, p. 1; (c) J. A. McCammon and S. C. Harvey, *Dynamics of Proteins and Nucleic Acids*, Cambridge University Press, Cambridge, 1987.

22. (a) R. J. Bartlett and J. F. Stanton, *Applications of Post-Hartree–Fock Methods: A Tutorial, Reviews in Computational Chemistry*, Vol. 5, K. B. Lipkowitz and D. B. Boyd, Eds., Wiley, New York, 1994, p. 65. (b) T. D. Crawford and H. F. Schaefer, III, *An Introduction to Coupled Cluster Theory for Computational Chemists, Reviews in Computational Chemistry*, Vol. 14, K. B. Lipkowitz and D. B. Boyd, Eds., Wiley, New York, 2000, p. 33. (c) C. Møller and M. S. Plesset, *Phys. Rev.*, **46**, 618 (1934).

23. N. L. Allinger, J. T. Fermann, W. D. Allen, and H. F. Schaefer III, *J. Chem. Phys.*, **106**, 5143 (1997).

24. J. Cioslowski, *Ab Initio Calculations on Large Molecules: Methodology and Applications, Reviews in Computational Chemistry*, Vol. 4, K. B. Lipkowitz and D. B. Boyd, Eds., Wiley, New York, 1993, p. 1. Also see J. Cioslowski, Electronic Structure Calculations on Fullerenes and Their Derivatives, Oxford University Press, Oxford, 1995.

25. A. L. L. East and W. D. Allen, *J. Chem. Phys.*, **99**, 4638 (1993).

26. (a) F. Pawlowski, P. Jorgensen, J. Olsen, F. Hegelund, T. Helgaker, J. Gauss, K. L. Bak, and J. F. Stanton, *J. Chem. Phys.*, **116**, 6482 (2002). (b) W. D. Allen, A. L. L. East, and A. G. Csaszar, in *Structures and Conformations of Non-Rigid Molecules*, J. Laane, M. Dakkouri, B. van der Veken, and H. Oberhammer, Eds., Kluwer, Dordrecht, The Netherlands, 1993, p. 343.

27. I. Ufimtsev and T. Martinez, *C&E News*, September 22, 2008, p. 88.

28. (a) R. G. Parr and W. Yang, *Density-Functional Theory of Atoms and Molecules*, Oxford University Press, New York, 1989. (b) L. J. Bartolotti and K. Flurchick, *An Introduction to Density Functional Theory, Reviews in Computational Chemistry*, Vol. 7, K. B. Lipkowitz and D. B. Boyd, Eds., Wiley, New York, 1996, p. 187. (c) F. M. Bickelhaupt and E. J. Baerends, *Kohn–Sham Density Functional Theory: Predicting and Understanding Chemistry, Reviews in Computational Chemistry*, Vol. 15, K. B. Lipkowitz and D. B. Boyd, Eds., Wiley, New York, 2000, p. 1. (d) A. St-Amant, *Density Functional Methods in Biomolecular Modeling, Reviews in Computational Chemistry*, Vol. 7, K. B. Lipkowitz and D. B. Boyd, Eds., Wiley, New York, 1996, p. 217.

29. Y. Zhao and D. G. Truhlar, *J. Chem. Phys.*, **125**, 194101 (2006), and references therein.

30. N. L. Allinger and J. T. Sprague, *J. Am. Chem. Soc.*, **95**, 3893 (1973).

31. (a) K. Machida, *Principles of Molecular Mechanics*, Kodansha and Wiley, co-publication, Tokyo and New York, 1999. (b) U. Burkert and N. L. Allinger, *Molecular Mechanics*, American Chemical Society, Washington, D.C., 1982. (c) A. R. Leach, *Molecular Modeling Principles and Applications*, Pearson Education, Essex, England, 1996. (d) A. K. Rappe and C. J. Casewit, *Molecular Mechanics Across Chemistry*, University Science Books, Sausalito, CA, 1997. (e) N. L. Allinger, J. R. Maple, and T. A. Halgren, in *Encyclopedia of Computational Chemistry (under* Force Fields), P. v. R. Schleyer, N. L. Allinger, T. Clark, J. Gasteiger, P. A. Kollman, H. F. Schaefer, III, and P. R. Schreiner, Eds., Wiley, UK, Vol. 2, 1998, pp. 1013–1035.

32. D. B. Boyd and K. B. Lipkowitz, *J. Chem. Ed.*, **59**, 269 (1982).

33. I. Pettersson and T. Liljefors, *Molecular Mechanics Calculated Conformational Energies of Organic Molecules: A Comparison of Force Fields, Reviews in Computational Chemistry*, Vol. 9, K. B. Lipkowitz and D. B. Boyd, Eds., Wiley, New York, 1996, p. 167.

34. D. H. Andrews, *Phys. Rev.*, **36**, 544 (1930).
35. (a) F. H. Westheimer and J. E. Mayer, *J. Chem. Phys.*, **14**, 733 (1946). (b) F. H. Westheimer, *J. Chem. Phys.*, **15**, 252 (1947). (c) F. H. Westheimer, in *Steric Effects in Organic Chemistry*, M. S. Newman, Ed., Wiley, New York, 1956.
36. (a) C. K. Ingold, *Structure and Mechanism in Organic Chemistry*, Cornell University Press, Ithaca, NY, 1953, pp. 403–412. (b) I. Dostrovsky, E. D. Hughes, and C. K. Ingold, *J. Chem. Soc.*, 173 (1946).
37. N. L. Allinger, *J. Am. Chem. Soc.*, **81**, 5727 (1959).
38. (a) J. B. Hendrickson, *J. Am. Chem. Soc.*, **83**, 4537 (1961). (b) J. B. Hendrickson, *J. Am. Chem. Soc.*, **86**, 4854 (1964).
39. (a) E. L. Eliel, N. L. Allinger, S. J. Angyal, and G. A. Morrison, *Conformational Analysis*, Wiley-Interscience: New York, 1965, p. 43. (b) M. Hanack, *Conformation Theory*, Academic, New York, 1965. (c) H. Dodziuk, *Modern Conformational Analysis: Elucidating Novel Exciting Molecular Structures*, VCH, New York, 1995.
40. K. B. Wiberg, *J. Am. Chem. Soc.*, **87**, 1070 (1965).
41. N. L. Allinger, *J. Am. Chem. Soc.*, **99**, 8127 (1977).

4

MOLECULAR MECHANICS OF ALKANES*

POTENTIAL ENERGY SURFACE

Let us begin with a simple diatomic molecule, for example, HCl. This molecule has three translational and two rotational degrees of freedom. It also has one degree of vibrational freedom, which is the bond stretching. Hooke's law (the harmonic approximation) says that the energy of stretching is given as in Eq. (4.1):

$$E = \frac{k}{2}(l - l_0)^2 \tag{4.1}$$

A better approximation would add higher terms (cubic, quartic, etc.), and a still better approximation would be the Morse function, but we will use the simple form for present purposes. Before computers, the harmonic approximation was almost always used in actual calculations. Higher approximations are almost always required to obtain

---

*The development in this chapter concentrates on the MM4 force field. We believe it is the most complete and accurate one currently available for alkanes. For additional information on this and other force fields, see Molecular Mechanics Programs in the Appendix.

Molecular Structure: Understanding Steric and Electronic Effects from Molecular Mechanics,
By Norman L. Allinger
Copyright © 2010 John Wiley & Sons, Inc.

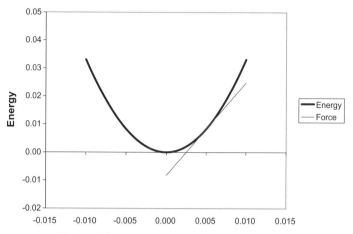

Figure 4.1. Energy vs. bond length (schematic).

chemical accuracy, but they were largely out of reach for hand calculations. With computers, the computational times are not increased much by better approximations, which are now widely used. It will be obvious how we would extend the calculation to higher approximations.

We know that for such a classical system (two weights joined by a spring), as the spring is stretched, the energy behaves as described by Eq. (4.1). The *force* required to stretch the spring is the derivative of the energy with respect to the stretching motion [Eq. (4.2)]:

$$\frac{dE}{dl} = F = k(l - l_0) \qquad (4.2)$$

Thus, the energy has a parabolic relationship to the bond length, and the force required to stretch the bond is the slope of that curve, which varies linearly with the bond length. The force is often referred to as the *energy gradient* (Fig. 4.1).

The second derivative of the energy (or the first derivative of the force) with respect to the bond length distance is given by Eq. (4.3):

$$\frac{d^2E}{dl^2} = k \qquad (4.3)$$

This quantity k is referred to as the *force constant*. If it is a large number, the bond is very strong and hard to stretch, but if it is a small number, the bond is weak and stretches easily. It is just a constant for a given bond for small displacements.

Such a linear molecule has what is sometimes referred to as a one-dimensional potential energy relative to distance. If we take the energy of the system as the ordinate, then the distance between the atoms can simply move over an interval, and a plot of the energy versus distance is as shown in Figure 4.1. This is said to be a one-

POTENTIAL ENERGY SURFACE 53

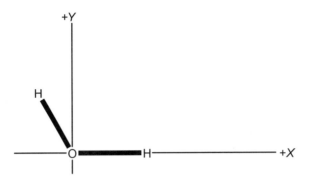

Figure 4.2. Water molecule.

dimensional figure because there is only one degree of freedom in the motion, back and forth along the X axis.

If we come to a somewhat more complicated molecule, such as water, we have three atoms attached, in this case in an angular manner. If we put the oxygen at the origin of a Cartesian coordinate system in two dimensions, and place one of the hydrogens along the X axis at the appropriate bond length distance, the other hydrogen will be found at a bond angle of approximately 105° from the first hydrogen. We may place it in the XY plane, about 15° to the left of the X axis, as in Figure 4.2. If we imagined the oxygen as our reference point, then stretching the hydrogen along the X axis leads to a one-dimensional figure (which can be represented by the harmonic approximation, or a series approximation, or a Morse curve), and similarly for the other hydrogen.

In addition, we may bend the (H–O–H) angle by moving the upper hydrogen in Figure 4.2 in an arc, keeping it at the same distance from the oxygen. As we bend the oxygen away from its normal bond angle by a small amount, we find that the energy increases according to the magnitude of the bending. The equation that describes this energy change looks like Hooke's law, but in this case involves an angle instead of a distance [Eq. (4.4)]:

$$E = \frac{k}{2}(\theta - \theta_0)^2 \qquad (4.4)$$

If we make a plot of the energy change described by Eq. (4.4), we obtain another one-dimensional curve, which again is a parabola for small bendings. But, if we bend the system as far as 180°, in the case of water, the molecule becomes linear and the energy reaches a maximum. If we continue moving the hydrogen further in the same direction, it goes back down to a symmetrical minimum, as shown in Figure 4.3. We have again the same molecule, which corresponds to the original molecule oriented differently, although in this case we reached the geometry by a bending motion instead of by a rotation. So we can imagine a "potential energy surface" for the water molecule (the X-Y plane in Fig. 4.2) in which the atoms are in the lowest energy position as indicated

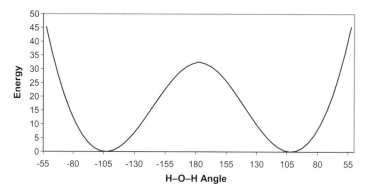

Figure 4.3. Variation of energy with angle for water.

in Figure 4.2, and the energy varies according to the motions of the atoms on that surface.

One needs to keep in mind, however, that one is treading on dangerous ground by applying molecular mechanics in cases such as this. The hydrogen nucleus has quite a small mass, and hence the hydrogens can cross the barrier indicated in Figure 4.3 without actually going up over the top of it. They can go through it in a process referred to as "tunneling." This is simply because the wave function describing the location of the hydrogen nucleus may have a maximum value on one side of the barrier, but it has a tail that extends through the barrier and out the other side. Hence, the hydrogen can go through the barrier, in effect. Although molecular mechanics can be used for a wide variety of problems involving molecules, there are some limitations.

Most molecules of interest contain many atoms, and we find it useful to talk about a "potential energy surface" in a polyatomic molecule. This is, of course, a multidimensional "surface." But it is convenient to mentally (and mathematically) isolate parts of these surfaces and think about them (or deal with them) one at a time.

If we want to describe a more general molecule, we can define the geometry in terms of the bond lengths and angles as we did with water, and we can calculate the energy from appropriate equations involving separately each bond and each angle. If the molecule under examination contains a linear connectivity arrangement of four (or more) atoms, such as the carbon atoms in butane, then there is another kind of freedom of motion (called an *internal coordinate* of the motion) which must be used to define the structure of the molecule. This coordinate is called the *torsional angle*. If we take *n*-butane as an example, the important torsional angle is defined by a rotation of the two ends of the molecule with respect to one another about the central bond. If we look at the central bond from an end-on view (a so-called *Newman projection*, where the ends of the molecule are projected on the plane that bisects and is perpendicular to the central C–C bond), we see various conformations as we change the torsion angle through a 360° rotation. The three conformations shown in Structure **1** correspond to energy minima and are the two gauche conformations (a *dl* pair, and the anti conformation).

POTENTIAL ENERGY SURFACE

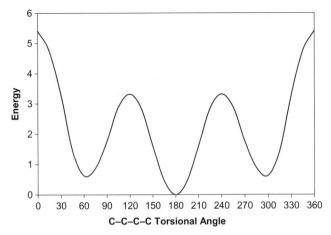

Figure 4.4. Variation of energy with torsion angle (ω) for butane.

1

These stable conformations are usually referred to as *conformers*. A *conformation* can be any three-dimensional arrangement, whereas a *conformer* is a stable (energy minimum) species. In the case of butane, as the torsion angle varies, so does the energy, as is shown in Figure 4.4.

The general equation that describes the variation of energy with torsion angle is conveniently written as a 3-term Fourier series as in Eq. (4.5):

$$E_\omega = \frac{V_1}{2}(1-\cos\omega) + \frac{V_2}{2}(1-\cos 2\omega) + \frac{V_3}{2}(1-\cos 3\omega) \tag{4.5}$$

For an alkane this equation has as its most important element a threefold cosine term. This leads to energy minima at approximately 60, 180, and 300° and to energy maxima in between, as in Figure 4.4. In the molecular mechanics formulation, a part of the extra energy in the energy maximum conformations comes from van der Waals repulsions between the atoms. The inherent threefold nature of the ethane-type barrier can be thought of as quantum mechanical (bonding/antibonding interactions between the bonds and/or substituents), and it is reproduced by the third term in Eq. (4.5). In molecular mechanics, additional onefold and twofold terms are added to the torsional equation in order to quantitatively reproduce the curve as shown for the central bond in butane in Figure 4.4, as determined by both experiment and high-level quantum mechanical calculations.[1] The V's in Eq. (4.5) are constants that are chosen to reproduce the known shape of this curve. If higher accuracy is wanted, the Fourier series can be continued. It is found, in general, that V_4 and V_6 terms are of small, marginally significant importance, and they are sometimes included. In MM4, a V_4 term is used in alkenes, and a V_6 term in alkanes, as might be expected from the symmetries of these molecules. These terms allow one to change somewhat the shape of the curve, without changing the torsion angles of the maxima and minima. This in turn allows a better fit to vibrational spectra, which depend on the exact shape of the curve near the energy minima. In ethane, for example, V_4 and V_5 terms cannot contribute anything meaningful because even if such terms are put into the equation, the symmetry of the molecule causes them to cancel out. Only V_3 and V_6 terms can actually contribute to the shape of the curve for this molecule. Similarly, only V_2 and V_4 terms can contribute meaningfully in ethylene. A V_1 term may contribute when the symmetry of a molecule is reduced by the presence of substituents, as, for example, when ethane is converted to butane. The V_5 term is a mathematical possibility but appears to have no physical significance in the first row of the periodic table.

The structure of a molecule is arguably its most important physical characteristic. In current nomenclature the *structure* of the molecule normally means the exact three-dimensional arrangement of the atoms that make up the molecule. One of the important goals of quantum mechanics in chemistry is to be able to correctly calculate structures of molecules. The structure of a molecule is that point on the potential energy surface where the energy of the system is at a minimum. In this case, the "potential surface" applies to all dimensions of the molecule. For molecules such as butane that have different conformations, there are different corresponding energy minima. The structure that has the lowest total energy is referred to as the *global energy minimum structure*.

Normally, in molecular mechanics one knows the connectivity of a molecule and wants to accurately determine its structure. This is done by picking an arbitrary possibility for the structure utilizing the connectivity and the rules of valence, and then systematically searching the energy surface to locate the minimum (or, in the case of more complicated molecules, minima). In principle, this is simple enough. We can calculate the energy of any particular structure from the bending, stretching, and torsion equations mentioned earlier, for each bond, angle, and torsion angle in the molecule, and sum them to get the total energy of the molecule. (There are other energies that also have to be added, as will be discussed later, but for now let us focus just on those mentioned.) If we can find points on the potential energy surface where that energy is a minimum, those points correspond to the structures of the conformers of the molecule. This energy minimization is carried out in practice by determining the derivatives of the energy of the molecule with respect to the atomic coordinates for the starting structure, and systematically moving the atoms in such a way as to minimize the energy. Thus, the problem is reduced to just a problem in mathematics, namely how to find the minimum points of a complicated function. There are a number of ways to solve this problem known to mathematicians, and some of them have been long known (since the time of Isaac

FORCE CONSTANT MATRIX

Newton). We will not go into the details of those procedures here. For that information one may consult any of several books on the subject that are available (see especially Ref. 2b and d.) We do want to discuss a little bit more about this subject, however.

FORCE CONSTANT MATRIX

To find the energy minima as described above, the standard mathematical procedures used involve solving a set of simultaneous equations, which is most conveniently done using matrix algebra. One normally will want what is called the **F** matrix describing a molecule. The **F** matrix is concerned with the potential energy of the molecule and is made up from the force constants that pertain to the internal distortions of the molecule.

The **F** matrix is a square array of numbers. The diagonal elements of the matrix are (in terms of internal coordinates) the force constants for stretching, bending, and torsion, which are described by equations given previously. The other (off-diagonal) matrix elements involve what are referred to as *cross terms*. One of the heaviest parts of the calculation in finding an accurate molecular structure by the method used in MM4 involves the diagonalization of this matrix.[2a,b] The computer time involved in this operation increases as the cube of the dimension of the matrix. Accordingly, much more time is required for larger molecules, and this procedure can become time consuming (by molecular mechanics standards). The calculations are always very fast compared to quantum mechanical calculations (by orders of magnitude).

To illustrate this with a concrete example, let us consider the butane molecule but neglect the hydrogens. Thus, we have just four identical atoms (carbon) connected together in a linear manner. For convenience, we will call this 4-atom structure dehydrobutane, Structure **2**. We can number these 1, 2, 3, 4, according to the connectivity.

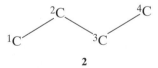

2

Diagonal Part

As a first approximation, we can describe dehydrobutane with a force constant matrix as shown in Eq. (4.6):

$$\mathbf{F} = \begin{pmatrix} k_{s12} & 0 & 0 & 0 & 0 & 0 \\ 0 & k_{s23} & 0 & 0 & 0 & 0 \\ 0 & 0 & k_{s34} & 0 & 0 & 0 \\ 0 & 0 & 0 & k_{\theta 123} & 0 & 0 \\ 0 & 0 & 0 & 0 & k_{\theta 234} & 0 \\ 0 & 0 & 0 & 0 & 0 & k_{\omega 1234} \end{pmatrix} \qquad (4.6)$$

Our matrix elements (force constants) in the **F** matrix will be first the stretching constants for bond 1–2, for bond 1–3, and for bond 1–4, which we will represent as k_{s12}, k_{s23}, and k_{s34}. The bending constants for the molecule will be represented by $k_{\theta 123}$ and $k_{\theta 234}$, and the torsion constant is represented by $k_{\omega 1234}$. The numerical values for constants such as these are normally determined experimentally by vibrational spectroscopy, as the values of these constants are primary determinants of the vibrational frequencies of the molecule.

Since our molecule contains just four atoms and is nonlinear, it will have three translational degrees of freedom, three rotational degrees of freedom (for overall rigid-body rotation), and $3N - 6$ ($3 \times 4 - 6 = 6$) internal degrees of freedom. These latter are conveniently defined by the six internal coordinates as shown on the diagonal of the force constant matrix. As previously indicated, these constants are often determined experimentally by examination of the vibrational spectra of the molecules. They can also be calculated by quantum mechanical methods. This matrix contains a first approximation to the potential energy description of the molecule. This matrix is a short-hand description of the key components of the equations (given earlier) that we have to solve to locate the structure (energy minimum) on the potential surface.

Vibrational Spectra

This will be a good point to digress slightly from the molecular mechanics usage of the force constant matrix, and talk a little bit about vibrational spectra. We would like to be able to calculate the vibrational spectra of molecules after we determine their structures. The model used for a molecule in the classical calculation is that of a number of masses (atoms) held together by springs (bonds). Such a system can vibrate, and will vibrate at only certain frequencies. If the system is vibrating, only these frequencies will be observed in the vibrational spectrum. A major difference between the classical model and the quantum mechanical molecule is that the classical model can stop vibrating, and in fact at 0 K would remain motionless. The actual (quantum mechanical) molecule at 0 K is not motionless, but is confined to its lowest vibrational state. The lowest vibrational state (ground state) has the lowest energy (zero-point energy) that the molecule can actually have. The classical viewpoint is adequate for most purposes in which we will be interested, so we will use that approach, with the quantization superimposed upon it. And we will normally restrict ourselves to the harmonic approximation in the discussion, although in principle the extension to higher approximations is straightforward. (In practice, it can be complicated, however.)

As long ago as 1910, Weniger[3] observed that aldehydes and ketones always showed an absorption frequency around $1750 \, \text{cm}^{-1}$ in the infrared spectrum, and that frequency became a diagnostic indication of the carbonyl functional group in qualitative organic analysis. If one has a molecule containing N atoms, it must have a total of $3N$ degrees of freedom. The molecule has three degrees of translational and three of rotational freedom. There are thus $3N - 6$ degrees of vibrational freedom. ($3N - 5$ for linear molecules.) A vibration for each of these frequencies (more or less coupled together) is normally found in the infrared. They also can be determined in the Raman spectrum, where typically the vibrations are the same, but the selection rules (which lead to strong, weak, forbidden, etc.) are different, which often provides additional useful information. Spectroscopists thus knew about, and utilized, the force constant matrix to calculate vibrational spectra, long before molecular mechanical or quantum mechanical calculations of structures were undertaken.

How does this work exactly? If we imagine a classically ideal system of weights connected together by springs, and the system is suspended in space, we can give it a whack and it will begin to vibrate. The vibrations can be factored into what are called "normal modes." And there will be $3N - 6$ of these normal modes. Hence, our molecule will show that number of bands in the infrared spectrum (with, of course, a number of exceptions that we will not go into at this point). After it had been observed that such bands can be seen in the infrared spectra of molecules, a classical model was used to try to understand what frequencies would be observed for any particular molecule. If we consider a simple diatomic molecule, where we have one atom of infinite mass, and an attached atom, the vibrational frequency is given by

$$v = \frac{1}{2\pi}\sqrt{\frac{k}{m}}$$

where v is the vibrational frequency, k is the force constant (the strength of the bond), and m is the mass of the attached atom. So the small (actually low-mass) atoms tend to give high stretching frequencies, while the large (heavy) atoms give low stretching frequencies. For a polyatomic molecule this formulation expands into what is called a *mass-weighted force constant matrix*, diagonalization of which gives the vibrational frequencies of the molecule. The theory behind vibrational spectra is well worked out. It is discussed in books on spectroscopy,[4] and the standard work is the book by Wilson, Decius, and Cross.[4a]

Thus, it was determined early on that if one had an experimental vibrational spectrum for a known molecule, one could formulate a force constant matrix that would describe this observed spectrum. In that way one could obtain information regarding the various bond strengths of bonds between different atoms, and similar information regarding bond and torsion angles.

The vibrational frequencies of the atoms in molecules do not behave independently, but they couple together. Generally speaking, frequencies that are similar couple rather strongly, and those that are very different couple only weakly. This leads to the idea of "group frequencies," and that specific functional groups within a molecule might be identified by specific frequencies. Unfortunately, this situation is muddied up by the fact that the coupling shifts the bands around, sometimes quite a bit. In molecules containing more than a few atoms, the large number of frequencies can lead to spectra too complicated to fully interpret by inspection.

The infrared spectrum covers the range from approximately 4000–500 cm^{-1} (2.5–20 μm). Still longer wavelengths are referred to as the *far infrared*, and shorter wavelengths lead through the near infrared and into the visible spectrum. The bonds that involve hydrogen at one end (such as C–H, O–H, N–H, etc.) tend to show stretching vibrations in the high-frequency region (3000–4000 cm^{-1}) because of the mass effect of the very light hydrogen. The range from about 600 to about 1600 cm^{-1} is often referred to as the *fingerprint region*. Like people, organic compounds have a "fingerprint" in this region by which they may be identified, with few exceptions. The spectral range from about 600 up to 1300 cm^{-1} is relatively crowded with bands, containing mainly bendings from first-row atoms (e.g., C–C–C) and heavy atom stretching vibrations. Stronger bonds (higher force constants) tend to show higher vibrational frequencies. Thus, triple bonds (such as cyanides or acetylenes) are found in the range of 2100–2260 cm^{-1}. Double bonds (C=O and C=C) are generally found in the range of 1620–1740 cm^{-1} or below that if conjugated. Single bonds are generally found at lower frequencies (1000–1400 cm^{-1}, and still lower for second-row and heavier atoms). In the double-bond region, carbonyl bands are especially obvious because they are very strong.

The intensities of the vibrational bands observed are proportional to the change in the dipole moment when that vibration occurs. What this means is that bonds that are vibrating across a center of symmetry (like the C=C stretch in acetylene) have zero change in dipole moment upon vibration, and hence they have zero intensity. Such bands are said to be *forbidden*, although sometimes they are weakly observed because of various second-order effects. Bonds like carbonyl, which are highly polar, tend to emphasize that polarity more when they vibrate, and hence give intense bands in a relatively vacant region of the spectrum. They can ordinarily be easily identified and accurately measured. Since they have frequencies that are rather different from whatever else is vibrating in the molecule, they couple slightly with other vibrational motions. These small couplings move the carbonyl band around, so that one can rather easily differentiate ketones from amides, or from esters, for example, and one can tell a cyclopentanone from a cyclohexanone, and so forth. Thus, a great deal of information can often be determined about a molecule from the presence of the carbonyl vibration and its exact frequency. Much useful information regarding infrared frequencies is given in books and manuals that are concerned with the use of infrared spectra in structure determination or interpretive spectroscopy.[5]

We want to make the point that the *force constants* (the quantities in the **F** matrix used for calculating the vibrational spectra, which correspond to the atomic motions in question and are symbolized by k) and the *molecular mechanics parameters* also symbolized by k, for stretching, bending, and torsion, are *not* identical. The force constants are the elements in the force constant matrix. But, in molecular mechanics these elements typically contain not only the parameters for the stretching, bending, or torsion that they are used to represent, but they also contain contributions from the van der Waals and electrostatic terms, which cause their numerical values to change somewhat. Thus, k as used in molecular mechanics to represent a *stretching parameter*, and k, as used in vibrational spectroscopy to represent the *stretching constant*, are normally represented by the same symbol. But they represent related, yet somewhat different, quantities. An important difference between the stretching parameter and the stretching constant is that the former is transferable from one molecule to another, whereas the latter is not (except perhaps as an approximation). That is, if one is looking at the stretching parameter for the stretch of the C–C bond in ethane, one uses an identical value for that parameter when one looks at the C–C bond in propane, or in cyclohexane, or in an alkane in general. The molecular mechanics parameters are *transferable* from one molecule to another. This is an assumption that is made, and it is found to work quite well. And it is fundamental to the design of molecular mechanics.

Since the symbolism in common usage does not usually differentiate which of these quantities is under discussion, the reader must always keep in mind whether molecular mechanics parameters, or force constants, are being discussed. In the past, much confusion has resulted in the literature from the fact that authors do not commonly differentiate these two cases. They know which one they are discussing, and they assume that the reader will also know what they are discussing.

So how does our description of the vibrational frequencies of the molecule as represented by a classical system of weights and springs come out when compared with experiment? It comes out pretty well. We can calculate vibrational frequencies to within about $25\,cm^{-1}$ with the MM4 program over a very broad selection of molecules. (This could be done more accurately with a well-defined subgroup of molecules, if desired.) One can also calculate these frequencies quantum mechanically (by calculating the energy minimum, and then two points close to the minimum, and approximating the potential surface there as a parabola, and then calculating the force constants and the spectrum as described for the molecular mechanics case). The quantum mechanical values obtained have to be scaled to account for systematic (basis set and correlation truncation) errors in the calculation. This method, properly used, can give results with an accuracy of about 30 wave numbers over a broad range of compounds (using the B3LYP/6-31G* basis). Both the quantum and the molecular mechanical methods use the harmonic approximation, where the systematic error from that approximation is scaled away so far as possible.

While the **F** matrix shown in Eq. (4.6) is a reasonably good approximation to a mechanical system involving four masses connected by springs, there is something more that needs to be added when we are talking about molecules. We will have to eventually arrive at something that looks similar to Eq. (4.6) for our molecule, but there are additional quantities that we need to introduce that differentiate the real molecule from a collection of only weights and springs. Two of these additional quantities are van der Waals interactions and electrostatic interactions. In the classical system, there are assumed to be no interactions between the atoms, apart from the springs. In the molecular mechanics models there are. If we are going to understand the molecule with this model, then we might think about the interaction between carbon 1 and carbon 4, as an example. There is no explicit distance-dependent term in the **F** matrix in Eq. (4.6) for this quantity. But there will always be such a distance-dependent term (van der Waals) with a molecule. And in a more general case, where C_1 and C_4 are not simply carbon atoms but are atoms in general, some or all of them may carry a partial positive or negative charge, so that there will be an electrostatic interaction between them. In the molecular case, there are additional equations that describe these additional forces. Hence, in the general molecular case, for example, k_{s12} must be the force constant for the change that results when the bond between atoms 1 and 2 stretches, so if the van der Waals and/or electrostatic forces change with that bond stretching, then those changes will also come into the matrix element at that location.

The vibrational frequencies calculated from a force constant matrix of the type represented here [Eq. (4.6)] will be a reasonable approximation, but they will not be very accurate for calculations involving molecules. This matrix can be used in studying other aspects of the potential energy of the molecule as well. But again, the results will not be very accurate. Why are these results approximately true but not accurate? One reason is that the different atoms "feel" one another in ways that are not completely accounted for by just the diagonal part of the force constant matrix. Rather, there are additional interactions that occur between the atoms in real molecules. When one hydrogen moves, its bonding electrons move with it to some extent, and the interactions between those electrons and the bonding electrons of the other hydrogen are thus changed. So a better description of the force constant matrix involves including these *off-diagonal* elements, all of which were set equal to zero in the matrix as written in Eq. (4.6), as a first approximation.

The vibrational motions of a molecule are always coupled to one another. Generally speaking, vibrations that have similar frequencies tend to couple strongly, and those that have very different frequencies couple weakly. We can understand this better with the help of a simple example, the water molecule, discussed earlier. Each of the hydrogens shows a stretching vibration with respect to the oxygen. If the oxygen were infinitely heavy, ideally both of the hydrogens would have the same vibrational frequency. As they vibrated, the oxygen would remain at rest. But the oxygen is not infinitely heavy, relative to the hydrogen. So when a hydrogen moves during the vibration, the oxygen also moves a little bit in the opposite direction. And the second hydrogen feels that little bit of motion of the oxygen, brought about by the first hydrogen. So these two O–H vibrations are coupled together. In fact, the two hydrogen motions coupled together are described as symmetric and antisymmetric vibrations of both hydrogens at the same time, according to the two structures shown on the left in Figure 4.5.

These two motions have similar but different frequencies. If we had an ideal mechanical system, simply weights and springs, then we could calculate accurately these frequencies. But for water such calculations do not come out quite right because of additional coupling. The coupling between the two hydrogen atoms (and their electrons) can be better represented by adding an additional equation into our system, as shown in Eq. (4.7):

$$E_{ss} = k_{ss}(l_1 - l_0)(l_2 - l_0) \qquad (4.7)$$

Note that Eq. (4.7) is similar to Eq. (4.1), but instead of stretching just one bond [as in Eq. (4.1)], two bonds are stretched at the same time. In the simplest approximation, this k_{ss} is simply taken to be zero. In a better approximation, it is given a value that enables the experimental spectrum to be accurately reproduced by the molecular

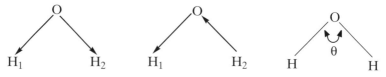

Figure 4.5. Vibrational motions in water.

FORCE CONSTANT MATRIX

mechanics force field. This quantity appears in the force constant matrix as an off-diagonal element, coupling together the two O–H stretchings (later).

There is also a bending motion, where the angle HOH (θ in the right-hand structure in Fig. 4.5) alternately increases and decreases. It is found that this motion is again qualitatively represented by a classical equation [Eq. (4.8)], but again, to properly fit the vibrational frequencies of water, we also need to include stretch–bend interaction equations [Eqs. (4.9) and (4.10)]:

$$E_\theta = \frac{k_\theta}{2}(\theta - \theta_0)^2 \tag{4.8}$$

$$E_{s(1)\theta} = k_{s\theta}(l_1 - l_0)(\theta - \theta_0) \tag{4.9}$$

$$E_{s(2)\theta} = k_{s\theta}(l_2 - l_0)(\theta - \theta_0) \tag{4.10}$$

Looking again at our dehydrobutane molecule, and the force constant matrix that describes it [Eq. (4.6)], we see that there are stretching interaction constants that might be expected between the three different stretching elements, and there is a bending interaction element expected between the two angles. There are also stretch–bend, stretch–torsion, and bend–torsion interactions possible between the appropriate elements. Note that each degree of freedom can interact with each other degree of freedom in the molecule, and this is true in general, regardless of the size of the molecule. Most of the direct interactions that are between atoms that are located further away from one another than 1,4 are small, however, and those terms are routinely set equal to zero. Those 1,5 and more distant interactions may show van der Waals and/or electrostatic interactions, however, in which case they may contribute off-diagonal terms to the force constant matrix. So for a large molecule, there can be a stretch–stretch interaction between two bonds that are far removed in space. One would think, and one finds, that such interactions are normally negligible (zero). The things that actually interact significantly in the molecule usually involve parts of the molecule that are close together, either because of their connectivity or because the molecule twists in such a way as to bring close together parts that are far away in connectivity (through bonds).

Off-Diagonal Part

So while our four-atom carbon system can be described approximately by Eq. (4.6), taking all of the off-diagonal elements to be zero, a better description is as given in Eq. (4.11), where the various interaction elements are included.

$$\mathbf{F} = \begin{pmatrix} k_{s12} & k_{(s12)(s23)} & k_{(s12)(s34)} & k_{(s12)(\theta123)} & k_{(s12)(\theta234)} & k_{(s12)(\omega1234)} \\ & k_{s23} & k_{(s23)(s34)} & k_{(s23)(\theta123)} & k_{(s23)(\theta234)} & k_{(s23)(\omega1234)} \\ & & k_{s34} & k_{(s34)(\theta123)} & k_{(s34)(\theta234)} & k_{(s34)(\omega1234)} \\ & & & k_{\theta123} & k_{(\theta123)(\theta234)} & k_{(\theta123)(\omega1234)} \\ & & & & k_{\theta234} & k_{(\theta234)(\omega1234)} \\ & & & & & k_{\omega1234} \end{pmatrix} \tag{4.11}$$

Note that this matrix (and similarly other **F** matrices) is symmetric about the diagonal. That is, the matrix element in the second column and first row is a stretch–stretch interaction between atoms 1–2 and 2–3. The matrix element in the first column of the second row (not shown) is exactly that same quantity. Hence, the upper triangle (above the diagonal) of the matrix is identical numerically and conceptually to the lower triangle (below the diagonal, which is not shown). This is the normal situation in mechanics, and hence we will normally write out and discuss only the upper triangle of the **F** matrix.

Evidently, it will take a much greater effort to determine all of the elements given in Eq. (4.11) than it did to determine those given in Eq. (4.6). To apply Eq. (4.6) to large molecules, the amounts of information needed and the calculation times increase as the number of internal degrees of freedom of the molecule (the number of diagonal elements). In Eq. (4.11), the amount of information needed increases as approximately the square of that number [actually as $(n^2 + n)/2$]. Hence, solving problems of this type becomes much more time consuming as the size of the molecule increases. Even worse, we have to know the values of all of these additional force constants.

The good news is that we know that in a large molecule parts of the molecule that are far apart are not likely to show much if any interaction with each other. So what is done in real life is to initially set all of the off-diagonal elements in the **F** matrix at zero and determine the actual values for only a relatively small number that are known to (or expected to) be important. Even so, the determination of what we might call a "chemically accurate" force constant matrix is many times more complicated than determining just a diagonal matrix. And why do we want to do all of that work? Because if we do not, the results that we obtain will probably not be of chemical accuracy. (Chemical accuracy might be thought of as something less than 1 kcal/mol in terms of energy and about 0.005 Å in terms of bond length, although it may vary quite a bit from one type of situation to another.) If one is calculating heats of formation, for example, numbers that are in error by 10 kcal/mol are ordinarily useless. One can often guess numbers with better accuracy than that. Numbers with errors of 1 kcal/mol are not as accurate as can be obtained by the best experiments, but they are more accurate than some that are obtained by experiments in difficult situations. Current day force fields typically allow energy calculations between conformations of ordinary organic molecules to within fractions of a kilocalorie/mole, for example. For those who work with larger, and especially very large molecules, it should be noted that both experimentally and computationally, the error expected in a heat of formation increases with increasing molecular weight. Experimentally, this is true because when one burns a sample, for example, one is burning a certain number of *grams* of compounds, and the number of *moles* involved becomes smaller as the molecular weight increases, so the smaller sample in molar terms leads to more error per mole. Computationally, the expected error is on a per bond basis. Hence, larger molecules have larger numbers of bonds and larger uncertainties in the computational energies.

Let us think again about the dehydrobutane molecule and its internal degrees of freedom. Evidently, there are three bond lengths that can stretch, two bond angles that can bend, and one torsion angle which that rotate. Thus, there are six internal degrees of freedom in the molecule, and, hence, we will need to solve six simultaneous equations to determine that point on the potential surface where the energy is simultaneously a minimum in each of those degrees of freedom. But, if we think about this a little further, we will realize that these six degrees of freedom are not necessarily totally independent. That is, they can interact, which will lead to additional equations. The stretching interaction between atoms 1 and 2 interacts with that between atoms 2 and 3, for example. Just as the two H–O stretches in water interact to yield a stretch–stretch interaction [Eq. (4.7)], so here the two C–C bonds can interact. This leads in Eq. (4.11) to the off-diagonal matrix element $k_{(s12)(s13)}$.

Thus, instead of having the square of one stretch, we have the product of the two stretches. And there can, in principle, be a cross term like this between each pair of the internal coordinates, so we may have bend–bend, stretch–bend, torsion–stretch, and so forth. Fortunately, many of these, most in the case of a large molecule, will turn out to have values of zero (or at least small enough so that zero is a good approximation). Hence, one usually begins by assuming that these cross terms are negligible (zero), unless there is some reason to think otherwise. Much of structural chemistry is connected with cases where we have compelling reasons to think otherwise (later).

It will be convenient to now discuss these equations in terms of the **F** matrix, which would then be used to actually solve the equations. We shall see later that it turns out that most of these troublesome cross terms result from the interaction of a lone pair of electrons with other nearby parts of the molecule. Since there are no lone pairs of electrons in hydrocarbons, this cross term problem is minimal in such compounds, and taking all of the cross terms equal to zero is a fair approximation. But a few of these terms are somewhat important, even in hydrocarbons. These few will be discussed here.

STRETCH–BEND EFFECT

Vibrational spectroscopy was a forerunner of molecular mechanics in the sense that spectroscopists used many of the basic methods used in molecular mechanics to calculate spectra, long before it was possible to calculate structures. It was early found that a purely diagonal force field was less accurate than desired, and that cross terms could improve the quality of the spectra calculated. One of these cross terms was a stretch–bend term. It has the form shown in Eq. (4.12):

$$E_{l_1\theta_{123}} = k_{l_1} k_{\theta_{123}} (\Delta l_1)(\Delta \theta_{123}) \tag{4.12}$$

The sign of the constant term is such that as the bond angle becomes smaller, the bonds tend to stretch. (The physical effect that leads to the requirement for this type of term will be discussed in the next section.) This term is particularly important in cyclic compounds. The C–C–C bond angles in cyclohexane normally have values a little larger than tetrahedral, just as do open-chain alkanes. But small rings, such as cyclopentane and cyclobutane, necessarily have these angles squeezed down smaller than tetrahedral. And the result is that carbon–carbon bonds in those rings are substantially longer than those in cyclohexane, or in normal alkanes. Many larger rings (cycloheptane, cyclooctane, and especially the medium rings such as cyclodecane) tend to have these bond angles opened to larger than normal values. That, in turn, causes these bonds to shrink in length, although in many cases the expansion resulting from the internal congestion in such compounds largely negates the results of the stretch–bend terms.

UREY–BRADLEY FORCE FIELD[6]

An alternative way that was found of allowing for the effect of the stretch–bend interaction in a molecule such as water or propane was not to add cross terms to the force field but rather to add what are called *Urey–Bradley terms*. In this type of force field one uses harmonic terms for bending and stretching as usual, but instead of adding a cross term, an additional term is used in which a harmonic distance potential is added between atoms that have a 1,3 relationship to each other, but with a much smaller force constant. Thus, for propane in the ground state, there would be some optimum distance between carbons 1 and 3. A harmonic potential, which would look exactly like that used for bond stretching (Eq. 4.1) with a smaller stretching constant, would be applied to the distance between atoms 1 and 3. What that means is that if the bending motion, say, of the molecule were such as to bring those two atoms closer together, then a 1,3 stretch would result, which tries to push them back further apart. So the result would be similar to that of the stretch–bend effect discussed previously.

It was noticed long ago that in simple molecules such as propane and isobutane, the C–C–C bond angles are larger than tetrahedral by 2–3°. This suggested that the methyl groups in those molecules are repelling one another (discussed further in the next section). The Urey–Bradley picture is also consistent with this viewpoint. This lets us understand (in part) why in cyclopentane, and even more cyclobutane, for example, the bond lengths are much longer than those in butane itself. (In order not to oversimplify the problem, it should be pointed out that an eclipsed bond is normally longer than the corresponding staggered bond (e.g., in ethane). Cyclopentane and cyclobutane, therefore, have their bonds elongated somewhat by this partial eclipsing. But that elongation is insufficient to explain the observed bond lengths. The inclusion of a stretch–bend interaction is also needed to account fully for the stretching that is observed.[7])

The results given by a Urey–Bradley force field are very similar, although not identical, to those given by a force field utilizing stretch–bend cross terms. One is likely to need many cross terms in the force field anyway, apart from the stretch–bend interactions, so it is more convenient for most purposes just to add stretch–bend terms here as well. The Urey–Bradley force field is consequently seldom used anymore, but it is mentioned here because it is conceptually useful.

VAN DER WAALS FORCES

These forces present a bigger problem in molecular mechanics than might be recognized at the outset. The van der Waals force has two parts, an attraction, which goes to zero at long range, and a much more serious repulsion, which comes in at a shorter

VAN DER WAALS FORCES

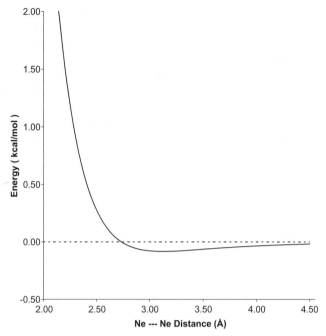

Figure 4.6. The van der Waals energy for the interaction between two neon atoms.

range. The van der Waals interaction energy between two neon atoms as a function of distance is shown in Figure 4.6.

These two forces together yield a curve that starts with an energy of zero at an infinite distance. As the atoms approach one another, the curve goes downward, to some minimum energy value (called the sum of the van der Waals radii of the interacting atoms), and then turns and goes steeply upward. The attractive part of the curve is due to electron correlation, and the energy varies rather strongly with distance. The leading term in this attraction has a $1/r^6$ dependence, and only the leading term is normally used in molecular mechanics. The repulsive term begins to take effect when the atoms come much closer together, and it has a much stronger distance dependence. In his early studies, Lennard-Jones used an r^{-12} dependence for this repulsion,[8] but that is now known to be somewhat too steep. That particular formulation has great mathematical convenience, however, which was the main reason for using that exact exponent. One formulation of the van der Waals potential, then, is given by Eq. (4.13):

$$E_{vdw} = -\frac{a}{r^6} + \frac{b}{r^{12}} \qquad (4.13)$$

If one were to use an equation of that form (the so-called *6/12 potential*), a number having a fractional value between 8 and 9 would be better[9] than 12. But the 6/12 potential is still commonly used, especially in studies of macromolecules, for two reasons.

The first reason is that the calculation is very much faster if the number 12 is used, rather than some alternative. And for very large molecules, the rate-determining step in calculating molecular energies and the energy derivatives needed to calculate molecular structures comes from the van der Waals (VDW) part of the calculation. The second reason is that as long as one has a number substantially larger than 6, the exact value really makes a difference only at very close distances that are almost never met with in molecular structures.

A formulation that is somewhat better is referred to as the *Buckingham potential*.[10] This is an exponential formulation for the repulsive part, and the same r^{-6} potential for the attractive part as shown in Eq. (4.14):

$$E_{\text{vdw}} = -\frac{a}{r^6} + be^{-cr} \tag{4.14}$$

The exponential formulation, as used in Slater orbitals, is known to be the best simple representation of the electron cloud about an atom. The MM2, MM3, and MM4 force fields use this Buckingham formulation (which is often referred to as an *exp6 potential*). The disadvantage for large molecules is that it requires a longer computing time than does a 6/12 potential.

A Buckingham version of the van der Waals potential can be formulated as in Eq. (4.15) (called the Hill equation[11]), as it was used in MM2:

$$V_{\text{VDW}} = 2.90 \times 10^5 \varepsilon e^{-12.50 r_0/r} - 2.25\varepsilon \left(\frac{r_0}{r}\right)^6 \tag{4.15}$$

where r is the distance between the two atoms involved, r_o is the sum of the van der Waals radii of the two atoms, and ε is proportional to the depth of the well. This form of the equation can be formulated with just two parameters, r_0 and ε. Note that in Eq. (4.15), the energy minimum is equal to 1.1ε. Also, the r_0 term for the specific case of hydrogen is calculated in a special way (later).

These van der Waals potentials are experimentally known and can be calculated for rare-gas atoms. But there is a serious problem when one wants to use these potentials in molecular mechanics to study molecules, wherein atoms are bonded to one another. The rare-gas atoms are spherical, but because of the redistribution of electrons upon bonding, atoms in molecules are only roughly spherical. The obvious approximation is to consider a molecule as an assemblage of a group of spherical atoms, where the van der Waals potential is centered at the nucleus for each atom. And this approximation appears to give reasonable results.

We need to decide at the outset of the van der Waals model of the molecule if we want to calculate these interactions between all of the atoms of the molecule, or should we exclude specific atomic pairs. In large molecules the interactions are sometimes set to zero at long distances just to save computational time, but there are a few cases where they are really zero (in the MM model). These are when the atoms are 1,2 (bound together), 1,3 (bound to a common atom), or in some force fields, 1,4 (interacting through a torsional potential). MM4 uses an ordinary VDW potential for atoms that are

1,4 as we have found no advantage to doing otherwise, but 1,2 and 1,3 are clearly special cases.

If atoms are bound together, quantum mechanics and experiment both show that the bond between them is well described by a Morse potential. In the MM model, the VDW interaction is considered to just be lumped into the Morse curve and does not need to be considered separately. (It is much weaker than the normal bonding interaction, and the distance dependence curve has the same general shape.)

If atoms are bound to a common atom, one would expect a van der Waals interaction between them. However, if this interaction were included explicitly, since the atoms are quite close together (2–3 Å), one would have to be careful about the exact shape of this curve to get a correct energy variation with the vibrational motion. A first approximation would be just to find the minimum energy point and add the Urey–Bradley term (the harmonic approximation). And this is a reasonable approximation. But the mathematics is more simple if one uses a stretch–bend term rather than a Urey–Bradley (UB) term. So this stretch–bend term is important, even in hydrocarbons, and it is easy to see why in the UB model. It is almost as easy to see why in the stretch–bend model. When we need a cross term in the **F** matrix, it is because the two diagonal elements are not completely independent. That is, they interact. That interaction can be well represented in the present case by just linear terms as in Eq. (4.7). (If the linear terms were to be inadequate by themselves, quadratic or higher terms could also be added, but these are not needed here or in most of molecular mechanics.)

Next, the problem arises of how do we evaluate the parameters in this equation? Each different kind of atom must have its own values for r_0 and ε_0. We know the values for the rare gases, so we have a pretty good approximation as to the values for the individual atoms, but there is still quite a bit of leeway. When we wrote the *Molecular Mechanics* book in 1982 we included some graphs that showed at that time the various van der Waals functions that were being widely used by various research groups.[12] The differences between these functions were quite considerable for various reasons. These problems were largely resolved over the following years, and the potentials in use today are much closer to one another than they were in 1982. They still differ somewhat, however, because they are (slightly) force field dependent. The MM2 potential was found to be somewhat too "hard," and the energy increased too fast at short distances. The MM2 constants given in Eq. (4.15) are, therefore, a little different in MM3 and MM4 (discussed later), and the latter yield somewhat softer van der Waals curves.

How do we evaluate the parameters needed for the van der Waals function for all of the various atoms on which we will wish to carry out molecular mechanics calculations? There are different ways that one might try to do this. What we will discuss here is the method that was used for evaluating these parameters for MM4. These gave the best values that we were able to determine at the time. We do wish to point out that all of the parameters in a force field are, to some extent at least, interdependent. Accordingly, one cannot transfer parameters from one force field to another with reliable results. The way a molecular mechanics force field is developed is to fit various pieces of information as well as possible with the parameters available in that force field. For each error or inaccuracy in any particular parameter, the nature of the fitting is such that the rest of the parameters in the force field will adjust in such a way as to yield an optimum structure. Hence, to some extent errors tend to cancel themselves out, but they do it in a way that means the parameters become interdependent, and hence a property of that particular force field. They may be transferred to a different force field as an approximation, but this is not a recommended procedure for reasons that should be apparent.

In the MM4 case, we decided that it was most important to first fit the van der Waals properties for the alkanes, and this was done in the following way. First, it was assumed that the carbon van der Waals potential would be centered at the nucleus of the carbon atom. But for hydrogen, we know (e.g., from crystallography and from ab initio calculations) that the electron density of a hydrogen atom in an alkane is not centered at the hydrogen nucleus but is pulled inward toward the carbon nucleus to which it is bound. Accordingly, we broke the calculations involving hydrogen atoms into two parts. The first part is the location of the hydrogen nucleus, and for that we wish to fit to microwave data that actually locate exactly where the nucleus is. For the second part of the van der Waals calculations involving hydrogen, we assumed that the electrons can be represented by a soft sphere, but that sphere was not centered at the hydrogen nucleus. Rather it was offset inward from the hydrogen nucleus toward the attached carbon, by an amount that was to be optimized empirically.[13] This gave us what was considered to be a desirable, if not a necessary, separation between the hydrogen nuclear position and the center of hydrogen van der Waals interaction, which comes largely from the electrons. (The offset is along the C–H bond, measured as the fraction of bond length outward from carbon, and has the value 0.940 in MM4.)

Then we have some key pieces of information, two of which are the van der Waals characteristics of helium and neon. (Larger rare gas atoms also become important when one considers atoms further down in the periodic table.) How closely can we approximate the van der Waals characteristics of hydrogen with those of helium? And how closely can we approximate the van der Waals properties of carbon with those of neon? We would expect that carbon and neon would be rather similar in their properties. We know that as we go across the periodic table from left to right, the van der Waals radii of the atoms become smaller,[14] and we know by approximately how much. The value of ε for carbon would be expected to be similar to that for neon. The polarizabilities of atoms decrease somewhat as we go to the right in the periodic table. The number of electrons may increase, but they are more tightly held. But we also know that there is a big difference in polarizability between the first-row elements and second-row elements. To a first approximation, the first-row elements are similar. On the other hand, one expects a sizable difference between hydrogen and helium.

There is a real problem here in the following way. We can measure the van der Waals properties of a molecule such as methane, for example. But we cannot individually measure the van der Waals properties for the hydrogen atoms and for the carbon atom in methane. And the numbers that we would like to use will vary somewhat, depending on the exact function chosen for the van der Waals interaction, and on the other approximations that have been previously outlined. And from this point, different workers managed to deduce rather different values for van der Waals functions, as discussed earlier.

Our approach was to use graphite to help fix the van der Waals quantities in question for carbon. Graphite has sheets of carbon atoms bound into hexagons, and the interaction between these sheets is, in our model, simply due to the van der Waals forces between those sheets. So as a first approximation, we took the van der Waals radius and the ε value of carbon to have the same values as those of neon. Then we adjusted these values in such a way as to fit the structure and compressibility of crystalline graphite. The numbers came out as one might expect in that the carbon in graphite has a somewhat larger van der Waals radius than does neon, and the ε value is similar.[15] The values chosen for the van der Waals parameters of nitrogen, oxygen, and fluorine were then obtained by interpolation between the values for carbon and neon. For the second-row atoms, the values were extrapolated to the left using the values for argon, and paralleling those of the first row. Values for the remainder of the periodic table were obtained by extrapolation downward and to the left. They are probably not very accurate, but they have proved to be adequate for the purposes for which they have so far been used.

There is one other important case, which is when two different kinds of atoms, say carbon and hydrogen, undergo a van der Waals interaction. It was assumed in the earliest work that the van der Waals radii for different atoms could simply be summed. And it was also assumed that one could use the root mean square ε values of the different atoms for the well depth. This way of calculating the epsilons seems to be a good approximation and is still regularly used. However, the additivities of the radii, while a reasonable approximation, seemed not to be quite right. At least in the MM4 force field, we find the approximation adequate for interactions involving first- and second-row atoms with each other, but for interactions involving first- or second-row atoms with hydrogen, the values so calculated need to be reduced somewhat.[16] If this were not done, the resulting repulsions were found to be disproportionately large at close distances, and they lead to some distortions in molecular structures and properties (later).

When all of the above is done, one can obtain what appear to be fairly good van der Waals potentials for carbon and hydrogen in alkanes and reasonable potentials for the upper right part of the periodic table. The van der Waals characteristics finally settled on for this part of the periodic table are shown in Table 4.1.

We should perhaps here briefly discuss a point that is often overlooked by those who are not familiar with molecular mechanics. There are actually in the literature two

TABLE 4.1. MM4 van der Waals Parameters[a]: Radius (Å), ε (kcal/mol)

Hydrogen 1.640/0.017							Helium 1.530/0.026
Lithium 2.55/0.007	Beryllium 2.23/0.010	Boron 2.15/0.014	Carbon 1.960/0.037	Nitrogen 1.860/0.054	Oxygen 1.760/0.60	Fluorine 1.660/0.067	Neon 1.560/0.074
			Silicon 2.290/0.140	Phosphorus 2.190/0.168	Sulfur 2.090/0.196	Chlorine 1.990/0.224	Argon 1.880/0.252

[a] MM4 manual.

separate definitions of the expression *van der Waals radius*. In molecular mechanics, if we bring together two neon atoms, say, the position where the energy curve reaches a minimum is referred to as the *sum of the two van der Waals radii of the neon atoms* (Fig. 4.6). This is one of the definitions, and it is clear, straightforward, and unambiguous.

The second definition of van der Waals radius is used, for example, by Pauling in his book.[14] In that definition, if we had two neon atoms, packed into a neon crystal, the distance between them would be twice the van der Waals radius of neon.

At a quick glance without much thought, one might think these two definitions would yield numbers for the van der Waals radius of neon that are the same, but they are not even very similar. The Pauling *distance of closest approach* should be referred to as that, and not as the sum of the van der Waals radii. The distance of closest approach is always smaller (actually, very much smaller) than the actual sum of the van der Waals radii. This can be seen in the following way.[17] Suppose we have two neon atoms (1 and 2) motionless and at the sum of the van der Waals radii, as in Figure 4.6. They are at their minimum energy position. Suppose we now add a third neon atom 3, in a straight line 1–2–3 so that 3 is away from 2 by the sum of their van der Waals radii. Atoms 1 and 2 are at their van der Waals minimum, and so are atoms 2 and 3. But there is also an additional attraction between atoms 1 and 3 (attraction because they are further apart than the sum of their van der Waals radii). Because of this further attraction, atoms 1 and 3 will tend to move together, and that will lower the energy of the system. That motion will continue until the attraction energy between 1 and 3 does not increase faster than the repulsion energies that are increasing between 1 and 2, plus 2 and 3. Hence, at the energy minimum for the three atom system, the atoms 1 and 2, and also 2 and 3, will be closer together than the sum of the van der Waals radii. They will be closer together by something like 0.4 Å, a nontrivial amount. If we add a fourth atom to the straight line of atoms for the system (1, 2, 3, 4), there will be attractions of atom 4 with atom 1 and with atom 2, and those attractions will cause atom 4 to approach the remaining atoms more closely, which will cause pairs 1 and 2, as well as 2 and 3, to compress even further. And so on. So that the distance of closest approach in a crystal is quite a lot smaller than the distance that is defined by the sum of the van der Waals radii, which is the definition used in molecular mechanics.

And this is a general phenomenon. The sum of the van der Waals radii as defined by the van der Waals curves is considerably larger than the distance of closest approach of two atoms in a crystal (the Pauling radius). One can imagine exceptions, due to large charges, or charge transfer (bonding) interactions. But, generally speaking, the distance of closest approach between two atoms in a crystal, when there are no extraneous effects, will be very much smaller than the corresponding sum of the van der Waals radii. (These Pauling radii are what are usually used to determine atomic sizes in space-filling models.) Unfortunately, Pauling referred to these distances of closest approach as *van der Waals radii*, and that misnomer is widely seen in the older literature, and is still commonly used by crystallographers. For comparison, Pauling gives the van der Waals radius of hydrogen as 1.2 Å, that of carbon as 1.85 Å, nitrogen 1.5 Å, oxygen 1.4 Å, fluorine 1.35 Å, and that of chlorine as 1.80 Å. One expects, and we find (Table 4.1), that these Pauling distances and van der Waals radii are approximately parallel to one another as we move across or down the periodic table, with the Pauling distances being somewhat smaller (0.1–0.4 Å). This distance of closest approach also suffers from the problem that it cannot be accurately defined because the same atoms approach to somewhat different distances in different crystals.

All of the above have been known for a long time.[14,17] If one wishes to pack alkanes into a crystal by molecular mechanics, the van der Waals radius required for hydrogen (depends a little bit upon the remainder of the force field, but MM4 is typical) for MM4 is 1.64 Å, much larger than the Pauling distance of closest approach (1.2 Å), as expected.

One additional point should be made here, regarding van der Waals radii, but particularly those of hydrogen. The latter are particularly important because most organic molecules are more or less covered by a "skin" of hydrogen atoms. The alkanes, of course, are the most extreme case, but a similar situation applies to typical organic molecules. The attractive interaction between two hydrogen atoms only amounts at most to about 0.02 kcal/mol, a trivial amount. But when two modest sized molecules come together, there are quite a large number of such interactions, and this energy can become quite important. The biochemists refer to something called the "hydrophobic effect." In proteins, for example, this refers to the tendency for proteins to try to twist into conformations where alkyl groups are adjacent to other alkyl groups, and away from the surrounding water solvent. The alkyl groups do attract each other to a significant extent, but much of the effect in biology results from the fact that alkyl groups also disrupt the hydrogen bonding in the water solvent. By removing the alkyl group from the water and packing it next to another alkyl group, the molecule not only gains the van der Waals attraction between alkyl groups, but the system also gains hydrogen bonding between the no longer disrupted water molecules in the solvent.

The actual van der Waals radius of an atom is both quite important and quite unimportant, depending on the context. If one looks at Figure 4.6 (neon atoms), one can see that near the van der Waals distance the energy curve is very flat (at about 3.1 Å in this particular case), and the energy is very small (0.08 kcal/mol). So if one moves the van der Waals radius, say 0.1 Å in either direction, nothing significant happens to the energy in the area of the van der Waals radius itself. However, the steep repulsive curve that starts upward at about 2.7 Å and continues up ever more steeply at shorter distances is quite dependent upon the value of the van der Waals distance. If the van der Waals distance is moved inward, then that steep curve would also move inward. This motion would have significant consequences, not only in cases of hydrogens approaching each other too closely with resulting geometric changes (later) but also in affecting quantities such as the heat of sublimation of a crystal.[7]

Dr. Miller's Nuclear Explosion

In the early days of molecular mechanics by computer (about 1966), Dr. Mary Ann Miller, a co-worker, came to me one day and said something to the effect that she was minimizing the energy of a molecule, and two of the hydrogen atoms within that molecule collapsed to form helium, and there was a nuclear explosion.[18] And I said "what???" But I looked at her printout, and indeed it was true. The two hydrogens came together to exactly the same point, that is, they formed a helium atom (although an unusual isotope, helium 2) and the energy of the system went to minus infinity. After a few moments of bewilderment, it became clear what had happened. If one uses the exponential form of the van der Waals equation, the exponential term goes to a large positive finite value as the distance between the atoms goes to zero. But the r^{-6} term goes to minus infinity! So indeed, if you push two hydrogens together too closely, they combine to form helium, and there is a tremendous release of energy (a nuclear explosion). Of course, we did not wish to have nuclear explosions in the laboratory all the time, so a fix was put into the program so that if the distance between two atoms gets too close, the program automatically switches over to a function that prevents this from happening (see Nuclear Explosion Preventer in the Appendix). It is perhaps interesting just how perceptive the program was regarding such matters and shows the type of hazard that one might face if one used the earlier program as a black box.

van der Waals Interactions between Nonidentical Atoms

The foregoing discussion involving van der Waals forces has been mainly limited to interactions between pairs of atoms in which the two atoms were identical (two hydrogen atoms, two neon atoms, etc.). We now need to consider the other important case, which is the van der Waals interactions between two atoms when they are not identical. (It is always assumed that van der Waals interactions in molecular mechanics can be adequately described by two-body interactions only. No three-body or higher interactions are used, and up until now at least, this has always been an adequate approximation.) We need especially to consider the interaction between carbon and hydrogen but also between any two different kinds of atoms present in a molecule. Because of the forms used for van der Waals curves, the early assumption was made that the van der Waals interaction between nonidentical atoms could be formulated in a simple way, as follows. For the van der Waals distances, if the two atoms are the same, we simply sum the two van der Waals radii to obtain the van der Waals distance. If the two atoms are not the same, we can similarly sum the van der Waals radii of those two atoms to get the van der Waals distance. This assumption is almost always made, and is a reasonable first approximation.

Our treatment of the potential well depth parameter (ε) was to determine it along with the radii from the neon/graphite data, along with the radius to fit physical properties (including second virial coefficients of gases, the crystal packing distances, and the heats of sublimation of crystals). As before, the values were modified slightly when one of the atoms was hydrogen. Thus Eq. (4.15) would be expanded to give Eq. (4.16) for the interaction describing two atoms a and b, and it becomes:

$$V_{\text{vdw}} = 1.84 \times 10^5 (\varepsilon_a \varepsilon_b)^{1/2} e^{-12.00(r_a + r_b)/r} - 2.25(\varepsilon_a \varepsilon_b)^{1/2} \left(\frac{r_a + r_b}{r} \right)^6 \quad (4.16)$$

This is the equation used in MM4. If either a or b (or both) is hydrogen, its position is measured from the offset corresponding to the electron density (discussed earlier). We continue to use this well depth approximation for interactions of all first-row and second-row atoms with one another, and it seems to work adequately. There is, however, a problem when the interaction is between carbon and hydrogen that was noted long ago.[19] The problem is that this approximation leads to internal repulsions in alkanes that are excessive, as shown by the fact that molecular structures calculated using this approximation have bond lengths that are increasingly too long as the structure becomes more congested. In particular, if this approximation were used for the series ethane, propane, isobutane, and neopentane, one would find that the calculated bond lengths in these molecules would increase much more as we go along the series than is found experimentally.[7] While there may be other ways to overcome this problem, most force fields long ago simply reduced the van der Waals repulsion between carbon and hydrogen, relative to what it would have been with the formulation previously discussed. Reducing the ε value is of little help, and would greatly change the heats of sublimation of crystals by changing the long-range attractions. In order to make the geometric changes of the size actually needed, one has to reduce the sum of the van der Waals radii between the carbon and the hydrogen to a value that is about 5% less than the sum of those radii, and the ε is increased slightly so as to leave the long-range r^{-6} energies unchanged. When this is done, one can calculate these bond lengths properly, plus the other van der Waals properties for alkanes including those of their crystals, both geometries and heats of sublimation. For a long time it was thought that this C–H interaction was just a special case, but after sufficient studies were carried out[16] with molecules containing other first-row atoms, it was found that the optimum value for nitrogen, oxygen, and fluorine, the sum of the van der Waals radii between hydrogen and any one of those atoms must also be reduced somewhat (it turns out to be about 15% in each of these cases), in order to fit various kinds of data available (particularly energy barriers in congested molecules) for molecules involving these atoms.

CONGESTED MOLECULES

However, there is still something here that requires a little more work. And this is the potential between hydrogen atoms that are very much closer together than the sum of the van der Waals radii. This situation occurs widely in congested molecules, so that it is important to establish the energy of the H–H van der Waals interactions at distances that are appreciably shorter than the sum of the van der Waals radii. And, of course, one does not want to do this with just one molecule and one distance (although one could) because of the inherent errors that will be involved in determining whatever close value for the distances used. Rather, one wants to use a number of different compounds, with different distances, and do a least-squares optimization. It is rather difficult to calculate the energy accurately for the van der Waals interaction between two hydrogens when they are quite close together. The problem is that the molecule relaxes very much, and that leads to changes in the interatomic distances. How much distortion occurs depends on the particular molecule one is talking about, and how many and what kind of degrees of freedom can be distorted to let the molecule relax. It will also obviously depend on other parameters used in the force field, especially those for bending. Quantum mechanical calculations were not of much help here because molecules that were large enough to really show the effects that we were looking for could have calculations carried out upon them only with relatively small basis sets and with uncertain accuracy.

Tetracyclododecane

This molecule, and a few of its relatives, have been studied with some care, in order to obtain accurate information on the van der Waals potentials between hydrogens that are located much more closely with respect to one another than the sum of the van der Waals radii.[16] The structure of norbornane is shown, together with a related molecule, tetracyclododecane (Structure **3**):

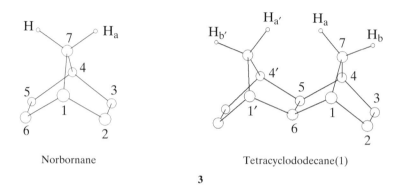

Norbornane

Tetracyclododecane(1)

3

Norbornane itself is, of course, a highly strained molecule. Both of these molecules have C_{2v} symmetry. For norbornane, the axis goes vertically through carbon 7, and the planes are defined by H–C_7–H_a, and perpendicular to that, C_1–C_7–C_4. This means that for norbornane, bonds and angles that look the same are the same. For example, the C–C–C angles at carbons 2, 3, 5, and 6 are identical, and so forth. For tetracyclododecane the axis is vertical, goes through the center of the C_5–C_6 bond, and midway between $H_{a'}$ and H_a. One plane includes $H_{b'}H_{a'}$, H_a, and H_b, and the plane perpendicular to that contains the C_5–C_6 bond and the midpoint between H_a and $H_{a'}$. In the following discussion we will discuss only one angle of an identical (symmetry-related) group.

In the tetracyclododecane molecule, two of the hydrogens on the bridges at the top of the molecule (H_a and $H_{a'}$) are much closer together than the sum of the van der Waals radii, and they are held in that position by the relatively rigid ring structure. The first point to make is that the symmetry of tetracyclododecane is C_{2v} (no imaginary vibrational frequencies). This means that H_a and $H_{a'}$ could have moved out of the symmetry plane they are in by one moving above and one moving below the plane of the paper, but they didn't do that. If the repulsion between H_a and $H_{a'}$ were increased sufficiently, they would do that, but here the repulsion between them is just not that great. Some pertinent bond lengths and angles for (1) are shown in Table 4.2 and corresponding data are given for norbornane for reference. Comparisons between these two molecules tell us something about the van der Waals interaction between closely spaced hydrogens and also give us a great deal of other interesting information.

CONGESTED MOLECULES

TABLE 4.2. Calculated and Experimental Geometries of Tetracyclododecane (1)[a]

Feature	MM4 1	MP2/6-31G** 1	Neutron Diff.[b] 1	MM4 Norbornane
7⋯7'	3.120	3.123	3.117	—
1–7–4	95.0	94.6	94.6	94.8
1–6–1'	120.4	120.4	119.8	—
1–7	1.537	1.537	1.539	1.537
H_a–7–H_b	106.8	107.8	107.6	109.4
H_a⋯$H_{a'}$	1.750	1.788	1.754	—

[a]Bond lengths are in ångstroms and angles in degrees.
[b]The values here were obtained by neutron diffraction on a derivative, discussed later.[20]

The structures of (1) found by crystallography[20] and by MM4[7] are shown in Table 4.2. The MM4 distance between the hydrogens (H_a and $H_{a'}$) is 1.750 Å, while the experimental distance is 1.754 (4) Å. The sum of the van der Waals radii (MM4) is 3.28 Å; thus, the closeness of approach of these two atoms is indeed quite small (53% of sum of the van der Waals radii). The MM4 energy of this van der Waals interaction is 2.58 kcal/mol. The molecule has undergone very substantial stretchings and bendings in order to reduce this energy to the observed point, as the stretching and bending energies near their minima do not increase very rapidly with distance, whereas the van der Waals repulsion that would exist in this molecule in the undeformed (Dreiding model) state is very high up on the van der Waals curve, where that curve is going practically straight up.

It is particularly interesting to see just how the molecule distorts here to minimize its energy. For clarity, the structures of norbornane and **(1)** are repeated in Structure **4**, and the MM4 bond lengths and angles are shown.

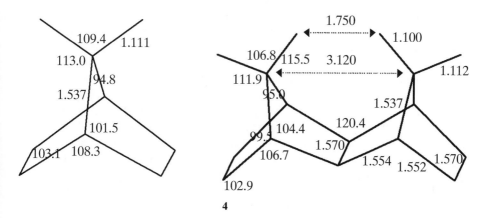

4

For those who have never thought much about this type of problem, it may seem that the resulting atomic distortions shown in 1 in Structure **4** are very odd. If we look at the hydrogens in (1) in Structure 3 that are subjected to the repulsion (H_a and $H_{a'}$), we note that the angle between the hydrogen and the inside of the skeleton is 115.5°, while the angle between the outside hydrogen ($H_{b'}$) and the skeleton is 111.9°. The corresponding angle is 113.0° in norbornane itself. Obviously, the H_a and $H_{a'}$ hydrogens bend away from one another, but only by an amount of 2.5°. And note that the CH_2 group undergoes a rocking motion. The repulsion moves only the inner hydrogen (H_a), but the outer one (H_b) moves with it in a cooperative way (by 1.1°) so as to minimize the bending energy from all of the angles at carbons 7 and 12. On the other hand, note the value of the interring bond angle (4–5–4′) at the bottom of the molecule. This angle is tertiary and would have a normal value of about 111°, but in this molecule it is open to 120.4°! So one angle far removed from the repulsion is distorted by roughly four times more than is the C–C–H angle where the offending hydrogen is actually involved.

The situation described is in fact completely normal. If one can distort a molecule in some degree of freedom to lower its energy, and if there are many such degrees of freedom, then what will be found in general is that the further the deformed angle in question is from the interaction, the greater the distortion will be. This is the opposite of what one might have intuitively expected. However, a little thought shows why this occurs, and why it is in fact a general phenomenon. If we bend the C_1–C_7–H_a bond angle at the top by say 1°, the hydrogens (H_a and $H_{a'}$) move apart by a certain number of angstroms. But if we bend the C_4–C_5–$C_{4'}$ angle at the bottom by 1°, the hydrogens are on a much longer lever arm, and their motion is much greater than it was when we bent the C–C–H bond directly. And this is, of course, a general phenomenon. The longer the lever arm, the more motion we will usually get away from the trouble spot, for a given amount of distortion at the angle in question. And so, generally speaking, distortions that occur in the molecule from steric repulsion are greater the further the distortion is from the center of that repulsion. If we look at the angular distortions of the C–C–C angles starting at the bottom of the molecule, and going up the "inside" of the bridge to the top, and compare the values with those of the parent norbornane molecule, we see that the central bond angle C_4–C_5–$C_{4'}$ is opened by about 9°, the next angle, at the bridgehead (C_5–C_4–C_7), is opened by 2.9°, and the C–C–H angle itself (4–7–H_a) by 2.5°, in accord with the stated principle. (Of course, the bending constants for C–C–C and C–C–H are not exactly the same, but this is a rule of thumb not strictly a law of nature.)

The many distortions discussed in the preceding paragraphs have very marked energetic consequences, of course. If one began with the tetracyclododecane molecule in a conformation as given by a Dreiding model, the hydrogens H_a and $H_{a'}$ would be very close together (about 1.1 Å), and the energy from the van der Waals repulsion would be enormous. So, of course, the molecule distorts to lower that energy as much as possible. Let's consider the energetics in norbornane and in tetracyclododecane. We can calculate the total energy for each of these molecules, and all of the component parts, from the molecular mechanics model. In norbornane, there is a considerable strain energy [which is well calculated, as the heat of formation of the molecule is well calculated by MM4 (Chapter 11)]. What we would like to know is the increase in strain in the tetracyclododecane, relative to two norbornane molecules, since if the latter were no more strainless than a Dreiding model, the strain energy in tetracyclododecane would be approximately twice that in norbornane. What we are interested in is the increase in the energy in tetracyclododecane, relative to two norbornane molecules. This is not an exact comparison but will give us an approximate idea of the strain energies.

The calculated molecular mechanics structure of a molecule is accompanied by information wherein the energy of the whole system is broken down into many parts, such as stretching, bending, van der Waals (which is subdivided into 1,4 van der Waals, and other van der Waals). What we see here is that after the tetracyclododecane has deformed into the actual calculated minimum energy structure, the energies for the important components relative to two norbornane molecules are as follows: bending, +3 kcal/mol, 1,4 van der Waals, −1.3 kcal/mol, other van der Waals, +2.9 kcal/mol, torsional, −1.5 kcal/mol. What this is telling us is that most of the energy has been taken into the bending degrees of freedom, consistent with the foregoing discussion. We divide the van der Waals energy into two parts, for convenience. The 1,4 van der Waals energy is between atoms that occupy that sort of relationship on a chain (such as a hydrogen on one end of ethane interacting with a hydrogen on the other end). These 1,4 van der Waals energies are quite large in most molecules, but they usually do not differ very much from one minimum energy conformation to another. The "other" van der Waals energy in the present case, when we compare the tetracyclododecane with norbornane times two, contains the large van der Waals repulsion term that we started with, plus other van der Waals repulsions that have been generated in the process of minimizing the first one. This total term amounts to +2.9 kcal/mol. It is also interesting that the torsional energy appears to have decreased somewhat. But this is an artifact because in two norbornane molecules there would be four eclipsed ethane units (2–3 and 5–6 in each), and in tetracyclododecane there are only three. An eclipsed ethane unit adds about 2 kcal to the torsional energy of the molecule (most of the other kilocalorie comes from increased van der Waals energy), so this torsional number results mainly from the fact that our comparison with two norbornanes is not exactly right.

So starting with a tetracyclododecane molecule that had a very large van der Waals repulsion in one place (something like 25 kcal/mol in the Dreiding model structure), we obtain a related structure in which that very large (total) van der Waals repulsion has been much reduced (to about 2.9 kcal/mol), at a cost of about 3.6 kcal in bending energy. (The actual calculated repulsion between H_a and $H_{a'}$ that remains is 2.58 kcal/mol.)

There is an additional significant quantity that we can calculate for such a disturbed molecule. We can calculate a formal "strain energy" for each of these molecules. For this we can use a "bond energy" type of scheme that follows along the lines developed over many years by a number of chemists, among whom the name Sidney Benson is best known. The details of this scheme will be presented in Chapter 11, but let us just say here that one can use some very simple molecules—ethane, propane, and isobutane in the present case—and from those component pieces one can calculate for a hydrocarbon such as norbornane or tetracyclododecane what the heat of formation would be if the molecule were "strainless." Using a molecular mechanics version of this Benson-type scheme, one can also calculate the actual heat of formation of the molecule of interest. The difference between these numbers gives us a *strain energy*. That is, how much strain exists in a particular molecule, relative to the basic strain-free fragments that make up the molecule. The numbers here come out that the strain energy of norbornane is 7.57 kcal/mol, quite a substantial amount, and not surprising in view of the five-membered rings, for example. For tetracyclododecane the strain energy is 21.13 kcal/mol. As a rough estimate, one expects approximately twice the norbornane strain energy in tetracyclododecane, just from the bending, stretching, and the like that go into making up the skeleton. And we expect roughly twice the norbornane strain because there are approximately twice as many things bent, stretched, and so forth as in the single norbornane molecule. The difference in strain between tetracyclododecane and two norbornanes then is 5.99 kcal/mol. This number is quite similar to the van der Waals repulsion remaining in the tetracyclododecane molecule, plus the additional bending that the molecule went through in order to minimize that repulsion.

Vibrational Motions of Compressed Hydrogens

There is one last point of interest to be discussed here. If two hydrogens are pushed very close together, well within the sum of their van der Waals radii as in tetracyclododecane, there is a large van der Waals repulsion between them. If we press those hydrogens together harder and harder, because the van der Waals part of the repulsive curve goes up so fast, we quickly come to the point where a hard push results in almost no further motion of the hydrogens toward each other. The hydrogens wish, of course, to avoid one another, and so the angles bend quite a lot as discussed, but the C–H bond lengths will also be compressed. Since there is rather little motion with compression, and much more motion with bending for the same expenditure of energy, the compression here will not be very large, but there will be some very perceptible effects. As the two hydrogens come together, so do the Morse potentials that would describe the vibrational motion of each of those hydrogens. The hydrogens do not ever coalesce (not withstanding the calculation by Mary Ann Miller), and they don't come anywhere near coalescing (under ordinary conditions). But rather, the Morse potentials are compressed, and the potential wells are much narrowed relative to what they were before the hydrogens came together. And the harder one presses the hydrogens together, the more those potential wells are narrowed. Hydrogens refuse to be localized for quantum mechanical reasons, so that as the well gets narrower, the vibrational energy levels move higher up to a wider part of the potential well so as to keep the hydrogens delocalized. This means that the frequencies of those vibrating hydrogens will also move higher. Hence the two hydrogens in tetracyclododecahedrane that are in close proximity are expected to show substantially higher than normal C–H stretching frequencies. So we can carry out a normal mode vibrational analysis on this molecule, and we should be able to identify the frequencies in question.

The first question we need to ask given that the frequencies will go higher is higher than what? The obvious model is norbornane. So we can carry out the calculations on norbornane, and find out what are the vibrational frequencies for the hydrogens and H_a and H_b. This is conveniently done with the MM4 program. One calculates the vibrational frequencies of the molecule, and one then looks at the C–H vibrational frequencies that are of interest, and that are in the 2800–3000 cm^{-1} range in ordinary hydrocarbons. There are 12 hydrogens, and hence 12 C–H bonds in norbornane, so we obtain 12 calculated frequencies in the range indicated. Some of these frequencies are degenerate, and some have zero calculated intensity from symmetry, and they overlap, so we probably won't really see that many, but we know the vibrational frequencies. Then we can put the molecule on the computer screen and examine the normal modes. Thus, we let the atoms vibrate for the different frequencies one by one, and we see which two of the frequencies correspond to the vibrations of atoms H_a and H_b. As indicated in our discussion of water (under Vibrational Spectra earlier in this chapter), in isolation these two hydrogens would vibrate at the same frequency. However, in fact, their motions are coupled so that one moves to a higher and one to a lower frequency. In norbornane these correspond to the antisymmetric frequency that is calculated to be at $2951\,cm^{-1}$ and the symmetric frequency at $2895\,cm^{-1}$. It is evident by inspection of the vibrations on the computer screen which frequencies are which.

We then repeat the calculation for tetracyclododecane, and this time there are 18 frequencies through which to search. There are actually going to be 4 frequencies that we will want to find here, since there are two hydrogens on each bridge, for a total of 4. These 4 frequencies have motions in which one geminal pair of hydrogens is moving in either a symmetric or an antisymmetric mode as described in the norbornane case, and the other pair can be simultaneously moving in a symmetric or an antisymmetric vibration with respect to the first one. The highest calculated frequency is at $3078\,cm^{-1}$, and it corresponds to the two hydrogens closest to each other (H_a and $H_{a'}$) vibrating in a symmetric mode, while the second frequency ($3028\,cm^{-1}$) is the antisymmetric mode. The two outside hydrogens H_b and $H_{b'}$, are coupled to each other and also to their geminal partners. The symmetric vibration is found at $2909\,cm^{-1}$ and the antisymmetric at $2907\,cm^{-1}$. It is of interest that the theory of all of this was worked out by Kivelson and co-workers as long ago as 1961.[21] They couldn't actually carry out the vibrational calculation at that time (it is much too large a calculation to do by hand), but they knew qualitatively what would happen.

Kivelson and co-workers[21] also noted that the shifting of frequencies to shorter wavelengths depended not only upon the distance between the hydrogens but also upon the angle at which they impacted one another. Thus, one can tell by a superficial examination of the infrared spectrum of the C–H stretching region of a molecule in which there are close approaches between hydrogens something about the distance between those hydrogens.

So this is pretty much what we expect. There are two hydrogens that are undergoing a serious mutual van der Waals repulsion, and those two hydrogens have the unusually high stretching frequencies $100–150\,cm^{-1}$ above the ordinary values of those in norbornane.

Other Very Short H⋯H Distances

While the distance between the closest hydrogens in tetracyclododecane is only $1.750\,Å$, there are other even shorter distances known in several molecules. In order to fix accurately the van der Waals repulsion between hydrogens, one desires data at quite short distances. While such data could be obtained by accurate quantum mechanical calculations on a molecule such as tetracyclododecane, one really would need calculations that were extrapolated to infinite basis set and correlation, and these are still out of reach for molecules of the size required to exhibit these phenomena. There are, however, quite a few experimental data available. For experimental techniques that will give us accurate numbers, we are limited to neutron diffraction for distances between hydrogens. There are now several of the latter available that were determined at low temperatures. The estimated standard deviations in the H⋯H distances are typically about 0.003 Å. There enters an additional uncertainty in these distances, however, from the results of the crystal packing forces. Since angular deformations of as large as 2° or 3° are possible, and known in some molecules, we have to be concerned as to what such deformations might do in the present case. Or, can we eliminate the possibility of such deformations? The best way to do this is to look at a series of compounds. They will pack in various lattices in which the forces that are impacting the H⋯H site are expected to be of different magnitudes and coming from different directions. It turns out that the data available in the literature consist of several molecules with different functional groups in different geometries, and hence different forces from the interactions of the molecules. Thus, while some specific deformations may occur in one crystal or another, an overall trend should be meaningful. Further, the measurements have all been made with relatively rigid compounds, cage structures that are unable to undergo significant torsional distortions, and that have larger than usual force constants for distortions because of those cage structures. And here molecular mechanics comparisons are helpful. If the molecular mechanics calculations on isolated molecules show the same progression of H⋯H distances in a series of compounds as are observed experimentally, there can hardly be significant lattice forces distorting the experimental data.

To carry out crystallographic studies, one first of all must obtain a suitable crystal. Hydrocarbons, generally speaking, do not form very good crystals. They tend to be waxy, disordered, and low melting. Hence, the crystallographer often studies interesting structures that have a fairly bulky polar group attached. One can compare calculations of the parent system of interest with the crystal structure, but, of course, there is the possibility that the substituent will deform the structure. The latter in fact is found to be the case with many of the molecules to be discussed here. Accordingly, the molecular mechanics calculations have been carried out in part on the exact same structures as were used for the diffraction studies. We have listed in Table 4.3 several structures of interest. Also are given the H–H distances from neutron diffraction and from MM4.

We may begin with the tetracyclododecane 2. The H–H distance found by neutron diffraction is 1.754 (4) Å in compound 2, which is compound 1 with a hydroxyl substituent added at carbon 2.[20] There is also available a neutron diffraction structure for the anhydride derivative, compound 3. The effect of the anhydride group is to repel the bottom side of the bicyclic system, which results in the top side parts being pushed closer together. Compound 3 has this H⋯H distance appreciably shortened, down to 1.713 (3) Å. There are also experimental structures available for compounds 4–6.[22] These have progressively shorter H⋯H distances, from 1.70 Å down to 1.629 (3) Å. The molecular mechanics calculated distances for this series of compounds may be compared with the experimental values. MM4 gives the following numbers for compounds 2–6 (Table 4.3): 1.756, 1.73, 1.712, 1.710, and 1.629 Å, respectively, and

CONGESTED MOLECULES

TABLE 4.3. Compounds with Short H–H Distances

		H_a–H_b		υ_{C-H}	
		Exp	MM4	Exp	MM4
2[a]		1.754(4)[20]	1.756	3052[21]	3075
					3026
					2910
					2908
3		1.713(3)[20]	(1.73)[c]	3078	3116
					3051
					2908
					2906
4		1.70[22]	1.712		3093
					3022
					2889
5[a]		1.76–1.79[24]	1.710	3048[22]	3086
				2963	3020
				2891	2889
6[b]		1.617(3)[22]	1.629	3119[22]	3179
		1.60(5)[25]			3074

[a]The substituent is a hydroxyl group.
[b]The substituent on the right is a benzoate. The left-most five-membered ring contains six chlorine atoms, a geminal pair on the lowest left carbon, and one on each of the other four carbons (represented by the larger attached spheres).
[c]The MM4 calculations cannot be carried out presently on compound 3 as MM4 parameters for anhydrides do not yet exist. The value is estimated from the MM3 value.[23]

close to the corresponding experimental numbers. The distance between the closest pair of hydrogens (compound 6) is only 1.617 Å from experiment, and the MM4 value (1.629 Å) differs from that by only 0.012 Å, which we regard as a reasonable agreement for a crystal/gas-phase comparison.

Finally, we can note the fact that each of the compounds that we have discussed in the foregoing in which there is a close H···H distance is expected to show a significant shift in the vibrational frequencies of those close hydrogens as observed in the infrared. These shifts were first noted by DeVries.[21b] The MM4 calculated frequencies for these hydrogens are given in Table 4.3, along with the experimental values. The MM4 values are noted to be qualitatively correct, but systematically too high.

ALKANES SUMMARY

A molecular mechanics force field should properly describe molecules in general, with experimental accuracy. It is an excellent tool for showing us what we do not know about chemistry.[2b] When the force field does not correctly calculate molecular properties, it indicates either that we simply have used an inadequate force field or the wrong parameters, or that we do not fully understand what is occurring.

The MM4 force field for alkanes does pretty much what we wanted it to do. It gives us to within "chemical accuracy" just about everything with which we have tried to deal. Are their further refinements that could be made? There are three that we know of, which are: (1) the butane barrier, (2) the tilt of methyl groups at the end of alkane chains, and (3) the accuracy of vibrational spectra.

1. After the MM4 development was begun, better calculations on the rotational barrier of butane were carried out,[1] and it was found that the cis barrier was actually higher by some 0.4 kcal/mol than had been recognized earlier. While this could easily be fixed, it would cause a multitude of small changes in the structures of hydrocarbons, and hence in organic molecules in general, so that it was not judged to be worthwhile at this time.

2. In a molecule such as propane, the MM4 force field causes the methyl groups to be tilted out away from one another because of van der Waals repulsions between the hydrogens on them (C–C–H_a is larger than C–C–H_s by 0.2°). Experimentally and by quantum mechanics, they actually tilt the other way (C–C–H_a is smaller than C–C–H_s by 1.0° from MP2/B). These angles of tilt involving hydrogens are rather small, and this problem could easily be fixed with a C–C–C–H torsion-bend term in the force constant matrix. We have looked at this problem,[26] and correcting it would be a large enough change to require reparameterization of the force field if one wants to include it in the proper way. As with item 1, this could be done, but relatively extensive studies have satisfied us that neither item 1 nor item 2 causes any error elsewhere that is large enough to warrant the effort that would be required for further refinement at present.

3. One item that does seem to be of perhaps more importance concerns the vibrational spectra. MM4 has a root-mean-square (rms) error of about 25 cm^{-1} in overall vibrational calculations for a test set of alkanes. (This corresponds to only 0.07 kcal/mol.) Spectroscopists commonly claim that they can measure vibrational frequencies down to 1 cm^{-1} without difficulty (in fairly small molecules), and numerous spectroscopic force fields in the literature indicate that they can often calculate these frequencies with average errors of 10 cm^{-1} or less. We have not worked on this problem very seriously. The MM4 force field is designed to calculate a great many different quantities

in a general way for organic molecules. This is a much more demanding criterion than being able to treat a handful of closely related molecules by a spectroscopic calculation only. Still, it seems likely that a better job could be done with vibrational frequencies in molecular mechanics. This is less straightforward than it can sound. While these frequencies may be measured with an accuracy of $1\,\text{cm}^{-1}$, one routinely finds in the literature that the same frequency as measured by different authors at different times may vary by $5-10\,\text{cm}^{-1}$. With molecules that are of modest size, there are frequently overlapping bands, overtones, and various other kinds of complications that make the problem more difficult than it may appear. Additionally, excluding rather small molecules, vibrational spectra are usually measured and reported in solution. The solvation of the molecules causes these frequencies to be moved about to some extent, even more so because infrared spectra are routinely determined in very concentrated solutions. Even if the solvent itself is inert, dimerization of polar molecules such as ketones in solution may cause significant shifts of some of their vibrational frequencies. The quantum mechanical calculation of these frequencies at the present level commonly used is similar in accuracy to the MM4 results. Hence significant improvements in force fields along this direction will probably have to be accompanied by better spectroscopic measurements than many of those reported in the current and older literature.

The MM4 frequencies are adequate for the calculation of a variety of properties, for example, entropies (Chapter 11). It appears that much of the calculational error in the vibrational frequencies stems from the couplings between frequencies. Thus, one frequency is calculated too low and another too high. But, if one is interested in a molecular property that uses a summation over the frequencies (e.g., entropy), most of the error cancels out.

Other improvements appear to be possible but marginal, for example, to improve moments of inertia, by improving bond lengths. It should perhaps be mentioned here that, generally speaking, the important quantity that is usually the most difficult to accurately calculate with a force field is the energy. The accurate data available experimentally for conformational energies is rather limited, but there is a large amount of heat of formation data, accompanied by other calorimetric data that have been tabulated in terms of thermodynamic functions. The usefulness of heats of formation and related data will be discussed in some detail in Chapter 11.

Our conclusion is that the MM4 force field for alkanes can indeed be improved somewhat, but the improvements would be small, and there appears to be no strong reason for making such improvements at this time. Force fields for organic molecules beyond alkanes are another matter, and will be discussed briefly in the following pages.

EXTENSION OF THE ALKANE FORCE FIELD

The alkanes constitute the most simple and fundamental class of organic molecules. They contain only two kinds of atoms, and only single bonds between those atoms. There are probably thousands, perhaps tens of thousands, of alkanes that are already known in the sense that they have been isolated and put into bottles. There are countless more that will be isolated later or that can be or have been investigated by computational studies. However, this class of molecules constitutes only a small fraction of organic molecules. The latter are, by definition, compounds that contain carbon, but in fact most of them contain one or more other elements in addition (besides hydrogen). Our molecular mechanics model can be extended in a general way to other kinds of organic molecules. This general extension can be considered in several steps. First, there is a wide variety of molecules that contain only carbon and hydrogen, but they also contain double or triple bonds. There are many more compounds of this type than are known for the alkanes themselves. However, the vast bulk of organic molecules contains in addition one or more atoms of other elements, and the most common and important of those would be oxygen, nitrogen, sulfur, phosphorus, and the halogens. These latter elements are incorporated into organic molecules in various ways, and the particular unit in which they are involved is referred to as a "functional group." Some functional groups are simple as, say, a bromine atom that replaces a hydrogen. Others are somewhat more complicated, such as the hydroxyl or amino groups, where we have hydrogens or additional groups attached to the heteroatom. Some are much more complicated, such as a sulfonyl chloride or a diphosphate.

We will consider some of these types of compounds in more or less detail in subsequent chapters of this book. We will give here a brief outline of how this presentation is arranged. First, we can consider the alkenes[27,28] (and also acetylenes[29,30] and allenes[30]). We approach these compounds with our molecular mechanics model in the same way we approach alkanes. From the molecular mechanics point of view, there is only one major difference between the alkenes and the alkanes (below). In the alkenes we still have only two elements to worry about, carbon and hydrogen. To a first approximation, we know the van der Waals properties of these atoms (although they might differ slightly from the same elements in alkanes in a better approximation), so to study an alkene at the lowest level, we need only determine the various parameters (the natural bond lengths, angles, stretching constants, torsion angles, etc.) that are associated with the double bond, similarly to what was done for alkanes. Parameters that were determined for the alkanes are, of course, carried over unchanged to the alkane parts of the alkenes. And thus we have a preliminary force field description of alkenes. We can compare the calculated geometries and energies, as well as the spectroscopic and other properties, for a set of alkenes with experimental and quantum mechanical information involving those same compounds, and see how well the model works. The answer is it works about as well as it did for the alkanes themselves. The alkene unit causes very little that is unexpected when it is inserted into an alkane molecule. And the same is found to be true for acetylenes and allenes. There is a great deal of interesting structural chemistry associated with these compounds to be sure, but it will not be discussed here in any detail. Leading references may be found in the studies cited in the References.

There is an important point to be mentioned here. Saturated hydrocarbons (alkanes) are treated in MM4 as very classical systems, and, in particular, they do not contain charge separation or dipole moments. This is not exactly correct (discussed in the following) but seems to be a good enough approximation for everything that we know to date. But when we introduce functional groups into the molecule, dipole moments, and their resulting induction of polarization into molecules, will become important. Even alkenes, which have relatively small dipole moments (mostly less than 1 D), require inclusion of charge separation in some form (bond dipoles or point charges) if they are to be accurately dealt with. And for more polar molecules, the situation becomes increasingly more complicated. Polarization adds a thick layer of complexity to molecular mechanics calculations, as will gradually unfold in what follows.[31]

Another level of complexity beyond alkenes as we further extend molecular mechanics calculations would be that of conjugated hydrocarbons. These are indeed a special case and something quite beyond what has been covered up to this point. Conjugated hydrocarbons will be covered in the next chapter.

After the molecular mechanics of the hydrocarbons is well in hand, we can consider the addition of functional groups to an alkane (or alkene, acetylene, etc.). Again, for each different functional group, we will have to know values for the natural bond lengths, angles, torsion barriers (both internal and external to the functional group), and the like. And these can all be worked out as previously described for alkane systems. At this point, however, things tend to get more complicated. If we insert a fluorine atom into an alkane, for example, we can try to treat that molecule by molecular mechanics as previously described in a straightforward manner. The question to ask is: Will this calculation give us the structure and properties of this molecule with chemical accuracy? The answer turns out to be no, it does not. We can deal fine with the bond length, angles, and the like involving the fluorine itself, but the *fluorine also induces significant changes into the alkane portion of the molecule.* For example, the C–C bond length of ethyl fluoride is shorter than the corresponding bond length in ethane itself by roughly 0.02 Å. And it is generally true that functional groups induce changes into the hydrocarbon part of the molecule. If we wish to obtain chemical accuracy from our calculation, there is still more to do.

This problem has been approached in three different ways. The simplest way is just to ignore it, and let the bond length of ethyl fluoride, for example, come out a little bit wrong. Alternatively, one can do the initial force field parameterization by including fluorides, hydrocarbons, and all other functional groups of interest together so that an average overall C–C bond length is obtained. This is simple and straightforward, and this approach can give a great deal of useful information.[32] It is not a very accurate way to treat the system, but it is a useful approximation because of its simplicity. If we want chemical accuracy, however, then we will have to deal with the functional groups themselves individually, and we will also have to take into account explicitly the different effects that are induced into a molecule when a specific functional group is attached to it. This has both its good and bad features. The good features are better accuracy and, importantly, better understanding as to what is actually going on. The bad feature is that it significantly complicates molecular mechanics. One cannot, for example, consider a carbonyl group to be the same kind of group in a ketone as it is in a carboxylic acid or in an anhydride, for example. Each of these groups really has to be treated as a separate entity, although in practice many of the parameters for one carbonyl group are common to all carbonyl groups. The behavior of the carbonyl group must be modified according to its environment as we go from one group to another. This should not surprise anyone because organic chemists have known for a couple of hundred years that a ketone has chemical properties different from those of an amide or an anhydride. So to a first approximation, a carbonyl group is a carbonyl group, but to a better approximation, a carboxylic acid differs from a ketone, which differs from an acid chloride, and so forth.

So then, can we really calculate structures and properties from molecular mechanics in a very general way for organic molecules, as we can with hydrocarbons? The answer is, yes, we can, but it involves a different level of approximation if we are going to require chemical accuracy in the results. As it turns out, the next level of approximation in molecular mechanics is to include what are commonly called "effects" in organic chemistry. That is, the *electronegativity effect*, the *Bohlmann effect*, the *anomeric effect*, and so forth. These things often involve off-diagonal matrix elements of major importance in the force constant matrix. And it is commonly found that things that we can take as *constants* for hydrocarbons become *functions* of other quantities, when we are dealing with the changes induced by heteroatoms in organic molecules. But as far as we know at present, these functions are indeed transferable from one molecule to another, just as the constants are transferable in hydrocarbons. And as far as we know, this can in fact be done, although there are a great many different functional groups. Therefore, there is a great deal of effort involved in the development of this model if one wants it to be "complete" or nearly so. As a fringe benefit from doing all of this, we can learn or better understand a great deal about some of the fundamentals of physical organic chemistry.

ALKENES

With one exception, the general mechanical model used in MM4 for alkanes carries over directly to alkenes. Of course, a double bond is shorter and stronger than a single bond, and therefore has a larger force constant and stretching frequency, and the basic geometry is trigonal rather than tetrahedral, for example. But the model is basically the same, only the numbers are different. The one important exception as far as MM4 is concerned is electrostatics. There has been discussion over the years about charges in alkanes. Clearly, a charge distribution occurs in molecules since we have positive nuclei surrounded by clouds of negative electrons. If the nuclei are the same, as in F_2, we suppose that there is no change in the charge distribution at the *atomic level*. That is, the charge of one fluorine is the same as the other. On the other hand, if we have a molecule such as HF, the fluorine is more electronegative, and we would expect that it would take some electron distribution away from the hydrogen, resulting in a charge distribution that has the hydrogen end of the molecule positive, and the fluorine end negative. This is all well described in detail by quantum mechanics. However, how do we incorporate it in a reasonable way in molecular mechanics at the atomic as opposed to the subatomic level? There are two straightforward possibilities. We can use point charges, that is, put a plus charge at the hydrogen nucleus and a minus charge at the fluorine nucleus in the case of HF, and, of course, these would be only partial charges, not a full electron charge, chosen so as to reproduce the dipole moment of the molecule. The other possibility is to represent this charge distribution by a point dipole placed on the bond between the two atoms. Both methods have been widely used, and both work about equally well. A number of studies have been made that show this to be true in different cases, and it seems to be true in general.[33] A better approximation would be to use both point charges at the nuclei and a point dipole in between. We chose to use point charges for ions and bond dipoles in molecules. This requires the calculation of charge–charge, charge–dipole, and dipole–dipole interactions in the general case, but only the last term is used for ordinary (neutral) molecules. More accurate, and more complicated, possibilities exist in terms of a general multipole expansion.[34]

To return to the present case, force fields have been devised that include a charge distribution in alkanes at the atomic level, and other force fields that are based on the assumption that the amount of charge distribution at the atomic level in alkanes is negligible for most purposes, and therefore electrostatics can be neglected in molecular mechanics calculations on alkanes. In our earliest molecular mechanics calculations[17] we concluded that since charges in alkanes were certainly small, we would omit them until there was evidence that they needed to be included. It is true that alkanes do have very small dipole moments, for example, propane $(0.083\,D)$[35] and isobutane $(0.132\,D)$.[36] If one wishes to reproduce these in molecular mechanics, then one would have to include electrostatics. But to our knowledge, this is the only improvement that electrostatics would make in the molecular mechanics of alkanes. The foreseeable improvement would be quite small, so we continue to omit electrostatics for alkanes.

On the other hand, unsaturated compounds such as alkenes do have significant dipole moments, and here we have included electrostatics in MM4 by assigning bond moments to C_{sp^2}–H and C_{sp^2}–C_{sp^3} bonds of 0.95 and 0.60 D, respectively. The resulting molecular dipole moments are pretty small for propene and methylacetylene, for example, 0.35 and 0.75 D, respectively. Dipole moments of this size for the most part have only a small effect on things that we wish to calculate in molecular mechanics. However, they may interact in a significant way with larger dipoles or charges that might be present in more complicated molecules, so we included them at the outset. Electrostatics does, of course, become quite important in molecules that have strongly polar groups. The α-haloketones (Chapter 7) are an intramolecular example where conformational equilibria are electrostatic and hence solvation dependent. Intermolecular electrostatic interactions are, of course, quite important for molecules in solution or in crystals (Chapter 10).

FUNCTIONAL GROUPS IN MOLECULAR MECHANICS

As outlined previously, when one inserts a functional group into a hydrocarbon, one not only has to deal with the functional group itself in molecular mechanics, but one also must deal with the various changes that may be introduced into the hydrocarbon portion of the molecule by the presence of the functional group. In an overview, these changes are induced into the molecule because what are often considered to be constants in molecular mechanics are actually functions of various other things. There are several different kinds of these functions, which are commonly called "effects" in organic chemistry. They have been fairly well studied, and they are reasonably well understood. This section will not discuss these in any detail but will simply alert the reader as to what is still needed in order to have a general and accurate molecular mechanics force field. The next chapter (Chapter 5) will discuss conjugated hydrocarbons, which are noticeably more complicated in molecular mechanics than hydrocarbons that do not contain conjugated systems. What is useful for conjugated hydrocarbons is also useful for, and can be easily extended to, conjugated molecular systems in general, including aromatic heterocycles. Later will come a series of sections, arranged in three chapters, that will discuss the *chemical effects* that are necessary to bring the molecular mechanics force field type of calculation up to the level of experimental accuracy for functionalized molecules.

REFERENCES

1. N. L. Allinger, J. T. Fermann, W. D. Allen, and H. F. Schaefer III, *J. Chem. Phys.*, **106**, 5143 (1997).
2. (a) K. Machida, *Principles of Molecular Mechanics*, Kodansha and Wiley, co-publication, Tokyo and New York, 1999. (b) U. Burkert and N. L. Allinger, *Molecular Mechanics*, American Chemical Society, Washington, D.C., 1982. (c) A. R. Leach, *Molecular Modeling Principles and Applications*, Pearson Education, Essex, England, 1996. (d) T. Schlick, *Optimization Methods in Computational Chemistry, Reviews in Computational Chemistry*, Vol. 3, K. B. Lipkowitz and D. B. Boyd, Eds., Wiley, New York, 1992, p. 1.
3. W. Weniger, *Phys. Rev.*, **31**, 388 (1910).
4. (a) E. B. Wilson, Jr., J. C. Decius, and P. C. Cross, *Molecular Vibrations, The Theory of Infrared and Raman Vibrational Spectra*, Dover, New York, 1955, 1980. (b) J. M. Hollas, *Modern Spectroscopy*, 2nd ed., Wiley, Chichester, UK, 1987, 1991.
5. Some good ones are: (a) P. Crews, J. Rodriguez, and M. Jaspars, *Organic Structure Analysis*, Oxford University Press, New York, 1998. (b) R. M. Silverstein, F. X. Webster, and D. J. Kiemle, *Spectrometric Identification of Organic Compounds*, 7th ed., Wiley, Hoboken, NJ, 2005. (c) J. B. Lambert, H. F. Shurvell, D. A. Lightner, and R. G. Cooks, *Organic Structural Spectroscopy*, Prentice Hall, Upper Saddle River, NJ, 1998. (d) D. L. Pavia, G. N. Lampman, and G. S. Kriz, *Introduction to Spectroscopy*, 3rd ed., Harcourt College Publishers, Fort Worth, TX, 2001.
6. H. C. Urey and C. A. Bradley, Jr., *Phys. Rev.*, **38**, 1969 (1931).
7. N. L. Allinger, K. Chen, and J.-H. Lii, *J. Comput. Chem.*, **17**, 642 (1996).
8. J. E. Lennard-Jones, *Proc. Roy. Soc. London, Ser. A*, **106**, 463 (1924).
9. A. Warshel and S. Lifson, *J. Chem. Phys.*, **53**, 582 (1970).
10. A. D. Buckingham and B. D. Utting, *Ann. Rev. Phys. Chem.*, **21**, 287 (1970).
11. T. L. Hill, *J. Chem. Phys.*, **16**, 399 (1948).
12. U. Burkert and N. L. Allinger, *Molecular Mechanics*, American Chemical Society, Washington, D.C., 1982, pp. 42–46.
13. (a) N. L. Allinger, M. T. Tribble, M. A. Miller, and D. H. Wertz, *J. Am. Chem. Soc.*, **93**, 1637 (1971). This offset appears to have first been used by (b) D. E. Williams, *J. Chem. Phys.*, **43**, 4424 (1965), and (c) R. F. Stewart, E. R. Davidson, and W. T. Simpson, *J. Chem. Phys.*, **42**, 3175 (1965).
14. L. Pauling, *The Nature of the Chemical Bond and The Structure of Molecules and Crystals*, 2nd ed., Cornell University Press, Ithaca, NY, 1948, p. 189.

15. J. T. Sprague and N. L. Allinger, *J. Comput. Chem.*, **1**, 257 (1980).
16. (a) N. L. Allinger and K. A. Durkin, *J. Comput. Chem.*, **21**, 1229 (2000). (b) T. A. Halgren, *J. Am. Chem. Soc.*, **114**, 7827 (1992).
17. N. L. Allinger, M. A. Miller, F. A. Van-Catledge and J. A. Hirsch, *J. Am. Chem. Soc.*, **89**, 4345 (1967).
18. M. A. Miller, unpublished results.
19. (a) N. L. Allinger, *J. Am. Chem. Soc.*, **99**, 8127 (1977). (b) E. M. Engler, J. D. Andose, and P. v. R. Schleyer, *J. Am. Chem. Soc.*, **95**, 8005 (1973).
20. O. Ermer and S. A. Mason, *J. Chem. Soc., Chem. Commun.*, 53 (1983).
21. (a) D. Kivelson, S. Winstein, P. Bruck, and R. L. Hansen, *J. Am. Chem. Soc.*, **83**, 2938 (1961). (b) L. deVries and P. R. Ryason, *J. Org. Chem.*, **26**, 621 (1961).
22. O. Ermer, S. A. Mason, F. A. L. Anet, and S. S. Miura, *J. Am. Chem. Soc.*, **107**, 2330 (1985).
23. N. L. Allinger, Y. H. Yuh, and J.-H. Lii, *J. Am. Chem. Soc.*, **111**, 8551 (1989).
24. P. R. Seidel, private commun. to F. A. L. Anet cited in Ref. 22.
25. F. A. L. Anet and A. H. Dekmezian, *J. Am. Chem. Soc.*, **101**, 5449 (1979). [NMR on Structure 6 in Table 4.3, where benzoate is replaced by acetate, 1.60(5)].
26. J.-H. Lii and K.-H. Chen, unpublished work.
27. N. L. Allinger, F. Li, and L. Yan, *J. Comput. Chem.*, **11**, 848 (1990).
28. N. Nevins, K. Chen, and N. L. Allinger, *J. Comput. Chem.*, **17**, 669 (1996).
29. E. Goldstein, B. Ma, J.-H. Li, and N. L. Allinger, *J. Phys. Org. Chem.*, **9**, 191 (1996).
30. N. L. Allinger and A. Pathiaseril, *J. Comput. Chem.*, **8**, 1225 (1987).
31. (a) S. W. Rick and S. J. Stuart, *Potentials and Algorithms for Incorporating Polarizability in Computer Simulations, Reviews in Computational Chemistry*, Vol. 18, K. B. Lipkowitz and D. B. Boyd, Eds., Wiley, Hoboken, NJ, 2002, p. 89. (b) B. Ma, J.-H. Lii, and N. L. Allinger, *J. Comput. Chem.*, **21**, 813 (2000).
32. (a) T. A. Halgren, *J. Comput. Chem.*, **17**, 490, 520, 553, 616 (1996). (b) T. A. Halgren and R. B. Nachbar, *J. Comput. Chem.*, **17**, 587 (1996).
33. (a) D. E. Williams, *J. Comput. Chem.*, **9**, 745 (1988). (b) D. E. Williams, *Net Atomic Charge and Multipole Models for the Ab Initio Molecular Electric Potential, Reviews in Computational Chemistry*, Vol. 2, K. B. Lipkowitz and D. B. Boyd, Eds., Wiley, New York, 1991, p. 219.
34. (a) J. Applequist, J. R. Carl, and K.-K. Fung, *J. Am. Chem. Soc.*, **94**, 2952 (1972). (b) L. Dosen-Micovic, D. Jeremic, and N. L. Allinger, *J. Am. Chem. Soc.*, **105**, 1716 (1983). (c) L. Dosen-Micovic, D. Jeremic, and N. L. Allinger, *J. Am. Chem. Soc.*, **105**, 1723 (1983). (d) L. Dosen-Micovic, S. Li, and N. L. Allinger, *J. Phys. Org. Chem.*, **4**, 467 (1991). (e) Also see discussion in Chapter 10.
35. R. Lide, Jr., *J. Chem. Phys.*, **33**, 1514 (1960).
36. D. R. Lide, Jr., and D. E. Mann, *J. Chem. Phys.*, **29**, 914 (1958).

5

CONJUGATED SYSTEMS

INTRODUCTION

Back as far as the mid-1800s chemists recognized that alkenes, simple conjugated hydrocarbons such as butadiene, and aromatic hydrocarbons such as benzene were best considered as three distinctly different classes of compounds. The principle characterization method used then was chemical reactivity. Whereas the alkanes are relatively inert, alkenes were much more reactive toward a large number of reagents. The conjugated hydrocarbons were even more highly reactive, while the aromatics, as represented by the parent compound benzene, were much less so. As quantum mechanical and molecular mechanical methods were developed to treat the structures of these compounds, it was early noted that physical properties of alkenes are pretty similar to alkanes in most respects. They do have dipole moments, so that (weak) electrostatic effects have to be considered with these compounds. (Actually, some alkanes have dipole moments too, but they are small and negligible for most purposes.) But, as far as chemistry is concerned, the electrostatics of alkenes is not very important. Alkenes can be dealt with, generally speaking, easily and pretty well by molecular mechanics, whereas adding a heteroatom into a molecule results in what from a molecular mechanics point of view is a much more complicated system.

Molecular Structure: Understanding Steric and Electronic Effects from Molecular Mechanics,
By Norman L. Allinger
Copyright © 2010 John Wiley & Sons, Inc.

INTRODUCTION

Conjugated hydrocarbons, or polyenes, tend to be highly reactive compounds. There is clearly something different between a polyene and a collection of alkenes, from the viewpoint of organic chemistry. And aromatic systems are another special case. The polyenes and aromatics together are termed as *conjugated systems*. From a chemical point of view, this categorization may seem bizarre because one has very reactive compounds (polyenes) and very unreactive compounds (aromatics) in the same class, while as compounds of intermediate reactivity, alkenes are a separate class. Certainly this is a somewhat arbitrary, largely historical, but nonetheless useful way of categorizing these compounds.

Why do we want to categorize the conjugated molecules as we do? Let us think of the three compounds ethylene, butadiene, and benzene, as representing these three groups of structures. Each of these structures has the carbon atoms hybridized sp^2/p, so that there is a σ system and a π system. Most of the chemical properties of interest have to do with the π system since the π electrons lie in orbitals that are of higher energy, and they are therefore the most reactive toward electrophiles. And correspondingly, the π^* orbitals are the empty orbitals of lowest energy and the most reactive toward nucleophiles. The σ system is more ordinary and alkane-like (unreactive).

In the early days of quantum mechanics, before computers, it was practically impossible to deal in any meaningful way with systems as large and complicated as butadiene and benzene. So the physicists and others investigating the quantum mechanics of such molecules followed what to chemists would have been the logical path and divided the molecule into σ and π components.[1-3] Ethylene contains a total of 16 electrons and 6 nuclei, while benzene contains 42 electrons and 12 nuclei. The former, in terms of the number of particles, represented a really sizable problem to hand calculation, whereas the latter represented an almost impossible problem, unless we were fortunate enough to have symmetry cancel out most of the work. That actually happens in the particular case of benzene but not in compounds as simple as substituted benzenes. But, if we look only at the π electrons, ethylene contains two, and benzene contains six, and both of these are reasonable problems to deal with for hand calculations.

It is often stated that in each of these molecules the σ system and the π system will be orthogonal and hence will not interact. This is true in the sense that their orbitals do not mix and recombine and become as complicated as they would if the systems were nonplanar, but electrons are still electrons, and the σ and π electrons certainly know about the presence of each other. Their interaction can be represented by various simplifying approximations, but it cannot be completely ignored. Since theoretical chemists dealt with this kind of problem for many years before computers became available, they had a lot of time to think of ways to make approximations so that they could calculate something meaningful, even though the direct solution of the Schrödinger equation for even small organic molecules was for many years viewed as impossibly complicated.

The idea of σ–π separation though is quite useful because it permits what would otherwise be a very complicated problem to be reduced in essentials to a relatively simple problem. And even having computers available for the heavy work, a great deal of understanding, and ideas that prove to be important for molecular mechanics, can be gleaned from an examination of the σ and π systems of a molecule separately. For a modern review of σ-π separation in conjugated systems, see the chapter by Jug et al.[4]

In ethylene we have two p orbitals, one on each carbon, that are aligned and interact to form a π bond. In butadiene we have that twice, for the terminal bonds, but, additionally, the two orbitals on the central carbons also overlap and interact, although less strongly than the two on the end. Hence, the *bond orders*,[2] and all dependent properties, in butadiene are different for the end bonds from what they are for the central bond.*
And in benzene, the bonds are all the same but different from either the central or the end bonds in butadiene or from that in ethylene. And, if we think in terms of bond orders, we can see that by conjugating together various kinds of double bonds, not only between carbons but between various elements, we can in fact find a whole continuum of bond orders, and hence a whole continuum of bond lengths and related properties. With molecular mechanics, the fundamental strength of the basic method lies in the fact that bond properties are transferable from a specific bond in one molecule to the same kind of bond in different molecules. But often that does not happen quite that simply between two unsaturated carbons when they are conjugated. In that case the properties are highly dependent upon their extent of conjugation. This depends in turn on to what else they are conjugated.

One simple molecular mechanics approach is to assign different parameters to ethylene, to the end bonds in butadiene, to the central bond of butadiene, and to benzene, and say that all conjugated carbon–carbon double bonds can be approximated by one or the other of those bond types. One can then treat any conjugated system with a set of generic parameters, as appropriate for each of those structural types. This can be made to work for many compounds that are well approximated by one or the other of those structural types, but obviously it is a poor approximation for intermediate cases. This approximation, which has been widely used in molecular mechanics, frequently increases the probable errors in the calculated bond lengths and energies by maybe a factor of 3 or more. It would seem that there is no accurate way to treat such systems, except by somehow invoking quantum mechanics. This is most easily done by explicitly taking into account the bond order between each pair of atoms in the π system. If that is done, then indeed ethylene, the two bonds in butadiene, and the benzene bond all come out to be substantially different, as do different kinds of bonds in other compounds, throughout the whole continuum. Hence, we have a basis for treating this entire

---

*The value of the square of the wave function (Ψ^2) at a point is a measure of the electron density at that point. If one sums the values of the squares of the wave functions over the occupied orbitals of an atom for each individual electron, one obtains the formal electronic charge at that atom. Then, if the nuclear charge is subtracted from that value, what remains is the *Mulliken charge* at that atom. If that atom is part of a molecule, one can similarly sum over the wave function contributions from that atom, minus the nuclear charge, and obtain the Mulliken charge at that atom in the molecule. If one instead considers two atoms (a and b), and looks at the product of the two atomic wave functions (summed over occupied orbitals) between those two atoms ($\Psi_a \Psi_b$), one has a measure of the electron density that is shared by the two atoms, which can be considered as the covalent bond. The resulting number is referred to as the *Mulliken bond order*, and it is a measure of the strength of the bond. It can also turn out to be a negative number, in which case there is an antibonding interaction between the two atoms. Many molecular properties are related to these bond orders. In particular, the bond lengths are linearly related to the bond orders for a given type of bond (carbon–carbon π bonds, in the present case).[2,5,6]

set of molecules, not with a set of *constants* for parameters but with a *function* that will allow us to deduce specific parameters for bonds based on their specific bond orders. It was found that this scheme can be applied in a general way to any conjugated compound, not just to –C=C– bonds.

This might be a good point to look ahead a little bit, with one respect to the overall development of molecular mechanics. Originally, it was assumed (as an approximation that would hopefully be accurate enough) that the parameters that go into molecular mechanics were constants and could be simply transferred from one molecule to another. Thus, for example, the C–H bond was taken to have a certain bond length value for l_0, which was presumed to be a constant, at least insofar as the carbon was sp^3 hybridized. (It was earlier determined that in compounds such as aldehydes, for example, the carbonyl carbon had a substantially longer C–H bond than is normal in hydrocarbons.) Similarly, it was assumed that the harmonic approximation would be adequate for most purposes, as that approximation had always been used for just about everything in hand calculations, and it was indeed pretty good most of the time. However, as molecular mechanics developed over the years, it became evident that some quantities (including the l_0 value for the length of a C–H bond) were not really constants but were in fact functions of other variables. For the C–H bond, for example, the Bohlmann effect (Chapter 7) deals with this very problem. However, historically, the first example recognized where a constant for a parameter had to be replaced by a function that involved the bond lengths of carbon–carbon bonds in conjugated systems. This case was well understood at least as early as 1960.[5,6] The way we have chosen to treat conjugated systems in MM4 follows the π-electron theory that was so well worked out by hand calculations long ago (mainly in the 1950s through 1970s), at the dawn of the computer age.

In general, then, the use of transferable constants as parameters in molecular mechanics is a first approximation. Often the constants are not really constants (although they may appear to be when one is looking at a limited set of structures), but rather they are functions of variables, the nature of which depend on the case at hand. Thus, the simple logic behind molecular mechanics remains intact, but the details are somewhat more complicated than originally envisioned, if one wishes to exploit the inherent accuracy and generality of the method.

The standard way of treating conjugated systems using the σ–π separation approximation in the 1930s and into the 1960s was the so-called *Hückel method*.[1–3] In this method, the electron–electron interactions are not *explicitly* considered. Rather, the positions of the nuclei are fixed, and the electrons move in the field of the nuclei. Much of the error that results from neglecting the interactions of the electrons with one another can be circumvented with proper adjustment of empirical parameters. These parameters are also adjusted to allow (approximately) for the interaction from the σ system, which is thus taken into account without specific calculations. This method is crude but does often give qualitative results that enable rationalization of many chemical phenomena of interest. It was a powerful and useful tool in its time. A better approximation is the Hartree–Fock or self-consistent field (SCF) method in which the electron–electron interactions are explicitly considered[7–10] (Self-Consistent Field Method in Chapter 3). The quantum mechanical calculations on the π system in this case are carried out in a much more proper way. In the precomputer era, the full Hartree–Fock method was still too complicated for treating organic molecules, so some further approximations were made. First, rather than include all of the electrons in the molecule, only the π system was explicitly considered. This greatly reduced the size of the problem that had to be solved and meant that calculations on much larger systems could therefore be carried out. Second, the two-electron integrals are broken down into 1-, 2-, 3-, and 4-centered, of which the first two normally have much larger values. Since the numerically smaller 3- and 4-centered are far more difficult to calculate, as well as much more numerous, they were omitted, and the errors thereby introduced were minimized by the use of empirical parameters. Again, this is an approximation. But this approximation is a rather good one for conjugated π systems. (It might be added that the approximation is not very good for ordinary molecules, only for π systems.) This general method was devised by Pariser and Parr, and by Pople, and was found to be a much better and more realistic approximation than the earlier Hückel method, and it was still simple enough to actually be used for hand calculations on small systems. It was readily extended to larger systems once computers were available. What became known as the Pariser–Parr–Pople (PPP) method was subsequently widely employed by theoretical chemists to study conjugated systems in the 1960s and 1970s.[11,12] It was quite useful for the study of energies, including electronic spectra, of conjugated molecules.[13]

When extended to saturated molecules, this PPP-type procedure was called the CNDO method (complete neglect of differential overlap, meaning in effect the neglect of 3- and 4-centered electronic integrals; Chapter 3). This neglect was corrected for insofar as possible by empirical adjustments in the values of the 1- and 2-centered integrals and ionization potentials used. This kind of method works well in many respects, but for the most part it gives results that are of less than chemical accuracy. This general approach is referred to as *semiempirical molecular orbital theory*, and at the lowest level it is not very accurate for ordinary molecules. The general accuracy of the method can be improved, however, if one includes a few carefully chosen types of 3- and 4-centered integrals in the calculation, and many such methods were subsequently developed and given names such as MINDO (for modified neglect of differential overlap), as was discussed earlier in Chapter 3. It was mainly these 3- and 4-centered integrals that held up the molecular applications of quantum mechanics for more than 30 years. If one uses Slater orbitals, these functions in general cannot be integrated directly. This problem was insurmountable in the days of hand calculations. The way around it as a practical matter was both the coming availability of the computer and the simultaneous development of Gaussian orbitals by McWeeny,[14] and independently by Boys,[15] and then further by Pople,[16] of methods for using a series of Gaussian orbitals to approximate a Slater orbital. All of this led to functions that could be rapidly integrated (by computers), which in turn allowed the construction of the Gaussian program.[17]

The SCF-type methods in general do not in themselves include electron correlation, and this is a significant limitation. The PPP method is an SCF method and, therefore, has this limitation. Another significant limitation is that the ordinary PPP method is only applicable to systems wherein the σ–π separation can be properly invoked, that is, in general, planar systems. Many conjugated molecules are, of course, nonplanar,

INTRODUCTION

so this was a significant limitation in early PPP calculations. It is, however, possible to get around this latter problem. When one wants to determine the geometry of a molecule, either by quantum mechanical or by molecular mechanical methods, one begins with a starting structure. This structure is ordinarily "guessed," in the sense that it is put together using known structural fragments. It is then systematically altered by iterative methods in such a way as to find the exact structure for which the energy value is a minimum. One scheme for optimizing the structure (minimizing the energy) of a general conjugated compound by molecular mechanics, in principle, would be to separately minimize the sum of the energies of the σ and π systems. But this approach was really limited to planar (or nearly planar) systems. If the system were nonplanar, there would be significant (and very complicated) σ–π cross terms that would have to be dealt with. Thus, a different approach was designed that allows us to circumvent this problem.[18,19]

What we wish to do is to use the PPP method and overcome somehow the planarity problem. We want to reduce this type of calculation to the usual molecular mechanics format, where we know specifically values for l_0 and the stretching parameter k_s (and other quantities derived therefrom) for each bond in the conjugated system. We do this in the following way.

We begin with the structure of interest that is to be optimized. If that structure is nonplanar, it is "planarized" by removing the direction cosine terms in the geometric structure that account for the nonplanarity. In other words, the molecule is just mathematically squashed down into a plane. All of the π–system calculations are then carried out on this planar form, where the σ–π separation is valid. From those calculations, we do not take energies directly but only bond orders. From these bond orders, we can then generate all of the properties of the bonds (since the stretching, bending, and torsional profiles are all related to the bond order), and eventually including the energies. So we do the SCF calculations on the *planar structure of the π system*, where the σ–π separation is valid. Then we determine the bond orders, and using those we then determine all of the molecular mechanics force parameters. Having all of the parameters, the problem is reduced to a purely classical molecular mechanical calculation, and the molecule can subsequently deform away from planarity during the *molecular mechanics* geometry optimization. When handled in this way, the planarity deformation presents no problem, either in theory or in practice. This is because we do not try to calculate the out-of-plane deformation quantum mechanically (which would be difficult), but we calculate it molecular mechanically (which is straightforward).

Hence, although the development of a force field to deal with nonplanar, as well as planar, conjugated systems was tedious, it was quite straightforward. And in the end, the force field accurately describes a wide variety (preferably all) of conjugated structures. That is, all of the bond lengths and angles, and especially the torsion angles, must come out of the force field calculations, agreeing with those that are known from experiment, and/or from good ab initio calculations. And, an even more stringent test is that the energies (heats of formation) of all of these molecules must also come out correctly. If all of those things can be fit simultaneously, then we have a good model for describing conjugated systems. In fitting the MM4 force field to experiment, a total of 111 heats of formation for a wide variety of unsaturated hydrocarbons were also simultaneously fit along with the structures. This will be discussed in some detail in Chapter 11 under heats of formation. Also, all of these compounds, plus others that do not have known heats of formation (125 in all), simultaneously had their geometries fit to approximately within experimental error. Previously unknown experimental errors were uncovered by further investigation in a few cases where molecular mechanics calculations did not agree with the experimental data.

STRUCTURES OF CONJUGATED HYDROCARBONS

In the Hartree–Fock method as applied to π systems, one assigns an ionization potential to the p orbital of each atom involved and then carries out an SCF calculation and obtains π-electron densities at each atom and bond orders between each atom pair. It is relatively easy to make a few improvements in this scheme. In particular, because of induction in the σ system, the π-charge densities at different atoms will usually change during the SCF iteration sequence. And so a procedure called the *variable electronegativity SCF* (VESCF) *method* was developed.[13] In this approximation, as the charge density changes, the shielding of the nuclear charge also changes, and so does the ionization potential. This leads to a better description of the π system and to better bond orders and dependent properties. It also allows for the effect of alkyl groups attached to a conjugated system and is important if one wishes to calculate electronic spectra (discussed later in this chapter).

The MM2 treatment of conjugated π systems was devised in 1987,[20] based on earlier work,[18] and included this VESCF calculation from the start. The method was subsequently refined in MM3[21] and further in MM4.[22] This VESCF calculation is important for calculating accurate electronic spectra of hydrocarbons because the locations of any attached alkyl groups onto the conjugated system leads to sizable shifts in the absorption frequencies. But, since it is a way for allowing for induction in the σ system, one might suspect that it would become even more important when one looks at heterocyclic conjugated systems, and that is indeed found to be the case.

In order to study conjugated molecules by molecular mechanics, generally speaking we have to carry out the same kinds of operations that we would have to carry out to study unconjugated molecules. The one significant difference is that in the conjugated case, we need to establish for a given type of conjugated bond (in the simplest case C=C) the details of the bond order–bond length relationship. This was done by picking a few standard molecules for which accurate structures were known (specifically ethylene, butadiene, and benzene for the hydrocarbon case) and determining the details of the relationship involved. Many previous studies, going back to the days of hand calculations, made use of the fact that there is a linear relationship between the length of a bond and its bond order.[6] That is, the strongest $C_i=C_j$ gave the shortest bond in the case of ethylene, where the π-bond order is 1.0. If we conjugate the double bond, and reduce its bond order, the length increases. And, at least to a very good approximation, the relationship is linear and is given by Eq. (5.1) in the MM4 program:

$$l_{0ij} = l'_{0ij} - b_{ij} P_{ij} \tag{5.1}$$

where P_{ij} is the π bond order between atoms i and j, and l'_{0ij} is the bond length required to fit ethylene when $P_{ij} = 1.0$, and b_{ij} is a constant chosen to give the proper slope to the line so that the proper values are also obtained for the bonds in butadiene and benzene. Note that the quantity determined here (l_{0ij}) is not an l but an l_0. That is, the actual value of l still has to be found by adding in the electrostatic and van der Waals interactions (which also takes into account effects from the σ system).

One can, of course, complicate the π-system calculation with continually improving approximations. However, there is a balance between the quality of the results desired and the amount of programming and computing time that will be required to obtain those results. For example, a method for including electron correlation is required to get good geometries for a few special kinds of conjugated systems discussed later in this chapter, for which a Møller–Plesset (MP2) perturbation calculation[17b,23] was used.[24] Generally speaking with MM4, if one has a molecule that is conjugated, versus one of similar size that is not, and one compares the computer time required to do the corresponding calculations, the addition of the Hartree–Fock calculations required for the π system may increase the computer time required by a factor of perhaps 2–5. If electron correlation (MP2) is needed, it would probably increase the time over that required for the Hartree–Fock calculation by another factor of perhaps 5–10. So the calculation time required depends to a significant extent on these additional complications. Nevertheless, the molecular mechanics method is still very fast compared to any full-scale quantum mechanical method of equivalent accuracy, and it remains so even for molecules containing π systems. The reason is simple. If one considers only π electrons, the quantum mechanical calculation required in molecular mechanics is dealing with only a small fraction of the total electrons in the system. Since the time requirements for accurate quantum mechanical calculations (including correlation) generally increase by something like the sixth power of the number of electrons included, the calculations rapidly become immensely time consuming as the number of electrons increases. The molecular mechanics procedure, invoking the σ–π separation, and using only 1- and 2-centered electronic integrals (ordinarily an adequate approximation for π systems), keeps the quantum mechanical part of the calculations very fast, and the accuracy can be pretty well maintained by guiding constraints provided by molecular mechanics. For a molecule containing say 8 atoms heavier than hydrogen, the molecular mechanics calculation will probably still be about 4 orders of magnitude faster than the full quantum mechanical calculation, to obtain structures of similar accuracy. Beyond that size the speed advantage of molecular mechanics continues to increase rapidly.

We would suspect, and we find, to use 1,3-butadiene as an example, that the torsion barrier about the central bond is much lower than that in ethylene. The torsion barriers about the end bonds of butadiene, on the other hand, are quite high, but still somewhat lower than that of ethylene. As the bond order is reduced from 1.0 (in ethylene) to a smaller number as in the bonds in butadiene (0.96 and 0.26 for the short and long bonds, respectively, the V_2 coefficient describing the barrier is also reduced. The V_2 term is, in fact, gradually replaced by a smaller V_3 term, which dominates if the bond order falls down near to zero, and the double bond goes over to a single bond. If we wish to have the force field describe not only ethylene, benzene, and butadiene, but also such things as 1,2-cyclohexadiene or cycloheptatriene (which are both nonplanar), we have to take all of these things into account.

But there is more. When we consider propene as a simple example, rotation of the methyl group has an energy profile that puts a hydrogen eclipsing the double bond at the energy minimum, and there is a three-fold barrier to that methyl rotation of about 1.5 kcal/mol. This type of term also comes into play with 1,2-cyclohexadiene because the ring prevents the alkyl fragments attached to the diene from locating themselves at the position where their torsional energies would be at minimum values. But the height of that barrier is also bond order dependent.[25] This leads to consequences that are not always foreseen. For example, consider the molecule 2-methylbutadiene, Structure **1** in the stable (anti) conformation. Is Structure **1a** the ground state, or is it **1b**?

Since the methyl group hydrogen tends to eclipse a double bond, and to be staggered with respect to a single bond in a ground state, in this case **1b** has the methyl hydrogen eclipsing the bond of high bond order, and the hydrogens are staggered with respect to the bond of low bond order, and that is the ground state. While the orientation of a methyl group in methylbutadiene may in itself be of limited interest, there are consequences of that fact that should be noted. That which applies to methyl groups, applies, generally speaking, to alkyl groups. This means that this effect may be expected to have consequences in the conformational behavior of more complicated molecules whenever a carbon with sp^3 hybridization is attached to a conjugated system. Further consequences of this situation will be discussed in Chapter 8, where it turns out to be important in carbohydrates.

Let us now consider the structures of some conjugated hydrocarbon molecules as they are calculated with the MM4 force field. Table 5.1 shows structural data on the simple molecules ethylene, benzene, and *trans*-butadiene. Thus, we see that we can first of all calculate to within experimental error the structures of our simple standard molecules (Table 5.1), including the rotational barriers of ethylene and butadiene, which are not included in the table.[22]

We then applied the force field from those molecules to several somewhat more complicated cases, such as 1,3-cyclohexadiene (Table 5.2), and, as expected, we were also able to calculate the structure of those molecules to within experimental error.

STRUCTURES OF CONJUGATED HYDROCARBONS

TABLE 5.1. Structures of Ethylene, Benzene, and *trans*-Butadiene[22]

	Bond	Expt.[a]	MM4	Δ
Ethylene	a	1.337 ± 0.001	1.337	0.000
Benzene	a	1.399 ± 0.001	1.397	−0.002
trans-Butadiene	a	1.344 ± 0.001	1.343	−0.001
	b	1.467 ± 0.001	1.469	+0.002
Angle	ab	122.9 ± 0.5	123.0	+0.1

[a]All structures have electron diffraction (ED), r_g bond lengths.

TABLE 5.2. Structure of 1,3-Cyclohexadiene[22]

Bond/Angle	Expt. (r_g)[a]	MM4	Δ (r_g)
a	1.352	1.346	−0.006
b	1.470	1.471	+0.001
c	1.525	1.511	−0.014
d	1.536	1.544	+0.008
Av	1.460	1.455	−0.005
ab	120.1	120.3	+0.2
ac	120.1	120.9	+0.8
cd	110.7	111.4	+0.7
aba'	18.3	13.0	−5.3

[a]The estimated errors in the experimental bond lengths are 0.010–0.020 Å.

This general procedure was followed to develop MM4 for the calculation of conjugated hydrocarbon systems. In real life the procedure was modified significantly in the following way. The above illustration utilized only a small number of compounds as standards and derived a force field that was then applied to several individual molecules. But, of course, there are experimental errors in all of the molecular structures that are determined, and there are also computational errors in the quantum mechanical structures, as has been previously discussed in Chapters 2 and 3. So what was actually done to derive the MM4 force field for conjugated systems[22,26,27] was to select a diverse set of about 125 experimental and calculated structures, and then fit the structures given by the force field to that diverse set, partly by least-squares methods. Vibrational spectra, heats of formation, and rotational barriers were also fit simultaneously. The overall fit of the MM4 calculations to the available data can be fairly described, we believe, as being of "chemical accuracy" and is similar to the accuracy obtained for alkanes[28] and alkenes[29] discussed earlier. Several examples of compounds that are of special interest have been selected and will be discussed in detail in the following sections.

in-[3⁴,¹⁰][7]Metacyclophane

in-[3^{4,10}][7]Metacyclophane

This molecule (**A**, Structure **2**) is of particular interest because the hydrogen that is held over the center of the benzene ring is really squashed down into the ring. This is the type of system that can give us information regarding the repulsive part of the van der Waals curve between hydrogen and carbon.*

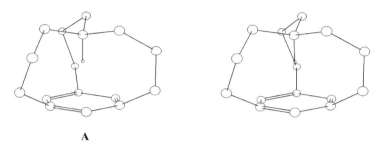

A

<u>Structure 2.</u> *in*-[3^{4,10}][7]Metacyclophane (stereographic structure with most hydrogens omitted for clarity).

If this repulsion is not well represented at the particular distance involved between the central hydrogen and the benzene carbons, that distance will be in error, and so will the strain energy of the system. There has been extensive discussion as to the nature of the interaction between a hydrogen attached to carbon and a π system, as in this molecule. Should this be considered a *hydrogen bond* or not?[30] Part of the definition of a hydrogen bond (discussed in Chapter 9) is that it holds the hydrogen more closely to the bonding atom than the sum of their van der Waals radii. In this case the deformation of the molecular structure, and the CH stretching frequency, both clearly indicate that these atoms are much closer together than the energy minimum distance of their van der Waals interaction. Of course, we expect that a hydrogen bond, like any other bond, will have an energy minimum and a strong repulsion if the two atoms come closer together than that minimum distance. And clearly, that is what we have here.

The structure of **A** is not known experimentally, so we will discuss instead the closely related trithiane **B**, shown in Structure **3**. The latter had its structure determined by X-ray crystallography at room temperature.[31] The positions of the sulfurs in **B** are indicated. The geometry of **B** differs somewhat from that of the hydrocarbon **A** [the angle at the sulfur in the upper right corner (105.0°) can be seen to be smaller than the corresponding angle at carbon in **A** (117.3°), for example, and the bonds from that sulfur are obviously longer than the corresponding bonds in **A**]. The X-ray structure **B** can be directly compared with the MM4 structure, for which the pertinent data are given in Table 5.3. Allowing for the experimental uncertainties (including a little distortion from crystal packing forces, Chapter 2), these structures are indistinguishable.

---

*Van der Waals forces are crucial quantities in molecular mechanics. The fitting of experimental data to derive a van der Waals curve in the case of hydrogen–hydrogen repulsions was discussed earlier (Chapter 4), and the same ideas were also used for the carbon–hydrogen case here, in an analogous manner.

STRUCTURES OF CONJUGATED HYDROCARBONS

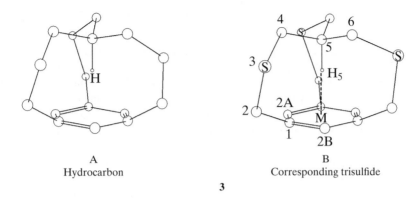

A
Hydrocarbon

B
Corresponding trisulfide

3

The ring distortions calculated and observed are evident (see Chapter 2 for some discussion of the problems inherent with X-ray structures). Note that the benzene ring and its attached atoms in a strainless aromatic system would be coplanar, and this would correspond to the angle that C_2 makes with the C_1–C_{2A}–C_{2B} plane being 180°. In fact, the X-ray value for this angle is 160.2°, and the corresponding MM4 value is 161.6, in reasonable agreement. Carbon 5 is held by the bridges at a distance that is much closer to the ring than the sum of the van der Waals radii, and hence there is a repulsion between C_1 and C_5. The ring carbons that are not attached to bridges can, however, pucker downward out of the plane of the bridged carbons, so the structure of the benzene ring is not planar, but it is a crown shape, with the bridged carbons upward in the figure, and closer to C_5, and the other carbons downward, and away from C_5. The distance is significant, although it does not really show in the drawings. The C_1/C_5

TABLE 5.3. Partial Geometry of 2,6,15-Trithia[3^{4,10}][7]metacyclophane (B)^a

Parameter	X ray[31]	MM4	Δ	E_{VDW}
H_5–M	1.69	1.72	+0.03	
C_5–M	2.78	2.81	+0.03	
C_4–C_5–H_5	105.4	105.1	−0.3	
C_4–C_5–C_6	113.2	113.4	+0.2	
S_3–C_4–C_5	116.1	115.3	−0.8	
C_2–S_3–C_4	106.3	105.0	−1.3	
C_1–C_2–S	110.2	112.0	+1.8	
C_1/C_5		3.12		0.20
C_2/C_5		3.17		0.15
C_1/H_5		2.20		1.02
C_2/H_5		2.24		0.83

^{a}Distances are in angstroms, angles in degrees, and the pairwise van der Waals energies are in kcal/mol. The point **M** is at the center of the aromatic ring.

distance is 3.12 Å, with a van der Waals repulsion of 0.20 kcal/mol. For C_2/C_5, the corresponding numbers are 3.17 Å and 0.15 kcal/mol. The distances between the hydrogen (H_5) and the ring carbons show similar behavior. The C_1/H_5 distance is 2.20 Å, with a repulsion of 1.02 kcal/mol, while the C_2/H_5 distance is 2.24 Å, and the repulsion is 0.83 kcal/mol. The distance between the squashed hydrogen and the middle of the benzene ring (**M**) is 1.69 Å by X ray and 1.72 Å by MM4. All of the other bond angles are, of course, strained by bending in the appropriate directions so as to minimize the effects of this repulsion. The bendings are of various sizes, as required by the shape of the energy minimum and its location on the potential surface. Chemists are familiar with Dreiding models, and other mechanical physical models, and their use to represent a benzene ring. With such models the benzene ring is pretty rigid. Actual benzene rings are a little more flexible than implied by such models. To bend one carbon of a benzene ring out of the plane sufficiently so as to twist the dihedral angle between it and the string of three attached carbons by 5°, for example, requires 1.23 kcal/mol of energy. A carbon attached to a benzene ring can be bent out of plane much more easily, 10° for 0.93 kcal/mol. With molecules such as the one being discussed here, there are many relatively small bendings that succeed in moving the offending hydrogen apart from the benzene ring by a sizable distance, and because the bendings are so much distributed, the total bending energy of the molecule is not nearly as great as one might suppose at first sight. For example, the tertiary C–C–C angle in isobutane is about 111.5°. The corresponding angles here (4–5–6) are opened somewhat, 113.2° for the specific angle shown by X ray (MM4, 113.4°).

It will be recalled from our earlier discussion about hydrogens that are highly compressed that the C–H stretching frequency for hydrogen 5 here would be expected to be unusually high (Chapter 4, Vibrational Motions of Compressed Hydrogens). In the cases discussed earlier, these high-frequency hydrogens had the option of vibrating in various off-axis directions so as to improve their van der Waals energies. Here the hydrogen is held by a threefold axis of symmetry and can only vibrate perpendicular to the mean plane of the benzene ring. The vibrational frequency turns out to be quite high experimentally (3325 cm^{-1}), relative to the analogous hydrogen in isobutane (2903 cm^{-1}). The MM4 value 3306 cm^{-1} agrees reasonably well with the experimental value.

AROMATIC COMPOUNDS

Back before the time of Michael Faraday (who discovered benzene in 1825[32]) it was known that a ubiquitous structural unit was found in natural compounds and seemed to be associated with the pleasant smells of many of those substances. The structure of the unit was not known at first, but its relative lack of reactivity was recognized early on. The structural unit could be seen to proceed unchanged through many series of reactions. Compounds that contained this unit were called *aromatic* because of their often fragrant odors, but this designation later became correctly and more usefully associated with their lack of reactivity. Subsequently, the structure of benzene was unraveled, and, indeed, its chemistry proved to be quite different from the usual chemistry of alkenes and alkanes. Thermochemical studies showed that benzene was significantly more stable than one might have expected from its structure, and hence *aromatic compounds* came to mean benzene-like compounds, characterized by stability or lack of chemical reactivity.[33]

While the thermochemical definition of an aromatic compound is commonly used and understood, there are two other definitions that have also been introduced. Sometimes it may be desired to decide whether or not a compound is aromatic, but it may not be that easy to obtain and interpret the desired thermochemical information. Hence, these subsidiary definitions. But these subsidiary definitions do not always lead to the same conclusion regarding the presence, absence, or degree of aromaticity as does the thermochemical definition. These two definitions will be briefly discussed in turn.

The first is the presence or absence of an electric ring current when the molecule is placed in a magnetic field.[34] If we have a cyclic polyene, or similar system, and place it in a magnetic field, the field may generate a current of electrons that will flow around the ring. If so, these moving electrons generate their own magnetic field, and this causes the chemical shifts of the protons attached to the ring to change very much from their normal values, as they would have in ordinary systems such as linear polyenes, for example, where no such current can be generated. This led to the suggestion in 1961[35] that a compound could be "described as aromatic if it will sustain an induced [π-electron] ring current." However, as will be discussed later in this chapter, while benzene and similar compounds do show large and usually very clear-cut chemical shifts from magnetically induced ring currents, such currents do not necessarily parallel thermochemical stability. Thus, the existence of such a ring current in a molecule is indicative of, but not proof of, the aromatic character of a molecule as defined in the classical (thermochemical) sense. It needs to be pointed out here, however, that there really is no completely agreed to definition of aromaticity.[33]

Another definition of aromaticity is sometimes used that is based on the bond lengths in the ring.[36] In benzene, the C=C bond lengths are all of equal length. In an ordinary polyene, alternating long and short bonds are observed. Hence "equivalent" bond lengths might be indicative of an aromatic ring. We use the word *equivalent* rather than equal because in naphthalene, for example, the bond lengths are certainly not equal, but they are equivalent in the sense that while they vary in π-bond order, and hence in bond length, none of the bonds have a bond order of near 0.4 or 0.9 as in butadiene. They are all in the intermediate range. And such systems are considered as aromatic. But again, equivalent bond lengths are an indication, not really a proof, of the existence of aromatic character in a ring. This point will be further discussed later.

Simple Benzenoid Compounds

To the organic chemists of the early 1900s and before, an aromatic compound was one that underwent substitution reactions, as opposed to the addition reactions of ordinary alkenes and polyenes. They also noted that if one had fused benzene rings in different arrangements, such compounds were by and large aromatic. Smaller members of this group included naphthalene, anthracene, phenanthrene, and coronene (Structure **4**):

Naphthalene   Anthracene   Coronene

Structure 4. Simple benzenoid compounds.

There are two general approaches to the solution of the Schrödinger equation for molecules, which are referred to as the *molecular orbital method*[17b] and the *valence bond method*.[37] In the first approximation, these two approaches give similar, but somewhat different, results. When refined, they approach the same limiting results. So these methods are just two ways of looking at the same problem. The molecular orbital method has been widely used in recent years, whereas the valence bond method has been used very little for quantitative calculations. However, the valence bond method has the advantage that the qualitative pictures used by organic chemists to describe molecules and their properties are more conveniently represented in that format. We will frequently use the valence bond pictures of these molecules here, and later throughout this book, to examine the structural problems of aromatic rings and organic molecules in general.

Benzene, and most aromatic compounds, usually undergo aromatic substitution reactions and not the addition reactions of ordinary alkenes. Phenanthrene (Structure 5), however, is different. When treated with bromine, it simply adds the bromine to the 9,10 double bond to yield 9,10-dibromophenanthrene. This unusual reactivity of phenanthrene can easily be understood from the valence bond viewpoint. In terms of valence bond pictures, we can describe the structure of the skeleton of this molecule to a first approximation with the complete set of five Kekule forms (resonance forms), as shown in Structure **5**:

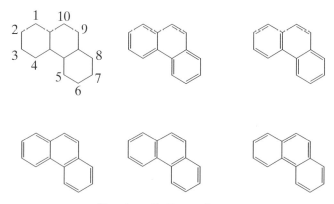

Structure 5. Phenanthrene.

If we look at these five forms, what do we see that is different about the 9,10 double bond, relative to the other bonds in the molecule? Note that the 9,10 bond is double in four out of the five Kekule forms. In contrast, the bond directly across the ring from the 9,10 bond is double in only one of the Kekule forms. Accordingly, we might conclude that the 9,10 double bond is 80% double, whereas the bond across the ring is only 20% double. And, if we similarly sum over the Kekule forms for each of the other bonds in the molecule, one by one, we will find that they all have a double-bond percentage that is in an intermediate range of less than 80% but more than 20%. What we are doing here is a qualitative solution to the Schrödinger equation using the valence bond method, and working our way through what amounts to the bond orders, and arriving at the conclusion that the 9,10 double bond has quite a high bond order, and hence should behave more like ethylene itself than like benzene. The remaining bonds in the molecule are much lower in bond order and more benzene-like. Hence, it should come as no surprise that the 9,10 double bond is the most reactive one, and it can sometimes undergo ordinary alkene addition reactions.

Double bonds are relatively short (about 1.34 Å), while single bonds between sp^2 carbons, tend to be much longer (about 1.45 Å). Conjugated C=C bonds typically span this range. The disagreement between the MM4 and X-ray structures of phenanthrene is relatively small and is believed to be due partly to experimental errors, partly to crystallographic distortions, and partly due to errors in the molecular mechanics calculations. It is perhaps of interest that when a similar comparison was originally made in 1971 using MM2 calculations[18] (which gave results similar to the MM4 calculations for this problem) and the then available X-ray structure,[38] there were two conspicuous discrepancies in bond lengths (of 0.040 and 0.050 Å). There was considerable speculation at the time regarding the short distance between the hydrogens attached to carbons 4 and 5, and whether or not the van der Waals repulsion between them might be sufficient so as to twist the phenanthrene rings out of plane, to minimize this repulsion.* Such distortions might, of course, also cause significant bond length changes. But the MM2 and MM4 calculations showed that this does not occur. There is indeed considerable repulsion between the 4 and 5 hydrogens, but they respond by in-plane bending, and the bond length distortions are not serious. After that time, a redetermination of the crystallographic structure was reported.[39] When compared to the new structure, the two discrepancies with the molecular mechanics structure were reduced to −0.015 and −0.004 Å, respectively. The older crystal structure was reported to have an ESD (estimated standard deviation) of 0.014 Å in the bond lengths, while the newer structure has the corresponding value reduced to 0.009 Å. In molecular mechanics we cannot

---

*The hydrogens at carbons 4 and 5 are each locked in rather tightly by the van der Waals repulsions between them and from the hydrogens on the other side (at positions 3 and 6). The (MM4) molecule is planar, with C_{2v} symmetry, and the lowest vibrational frequency is 86 cm^{-1} (real). The van der Waals repulsion between the hydrogens at 4 and 5 amounts to 0.98 kcal/mol, at a distance of 2.02 Å. (The neutron diffraction value at room temperature is 2.04 Å.[39]) But the hydrogen at 3 is only 2.44 Å from the hydrogen at 4, with a van der Waals repulsion of 0.26 kcal/mol. Bending the hydrogen at 4 away from that at 5 would involve pushing it closer to that at 3. And one would have to move the hydrogen at 5 toward that at 6 simultaneously. The bond angles involving the hydrogen at C4 are 117.5° in the direction of C3, and 121.4° in the direction of C5. The summation of the bending and van der Waals energies leads to the geometry described.

determine a comparable number because the calculations do not overdetermine the structure and allow the use of statistical methods to determine ESDs. But from a comparison of the accuracy of the results of numerous related molecules, we believe the corresponding molecular mechanics ESD would be about 0.005 Å. So the phenanthrene error was primarily in the earlier X-ray structure not in the molecular mechanics structure.

The MM4 program calculates the structures of these and many other simple benzenoid compounds to within experimental error, and of equal importance it calculates correctly their heats of formation (Chapter 11). This is not, of course, any great surprise since similar calculations have been made repeatedly in the past and, subsequently, with increasingly better approximations.[6,18–22] The bond order–bond length relationship idea proposed by Dewar works quite well when incorporated into more sophisticated molecular mechanics procedures.

Note now the molecule coronene, shown in Structure **4**. This molecule is, structurally, thermodynamically (Chapter 11) and computationally just another typical aromatic ring system. However, it leads us into more complicated and more interesting examples. While coronene might be called an "ordinary" aromatic hydrocarbon, the central six-membered ring in coronene can be replaced by a five-membered ring, and one of the peripheral six-membered rings can be removed, yielding the molecule that has the name *corannulene*. This molecule is more interesting in the sense that it is nonplanar.

Corannulene

The molecule is shown as Structure **6d**:

Structure 6. Corannulene (hydrogens omitted for clarity).

AROMATIC COMPOUNDS 109

Because of the five-membered ring in the center of the molecule, the planar form is not the ground-state structure. Rather, the molecule is deformed into the shape of an umbrella without a handle (**f**). The planar form (**g**) is the transition state in between two umbrellas (**f** and **h**). These three structures correspond to inverting the molecule. Calculation of the barrier to inversion depends upon having, among other things, the correct energies for the conjugation in the planar and puckered systems, along with appropriate bending and stretching constants for the C–C and C–H bonds, some of which depend on this conjugation. The energy barrier[40] of 11–12 kcal/mol experimentally and 14.3 kcal/mol by MP2/6-31G* is well calculated to be 11.2 kcal/mol by MM4.[22]

If we fuse an additional cyclopentane ring into the system (Structure **6e**) we would expect that the barrier for this inversion would be raised considerably because that ring also has to be distorted in going through the central planar structure of the corresponding transition state. We note that the energy is indeed raised very much, to approximately 26 kcal/mol experimentally[40] and 25.5 kcal/mol by MM4.[22]

It is reassuring that barriers such as these for corannulene and its cyclopentene derivative can be well calculated because it shows that the σ–π separation approximations employed in MM4 work well.

C_{60} Fullerene

If one starts with coronene and attaches additional six-membered rings onto that system, larger and larger planar aromatic compounds can be obtained that ultimately would lead to a large sheet of fused aromatic rings. Such a sheet is composed of pure carbon and is one atom thick. This substance is a known compound, called *graphene*. It is the strongest substance known.[41] When sheets of graphene are stacked one on top of the other, they yield graphite. And indeed, graphene can be produced by peeling it off of graphite. It appears that graphene has a great potential future use in computer construction (see Chapter 3, under Brief History of Computers).

Now imagine beginning again with coronene and adding more rings to the system. But this time we will mix in a few five-membered rings, as in corannulene. In this case, the system becomes nonplanar. Interestingly, if just a few five-membered rings are put into a large six-membered ring structure in just the right way, one can obtain a structure that is approximately spherical. A soccer ball (in most of the world outside of the United States it would be a football) has such a geometry, which in that case is made by sewing together hexagonal and pentagonal pieces of leather. The corresponding all-carbon structure is shown next to a sketch of a soccer ball. The molecule has the name C_{60} *fullerene*, and it can be prepared by heating graphite in an inert atmosphere (Structure **7**).[42]

It has the formula C_{60} and represents the third structure of carbon to be known (in addition to graphite and the diamond). Fullerene was discovered in 1985,[43] and a multitude of related compounds are now known. C_{60} fullerene represents a somewhat extreme example of an aromatic hydrocarbon molecule, both in size, and in its deviation from planarity.

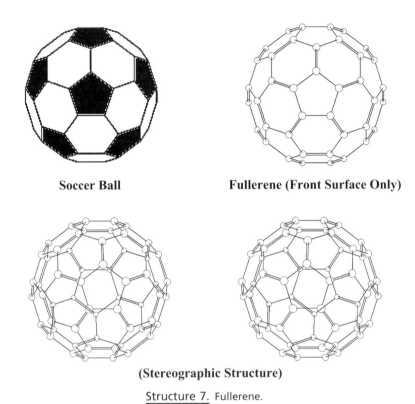

Soccer Ball **Fullerene (Front Surface Only)**

(Stereographic Structure)

Structure 7. Fullerene.

As it was to turn out, although C_{60} fullerene is a unique compound, innumerable related compounds of pure carbon have also now been discovered in which the overall structure of the molecule may be spherical, elliptical, or any of a multitude of prismatic structures, both smaller and larger than C_{60}.[44,45] Especially interesting to material science for their physical properties are various rod-shaped or tubelike related molecules.[44b] We will discuss here only the C_{60} example.

Theoreticians were thinking about C_{60} fullerene for a long time before it was discovered. The molecule was first discussed by Osawa[46] in 1970, and Hückel calculations on it were published by Bochvar and Gal'pern in 1973[47,48] and by Davidson in 1980.[49] These calculations clearly indicated the strong aromaticity expected for the molecule. A concise review of this early work is given by Kroto, Allaf, and Balm.[50]

This compound presents some challenges to structural chemistry. The ^{13}C NMR spectrum shows a single line, so all of the carbon atoms are equivalent. An ordinary X-ray structure for the molecule cannot be obtained because the molecule is rotating and hence disordered in the solid phase.[51] Since the molecule is essentially a sphere, this is not surprising. An X-ray structure was obtained for an osmium derivative of the fullerene, however.[52] Even more useful, a radial distribution function similar to what is normally obtained from electron diffraction experiments (Chapter 2) was obtained for the solid fullerene by both X-ray and neutron diffraction.[53] Because of the high symmetry of the molecule, the radial distribution function is sufficiently well resolved to establish the molecular structure rather accurately.

Because of the size of C_{60} fullerene, it also presents something of a challenge for computational chemistry.[45] Several different semiempirical schemes have been applied to the molecule to calculate the structure, the heat of formation, and heat capacities. These quantities are given approximately by semiempirical methods, but they are given more accurately by MM3,[45,54] and it is believed still more accurately by MM4. This compound was never used for parameterization of those force fields, and because of its size, it represents quite a large extrapolation from anything that was used. The structural features of the molecule are necessarily constrained by the high symmetry (point group I_h). The bond angles are all fixed by symmetry and the bond lengths. There are only two different types of bonds in the molecule. One is central to two fused six-membered rings, and the other is central to a five to six ring fusion. From the neutron diffraction study[53] these have lengths of 1.39 and 1.46 Å, respectively. The MM3 values are 1.390 and 1.453 Å, while the MM4 values are 1.383 and 1.448 Å. The agreement between MM3/MM4 and experiment is quite good, and this type of calculation has been highly useful for studies of compounds in this class.[44,45] One might wonder why the bond lengths around the five-membered ring are so much longer than those around the six-membered ring. There are minor contributions to these bond lengths from such things as stretch–bend effects, but the real difference here comes from the bond orders between the two types of bonds. The bonds that are fused between a five- and six-membered ring have relatively small bond orders (0.40), while those that are not so located have much higher bond orders (0.73). These numbers are reminiscent of those in butadiene. The bond orders in corannulene are not at all like that, so the fullerene bond orders are not only a function of ring size (five versus six), but rather they are a function of the conjugation of the whole molecule.[55]

It might be mentioned that there is quite a large empty space in the interior of the fullerene molecule. It has proven possible to trap various rare-gas atoms and small molecules within the fullerene interior, and the subject has been studied in some detail.[44,45] Helium and argon, for example, can be inserted into the fullerene cage at high temperatures and pressures. The resulting structures are referred to as *endohedral complexes*,* and they can be identified by mass spectrometry.[56]

The heat of formation of fullerene would also be a desirable quantity to know experimentally, as a test for the accuracy of various theoretical calculations, as well as studies on aromaticity. Some thermodynamic properties of fullerene will be discussed in Chapter 11.

---

*It may be of interest that the helium–fullerene complex is in fact a natural product. It has been detected in distant stars, where it has existed in large tonnage amounts for millions of years. This recalls to mind the fact that helium was discovered in the sun (by the presence of intense lines in the visible spectrum). This fact was indicated by the name chosen for it, which is from the Greek word *helios*, meaning sun. It was only later discovered to exist in Earth's crust.

AROMATICITY[33]

Aromaticity is a fundamental topic in organic chemistry. Some molecules such as benzene cannot be adequately written as a single valence bond structure. Rather, they have to be written in terms of multiple structures. This difference relative to most organic molecules yields properties, and indeed an entire chemistry, that is different from what might have been anticipated.

That benzene has different chemical properties from ordinary polyenes was well known to the early chemists. Polyenes, or even alkenes, react with a variety of nucleophiles, electrophiles, and free radicals, under various circumstances, to yield addition products. For example, bromine reacts more or less instantaneously with alkenes and polyenes, which is readily apparent when one drips bromine into an alkene solution, and the color disappears. The product in the simplest case will be simply a 1,2-dibromide. Benzene, on the other hand, requires prolonged refluxing with bromine, in the presence of a catalyst, to convert it to bromobenzene. And the latter reaction is not a simple bromine addition, it is a substitution of one hydrogen by one bromine. And a variety of compounds related to benzene (such as those discussed previously in this chapter) similarly are aromatic compounds, and chemically quite different from alkenes and polyenes. Why is this?

Hückel studied benzene and other monocyclic polyenes by solving the Schrödinger equation in a very simplified way. The π system of ethylene is the starting point. The energy of a 2p electron on carbon is our zero energy. When the two carbons are joined to give ethylene, two of these p orbitals are combined to form two molecular orbitals, bonding and antibonding, respectively. The bonding orbital has its energy lowered from that on the isolated atom by some amount, called β, and the antibonding orbital energy is similarly raised (to $-\beta$), as in Figure 5.1. In the ground state the electrons both go into the bonding orbital, paired, as in Figure 5.1. The interaction between the electrons is neglected, and the energy of the π system is then 2β (1β from each electron).

Extending the Hückel method to benzene, it is considered to have a regular hexagonal structure, planar with equal bond lengths. (The difference in the benzene bond lengths from that in ethylene is neglected.) The carbon nuclei become just unit positive charges, and the π system is considered as with six electrons moving in the field of

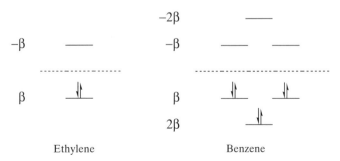

Figure 5.1. Hückel π orbital energies for benzene and ethylene.

AROMATICITY

those charges. The interaction of the electrons with one another is neglected, as previously. What Hückel found by solving the Schrödinger equation for benzene was that the six p orbitals combine to form a π system of six molecular orbitals, which have energies as indicated in Figure 5.1.

The π orbitals of benzene have energies that come as a lowest one and a highest one, and then there are two degenerate pairs of orbitals in between, with energies as shown. The six π electrons occupy the three lowest π orbitals in the ground state of benzene. It can be seen that relative to the zero line (which is the energy of an electron in each of the six separate p orbitals that we started with) the system has been stabilized by a considerable amount. We have two electrons in the lowest orbital, each with an energy of 2β, plus four with an energy of β, for a total of 8β of π bonding stabilization. For ethylene, consider the energy of 2β for comparison. Then three separate ethylenes taken together would have an energy of 6β. So the benzene molecule is more stable than three ethylenes by an amount $8\beta - 6\beta = 2\beta$. This 2β is referred to as the *resonance energy* of benzene. Note also that the benzene π system is a closed shell, analogous to the closed shells found in the rare-gas atoms. Historically, this simple qualitative explanation of the stability of benzene relative to the three constituent ethylene units was the starting point of understanding aromaticity.

One can write two Kekule forms for benzene, and these appear to be related to the stability of the molecule. But one can also write two Kekule forms for cyclobutadiene, and for cyclooctatetraene (Fig. 5.2), so are they not going to be similarly stable?

Chemists tried for a long time to synthesize cyclobutadiene, and endless failures are recorded in the literature. It was eventually synthesized[57] and is unstable in the sense that it cannot be isolated and put in a bottle. It can, however, be trapped in various ways.[58] But it was evident long ago that cyclobutadiene is not a stable molecule such as benzene. Cyclooctatetraene was synthesized by Willstätter and Waser in 1911,[59a] but it was highly reactive, such as a polyene, and certainly not stable such as benzene. So the Kekule forms written for these molecules seem to be misleading. Cyclobutadiene and cyclooctatetraene do not have any particular stability and are not at all like benzene. Why is that?

When Hückel solved the Schrödinger equation using his previously described method, he found that the energy levels of the π orbitals of benzene were as previously indicated in Figure 5.1. He applied the same method and approximations to solve the Schrödinger equation for cyclobutadiene and for cyclooctatetraene. To take cyclobutadiene first, there is one quite stable molecular orbital and one very unstable orbital, and then a degenerate pair of orbitals with an energy of zero β, as shown in Figure 5.3.

This means that with four π electrons, two would go into the lowest orbital with a total energy of 4β, and the other two (following Hund's rule), would form a triplet, with one electron going into each of the zero β orbitals, with unpaired spins. Thus, the

Figure 5.2. Kekule forms for benzene, cyclobutadiene, and cyclooctatetraene.

114 CONJUGATED SYSTEMS

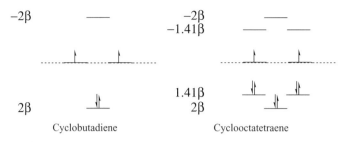

Figure 5.3. Hückel orbital energies for cyclotutadiene and cyclooctatetraene.

total energy of the π system would be four β, which shows no stability relative to two ethylene molecules (also 4β). Thus the resonance energy is zero, and the ground state would be a diradical. Clearly, this is not an ordinary closed-shell molecule, and it is not expected to be a stable system.

If we now turn to cyclooctatetraene (again assumed to be regular, planar with equal bond lengths), we have a system with eight π electrons, and the orbital energies were found to be as shown (Fig. 5.3). The electrons fill these orbitals in such a way as to again give an unstable diradical, with a resonance energy of zero.

Hückel was able to derive general equations that showed that for benzene, and for annulenes* that contain $(2 + 4n)$ π electrons, one has a system that has a stable closed shell (aromatic). But for a π system that contains $4n$ π electrons, the system is always predicted to be an unstable diradical (always assuming that the molecules are planar, regular polygons). These predictions were very tantalizing. They suggested a total of $(4n + 2)$ π electrons in a regular planar polygon would give an aromatic ring system, but $(4n)$ electrons would not. This fact was later called Hückel's $(4n + 2)$ rule for aromaticity. The simple prediction ultimately resulted in a flurry of synthetic activity, where a variety of compounds were prepared (or where it was attempted to prepare them) to bear out, or refute, Hückel's early suggestions.

The next important step in developing the theory of aromatic systems was to extend the solution of the Schrödinger equation for the π systems of the annulenes from the Hückel approximation to the SCF approximation (Chapter 3). In the latter, the interactions between the electrons are explicitly taken into account in an approximate way. The geometric results with benzene were similar to those from the Hückel method, but for cyclobutadiene, it was found that when the bond lengths were allowed to vary, the most stable geometry did not have bonds of equal length; rather they were alternating long and short, like those in a single Kekule structure. And importantly, the pair of π orbitals that were degenerate and nonbonding (zero energy) in the Hückel treatment now were split apart, one becoming a lower energy (bonding) orbital and the other a higher energy (antibonding) orbital. Here the electrons could both go into the lower energy orbital and form a closed-shell ground state. Nonetheless, when the energies of the annulenes as calculated for the π systems by the SCF method were examined (and

---

*Cyclic polyenes are called *annulenes*, where a prefix (number) can be added to indicate ring size. Thus, benzene and cyclooctatetraene are [6]annulene and [8]annulene, respectively.

AROMATICITY

TABLE 5.4. Dewar Resonance Energies and Bond Lengths in Annulenes[6b]

[n]Annulene	Resonance Energy	Bond Length	
		Min.	Max.
4	−18.0	1.34	1.51
6	20.0	1.40	1.40
8	−2.5	1.35	1.47
10	7.8	1.37	1.44
12	1.8	1.35	1.46
14	3.5	1.35	1.46
16	2.8	1.35	1.46
18	3.0	1.35	1.46
20	2.8	1.35	1.46
22	2.8	1.35	1.46

in this case actual numerical values for the π energies can be obtained), the results showed that, relative to open-chain polyenes, the difference between the $4n$ and $4n + 2$ electrons was maintained only in part. These resonance energies (specifically, those now called *Dewar resonance energies*[60]) are given as a function of ring size in Table 5.4.

Note that cyclobutadiene and cyclooctatetraene have negative (unfavorable) resonance energies (antiaromatic), while benzene is stabilized by a large positive resonance energy. [10]Annulene was calculated here to have a significant positive resonance energy, 7.8 kcal/mol. For still larger polyenes, Hückel's rule was not really borne out. The resonance energies from [12]annulene and larger are all about the same. The number 2.8 kcal/mol in Table 5.4 has that value because of where the zero-point reference is set, and the likely error that one might expect in these calculations is probably 2 or 3 kcal/mol. In other words, the resonance energies of the [12] and larger annulenes are zero here, to within our ability to calculate them. Note also the calculated bond lengths (also given in Table 5.4), which are equal in benzene but alternating for everything else. The alternation is less severe, but not by much, for [10]annulene.

The [4], [6], and [8]annulenes have resonance energies in Table 5.4 as qualitatively expected. The [10]–[16]annulenes are hard to relate to experiment because the real molecules do not have stable planar structures. The [18]annulene (which is very close to planar, see later) has a resonance energy of zero, both by calculation and by experiment (to within the accuracy of the methods), and it is now reasonably certain that larger annulenes will similarly have resonance energies of zero.

Cyclooctatetraene[58]

The structure of cyclooctatetraene is well known to be that of a tub.[61] This allows the C–C–C bond angles to be about 122° and near to the normal value for a polyene. There is a conjugation around the ring, which would be more effective if the ring were planar. But since the molecule has 8 π electrons, it does not obey Hückel's rule for aromaticity, and there would be little to be gained from this source by such planarity. However, the planar ring would have bond angles of 135°, so there would be a considerable bending energy for such a structure. When the molecule is in the tub conformation, the double bonds, of course, each prefer to be co-planar. This means that they are located as in Structure **8a**, at position 1,2 rather than at position 2–3, where they would be twisted. So the ground state of cyclooctatetraene is Structure **8a**. This structure can flatten out (Structure **8b**) and invert to give Structure **8c**.

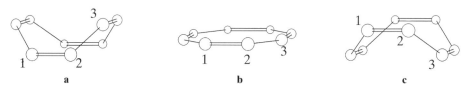

Structure 8. Cyclooctatetraene.

There is a barrier to inversion here that was measured[62] by some clever NMR methods at low temperatures. The free energy barrier from (a) to (c) (ring inversion) has an experimental value of 13.7 kcal/mol and the MM4 value is 13.5 kcal/mol, both at 263 K.

[10]–[16]Annulenes

Some of these should be aromatic and some not, according to Hückel's rule. But none of them can exist as planar regular polygon-type structures, and so we cannot put this rule to an experimental test with these compounds directly. They and many related compounds, however, can and have been prepared, and studied,[58] and it seems generally agreed that Hückel's rule is not well borne out. Rather, the resonance energies that can be inferred are more analogous to those in Table 5.4 from the SCF calculations. For [18] and larger annulenes on the other hand, planar conformations become possible. The molecules do not, of course, exist as regular polygons, but rather they fold in such a way as to maintain bond angles of approximately 120°, along with near planarity. These compounds are all quite reactive, similar to ordinary open-chain polyenes. We will discuss below one of these molecules ([18]annulene) in considerable detail, as that one appears to suffice to tell us about the higher annulenes.

[18]Annulene

The unusual properties of benzene and its Kekule structure go back to the beginnings of organic chemistry.[63] In 1931 Hückel published his famous $4n + 2$ rule for aromaticity,[1] which both rationalized the observed facts and also suggested that there would be a large series of hitherto unknown aromatic compounds possible. However, to be aromatic they had to be planar, or nearly so, which meant that actually [18]annulene was the next member of the series that was realistically a physical possibility. It was accordingly of great interest to chemists to see if in fact this compound was "aromatic."

AROMATICITY

Mislow proposed in 1952[64] that the internal hydrogens in the [18]annulene ring system would interfere with one another (from van der Waals interactions) to such an extent, that the molecule would probably not be planar (as is clearly the case with [10]annulene). The compound was synthesized in 1962 by Sondheimer and co-workers[65] and has been subjected to extensive studies subsequently. In spite of numerous studies by a variety of physical methods, and extensive computations as well, it can only be said at this writing that the structure is still not completely agreed upon. A brief exposition of the situation here up to the present time follows.

There are two principle structures that are possible for this molecule, and then there are some additional perturbations that are possible. But let us consider the general problem and the two principle structures first.

One can easily write a general structure for an 18-membered ring, which must be close to planar, and must contain a minimal amount of unfavorable van der Waals interactions. Let us initially assume the molecule is planar. There are two obvious possibilities, one of which would have D_{6h} symmetry and the other would have D_{3h} symmetry, as shown in Structure **9**.

[18]Annulene D_{6h} D_{3h}

9

The important point here is that the D_{6h} structure has equal bond lengths, such as benzene, and like one would anticipate for an aromatic compound, or alternatively, it might have alternating long and short bonds and be, in a structural sense at least, a D_{3h} polyene. There is also the possibility that the hydrogens in the interior show such unfavorable van der Waals interactions that they move up and down in an alternating fashion, out of the plane of the molecule, so as to minimize that repulsion, as suggested by Mislow. It turns out that they do this, but to a very minor amount. The resulting structures then become either D_6 or D_3, respectively.

So which is the structure? The first detailed theoretical studies on the structure of [18]annulene were carried out by Longuet-Higgins and Salem.[66] They predicted that there would be a substantial π electronic ring current for this molecule, resulting in chemical shifts of the inner and outer protons as a function of the degree of bond alternation. The observed NMR spectrum[67,68] indicated that, insofar as this approximate treatment is valid, bond alternation is present in [18]annulene.

An X-ray crystallographic study of the [18]annulene molecule was reported by Bregman et al. in 1965.[69] The X-ray work showed that the 12 inner bonds in the molecule had a mean length of 1.382 ± 0.003 Å, and the 6 outer bonds of the molecule had a mean length of 1.419 ± 0.004 Å. These bonds are somewhat different in length, but they are not alternating, and the differences in length are only about 0.04 Å, which is small compared to the 0.10 Å difference found in butadiene. The crystallographers worried some about this difference and also about the skeletal bond angles, which differ somewhat from 120°. Molecular mechanics says that these observed differences are just ordinary cis–trans differences due to the σ system. They are well reproduced by MM3 and MM4. The experimental deviations from planarity of the carbons was ±0.085 Å. Since this is not very much, the crystallographers also suggested that the isolated molecule might be planar, and the puckering could be due to crystal packing forces.

Other studies followed, including an analysis of the observed[70] ultraviolet spectrum by Van Catledge and Allinger,[71] that led these latter authors to state without equivocation "the molecule ([18]annulene) does not have this structure [the X-ray structure, approximately D_6] in solution." It is, of course, always possible that the structure in the crystal may differ from that in solution and/or from that in the gas phase. Conformational structures sometimes do differ in the crystal, relative to the other phases, but the kind of phase-dependent structural difference that would be required here is not known experimentally to occur in any other case, to the knowledge of the present author.

Thus, the situation up until the early 1970s was that there were two experimental structures available for [18]annulene, one approximating D_6 (equivalent bond lengths) and the other approximating D_3 (alternating bond lengths). The former was rather firmly established in the crystal (by X ray), and the latter was rather firmly established in solution and in the gas phase from the ultraviolet spectrum (by calculation and experiment). And the situation remained like that for quite a long time. Further studies involving thermodynamics, NMR, and crystallography were subsequently carried out to attempt to resolve this problem. These topics will be discussed in more detail in subsequent chapters, and hence a discussion of the final outcome of the [18]annulene problem will be postponed to Chapters 10 and 11.

Triquinacene (Homoaromaticity)

A conjugated polyene is represented by butadiene, which has two double bonds separated by a single bond. This puts the two double bonds fairly close together in space and allows an interaction of the two π systems, which in a sense are independent of the σ system, or the single bond. Hence one might ask the question: Suppose one puts two double bonds close together in space, but not bonded together via a σ system, will they interact as a conjugated system? Qualitative molecular orbital theory says yes, of course, they will interact, and the strength of the interaction depends on how close together in space they are, and on how much they overlap. And this prompted chemists to search for compounds that had double bonds that were not attached, but were close together in space. We might start by thinking about the compound 1,4-pentadiene:

$$CH_2=CH-CH_2-CH=CH_2 \quad \text{or}$$

AROMATICITY

We can ask is there any conjugation to speak of between the two double bonds? There really isn't, but if we could somehow force the bonds closer together, can we create some? If we can, we might call this *homoconjugation*.

The compound triquinacene was an early candidate as a compound that might show this effect (Structure **10**). In this particular case the effect could be referred to as *homoaromaticity*.[72]

Structure 10. Triquinacene (hydrogens omitted for clarity).

The double-bonded carbons that are adjacent to one another, but in different bonds, are approximately 2.54 Å apart in this molecule. This is much further apart than are the central atoms in butadiene (1.45 Å), but it is much closer together than nonbonded carbon atoms normally come (the sum of the MM4 van der Waals radii is 3.92 Å). Also, the dome-like shape of the molecule causes a larger overlap of the bottom sides of the p orbitals than would be found if the system were planar. [When 2p orbitals overlap at a given distance, the overlap is much greater if they are pointed at each other (σ overlap) than if they are parallel to each other (π overlap. The geometry here leads to a considerable portion of σ overlap.] The molecule was prepared and first studied in 1964.[73]

The usual way to determine the aromaticity of a compound is to determine its heat of formation, and compare this with an idealized calculated value, where there is no resonance. The actual path followed in the case of triquinacene was much more convoluted. To describe this history, we first need to say a little bit about thermochemistry and heats of formation. These topics will be discussed in more detail in Chapter 11, we will just give a brief outline here. The ideas are quite straightforward in principle, and the difficulty comes at the practical level. It is easy to imagine doing these things with high precision, but that is not so easy experimentally.

Consider the hydrogenation of triquinacene to its perhydro (actually hexahydro) reduction product. There are three double bonds in the starting material, so triquinacene(1) can be hydrogenated to dihydrotriquinacene(2) which can be hydrogenated again to tetrahydrotriquinacene(3) which can finally be hydrogenated to hexahydrotriquinacene(4). The most straightforward way in principle to determine the heats of formation of these compounds would be to directly measure the heat of combustion of each separate compound. But this requires substantial samples of compounds of quite high purity, and just obtaining the compounds can be difficult in practice. An alternative method would be to measure the heats of hydrogenation directly for each of the compounds (1), (2), and (3). If we then have the heat of combustion for any one compound in the series, we can convert the heats of hydrogenation into heats of formation. The numbers obtained from this approach are outlined in Table 5.5. We have starting with triquinacene, compound (1) going to (2), going to (3), going to (4). We would like to know the heat of formation of each of these compounds, particularly for compound (1).

TABLE 5.5. Heats of Formation of Triquinacene Products (kcal/mol)

	Triquinacene (1)	$\rightarrow$	Dihydro (2)	$\rightarrow$	Tetrahydro (3)	$\rightarrow$	Hexahydro (4)
Exper. H_f°	53.5 ± 1.0		30.5 ± 1.0		3.0 ± 1.0		−24.5 ± 0.86
ΔH_{H_2}		23.0		27.5		27.5	
MM4[a]	59.07		31.37		3.76		−24.01

Historically, the hexahydro derivative (4) had its heat of formation determined by heat of combustion.[74a] The measured value was −24.5 ± 0.86 kcal/mol, as listed in the Table 5.5. The heat of hydrogenation of each of the compounds (1), (2), and (3) to compound (4) were then each separately determined.[74b] These gave the values for the heats of hydrogenation (ΔH_{H_2}) in line 2 in Table 5.5. And from these values, the heats of formation of compounds (1), (2), and (3) were determined.[74b]

Now if we just look at the differences in the experimental heats of hydrogenation themselves, we see that the hydrogenation of triquinacene to the dihydro derivative is 23.0 kcal/mol, while the hydrogenation of the dihydro to the tetrahydro, and the hydrogenation of tetrahydro to hexahydro compound are each 27.5 kcal/mol. This immediately shows that much less energy is obtained in the first step, than in the latter two steps. This is quite analogous to what happens in the hydrogenation of benzene, where, because of the aromaticity the heat of hydrogenation of the first double bond is much smaller than the heats of hydrogenation of the other two bonds. One can similarly conclude from these numbers that the triquinancene is stabilized by homoconjugation by about 4.5 kcal/mol, relative to the other double bonds. But it turns out that the problem is actually more complicated than this.[75a] Dihydrotriquinacene(2) can relax its saturated cyclopentane ring by twisting, which yields a C_1 symmetry for the molecule. Compound 3 can similarly relax. Hexahydrotriquinacene(4) can similarly twist into a C_3 conformation, where each of the cyclopentane rings can relax (twist) into a lower energy half-chair. So interpreting this data is less straightforward than it appears.

In the meantime, computational chemists were at work using *ab initio*, semi-empirical, and molecular mechanics methods, to try to determine the aromaticity, if any, in triquinacene, and to some extent they also looked at these hydrogenation products as one way to obtain information on triquinacene. Importantly, Miller et al.,[75b] using *ab initio* theory, came to the conclusion that the heat of formation of triquinacene was not 53.5 kcal/mol, the experimental value, but was rather in the range of 56.6–57.0 kcal/mol (including zero-point and thermal energies). If correct, this would mean that triquinacene itself had no significant aromaticity or conjugation, and for some unknown reason there was an error in the experimental results. Houk offered an alternate interpretation of these data based on MM3 calculations.[75a] But our interpretation is that these calculations, and those involving cyclopentane rings in general with MM3, were less accurate than one would have liked. From current information, there is one useful piece of data from the MM3 calculations. These MM3 calculations can be carried out in two different ways. One can treat the alkenes as just that, unconjugated, simple alkenes. Or one can do the calculation considering the three alkenes as a conjugated system, where MM3 will automatically calculate the reduced overlap between the double bonds. The MM3 calculation for a conjugated system utilizes double-zeta π orbitals which should be an adequate representation. When the calculations are done both ways and compared, any errors in the cyclopentane part of the ring cancel out, because they are the same in both calculations. Thus one obtains the energy of the conjugated system, relative to the energy of three alkenes. And this number should be reasonably accurate. This number comes out to be 0.52 kcal/mol in the direction that triquinacene is stabilized by that amount.[75a,c] The accuracy of this value is not known, but would be expected to be of the order of 0.5 kacl/mol. Thus we cannot say for sure from the MM3 results that triquinacene has a resonance energy of 0.5 kcal/mol, but we can say with reasonable certainty that its resonance energy is considerably smaller than the 4.5 kcal number found experimentally.

The heats of formation are, however, more accurately calculated with MM4, where this cyclopentane problem has been removed, and MM4 properly handles the zero-point and thermal energies by statistical mechanics. The corresponding heats of formation calculated by MM4, treating these molecules as simple alkenes (unconjugated) are also listed in Table 5.5. First note that the hexahydro compound, which has its heat of formation determined by a heat of combustion, is fit to well within the stated experimental error. The dihydro and tetrahydro values also fit within the stated experimental errors of the hydrogenation calculations (Table 5.5). The only heat of formation that is seriously different from the experimental values is that for triquinacene. Here the MM4 value,[75c] if there were no conjugation, is 59.07 kcal/mol, about 5 kcals higher than the experimental value.

Finally, in 1998 Verevkin and co-workers determined the heat of combustion of triquinacene itself, and from that its heat of formation.[75d] Their value is 57.51 ± 0.70 kcal/mol. This value, of course, includes whatever conjugation energy is present in the actual triquinacene molecule. The MM4 value does not. The difference between the MM4 value and the experimental value is 1.56 kcal/mol. Thus one might take this to be an estimate of the resonance energy of triquinacene. However, the experimental value has a stated probable error of 0.70 kcal/mol. The accuracy of the MM4 number is not known, but is estimated to be 0.5 kcal/mol. That means that this 1.56 kcal/mol resonance energy is not really much outside of the combined errors expected in the number, but it suggests that the number must be fairly small, probably less than 2 kcal/mol. If we calculate the MM4 resonance energy of trequinacene as we did previously with MM3, treating the molecule as a conjugated system lowers the energy by only 0.32 kcal/mol, relative to the simple alkene calculation.

Our conclusion from the MM3/MM4 results is that the homoaromatic resonance energy of triquinacene is 0.4 ± 1.0 kcol/mol. It seems likely that there is a small resonance energy, but it is too small to specify with any confidence.

There are other ways to decide if a compound is "aromatic," of which the magnetic ring current from NMR, and various other spectroscopic methods were all examined.[76] From each of these methods, the result was the same, namely any homoaromatic resonance present in triquinacene was too small to measure.

Thus, there may be a nonbonded conjugated interaction or homoaromaticity in triquinacene, but it is quite small because the distance between the double bonds is just too great. This has prompted chemists to try to prepare additional molecules that would show this interaction, but where the double bonds were closer together, the overlap was greater, and the interaction would be much stronger. This led to speculation over the years, and a number of compounds were suggested[76] as potential structures that could be prepared that might show homoaromaticity.

These structures were chosen so as to push double bonds of the

type closer together in various ways in cyclic molecules. Calculations showed, however, that the interaction would be very small for each of the structures suggested so far, except for Structure **11** and closely related (substituted) compounds.

Structure 11. Stereographic projection (hydrogens omitted for clarity).

The four-membered rings in this molecule pull the double bonds much closer to one another than they are in triquinacene, about 1.87 Å compared to 2.54 Å. At this distance there is quite appreciable overlap between the orbitals of the non-bonded carbons, and homoaromatic interactions seem much more likely.

One can compare the heats of formation of the molecule calculated with MM3 as previously described, namely as a conjugated system, or as a tri-alkene system. The energy is lower for the conjugated system by 5.8 kcal/mol. This energy difference is approximately 10 times that for triquinacene, and about one-third of that observed for benzene (17.8 kcal/mol). Importantly, sizable magnetic susceptibility exaltations are also calculated for this molecule.[76] We are looking forward to seeing an eventual publication on the synthesis and experimental study of this compound, or a derivative.

ELECTRONIC SPECTRA[77]

Conjugated hydrocarbons, and particularly the aromatic ones, tend to have somewhat complicated electronic spectra that can ordinarily be observed in the ultraviolet. Historically, ultraviolet spectra were measured and used in chemistry in the 1930s, long before infrared or NMR spectra became available. Since several of the occupied π orbitals are, generally speaking, the filled orbitals of highest energy in the molecule, and several of the π^* orbitals are the unoccupied orbitals of lowest energy, it is normally transitions between these orbitals that are seen in the observable ultraviolet region, from about 200 to 400 nm, and occasionally extending into the visible region. For larger molecules there may be several observable transitions. Below about 200 nm the σ system absorbs a great deal of the impacting radiation, and such a spectrum becomes pretty much a generally uninformative continuum.

The use of an ultraviolet spectrum for the study of a simple conjugated polyene does not usually give very much information. Nonetheless, such spectra were routinely recorded for all new compounds at that time because so little information was available that one had to use whatever one could obtain. Something about the substitution pattern of alkyl groups on the polyene could be deduced from the frequency of the ultraviolet absorption. This kind of information was summarized by what were referred to as "Woodward's rules."[78] These rules were subsequently extended to conjugated ketones as well. These were simple additivity rules that enabled one to predict for relatively simple compounds at what wavelengths the ultraviolet absorptions were to be expected. This information was helpful in identifying unknown compounds. Quantum mechanics was applied to studies of the π structures of such compounds in the 1960s, and the underlying basis for Woodward's rules was rationalized in that way.[79] The variable electronegative self-consistent field (VESCF) method was a key to understanding Woodward's rules. In this method the carbon atoms forming the π system have ionization potentials that are affected by the attachment of alkyl groups or other substituents. The orbital exponents for the π orbitals of the carbon atoms are functions of these ionization potentials,[13] and they change as well. In addition, as the SCF interations proceed, the electron densities at the different atoms usually change, and these changes also cause further changes in both the ionization potentials and the orbital exponents. The result of all of this is a shifting of the π and π* energy levels of the system, which yields calculated shifts in the ultraviolet spectra. Such VESCF calculations agreed well with experiment for the electronic spectra of planar hydrocarbons[80] and provided strong evidence for the structure of [18]annulene as discussed earlier in this chapter. They also enabled the prediction of electronic spectra for conjugated heterocyclic compounds.[81,82] Later versions of the MM2 program (1977)[83] were capable of carrying out such calculations.[84,85] They have not, however, been extended to more modern computer programs to the author's knowledge.

Aromatic systems tend to show more complicated spectra than simple polyenes do from the π system transitions. Usually, several different transitions can be seen. They come at various frequencies in the ultraviolet, and also with widely varying oscillator strengths (intensities), and they frequently show fine structure due to concurrent vibrational motions (called vibronic transitions). From the differences in the frequencies of these vibronic bands, one can often identify the infrared transitions associated with them. These are sometimes allowed, and sometimes forbidden, by various selection rules. Accordingly, one can occasionally get a great deal of information from the ultraviolet spectra of aromatic compounds. C_{60} fullerene, for example, contains quite a complicated electronic spectrum, which has been recorded over the range of 190–700 nm. There are no less than nine observed electronic transitions in the range below 410 nm, and numerous transitions above that which are all forbidden and hence relatively weak. They are, however, of sufficient intensity to give the compound a strong magenta color.[50] An understanding of the electronic spectrum of this molecule has proved quite a challenge to spectroscopists.[86]

If there are atoms containing lone pairs that are also conjugated with double bonds or aromatic systems, then one can have not only $\pi \rightarrow \pi^*$ transitions but also corresponding $n \rightarrow \pi^*$ transitions. Since the n orbital usually lies at a higher energy than the π orbitals, these transitions are ordinarily observed at longer wavelengths than the $\pi \rightarrow \pi^*$ transitions. These $n \rightarrow \pi^*$ transitions are often weak, but observable, and are commonly seen in molecules such as ketones or pyridines. The $n \rightarrow \pi^*$ transitions are useful for a wide variety of studies in molecular structure. Carbonyl transitions, for example, usually occur in the near ultraviolet and are hence easily accessible in an otherwise empty region of the spectrum. They are responsible for optical rotatory dispersion and circular dichroism spectra,[87] which in turn are often used for (among other things) the assignment of absolute configuration in organic molecules. Sometimes the frequency range spanned by the transitions of the π system becomes sufficiently large, as the conjugated system becomes sufficiently large, that some of these frequencies move from the ultraviolet up into the visible part of the spectrum. These transitions frequently lead to the colors of dyes and related compounds.

STRUCTURES OF CONJUGATED HETEROCYCLES

The previously discussed calculations on conjugated hydrocarbons taught us a few points that will be pertinent here. First, one has to use a quantum mechanical calculation at some point to describe the π systems of these molecules in molecular mechanics. Other methods in general fall far short of yielding chemical accuracy for either structures or energies. The σ–π separation approximation is essential, and along with the VESCF approximation it allows us to calculate molecular structures and properties with an accuracy (for the most part) similar to that obtained for ordinary unconjugated hydrocarbons. Thus, we can expect that when computational objectives are further extended to include conjugated heterocyclic molecules more completely, it will be necessary to deal with the π systems in a manner similar to that used previously for the hydrocarbons. When the alkane calculations were extended to include heteroatoms, all sorts of complications developed because of the various "effects" that are popular in organic chemistry (see following chapters). We can anticipate that the detailed extension of force fields from aromatic hydrocarbons to heterocycles will be accompanied by similar complications if chemical accuracy is desired.

The parent heterocycles—pyridine, pyrrole, furan, and thiophene—proved to be straightforward to deal with, following the scheme outlined above.[81,85] Many other molecules of increasing complexity, containing two or three heteroatoms, and up to the size of porphin, were treated. These molecules were found to be unexceptional as far as molecular mechanics is concerned, and calculations on them pretty much yielded the expected results. However, when the nucleic acid bases were examined, unexpected difficulties arose.

From a biological point of view, there are a few relatively simple nitrogen heterocycles that occur in DNA (deoxyribonucleic acid) and RNA (ribonucteic acid), and

STRUCTURES OF CONJUGATED HETEROCYCLES

these molecules are consequently of exceptional importance in biology and, hence, in chemistry. These include especially adenine, guanine, cytosine, and uracil (Structure **12**):

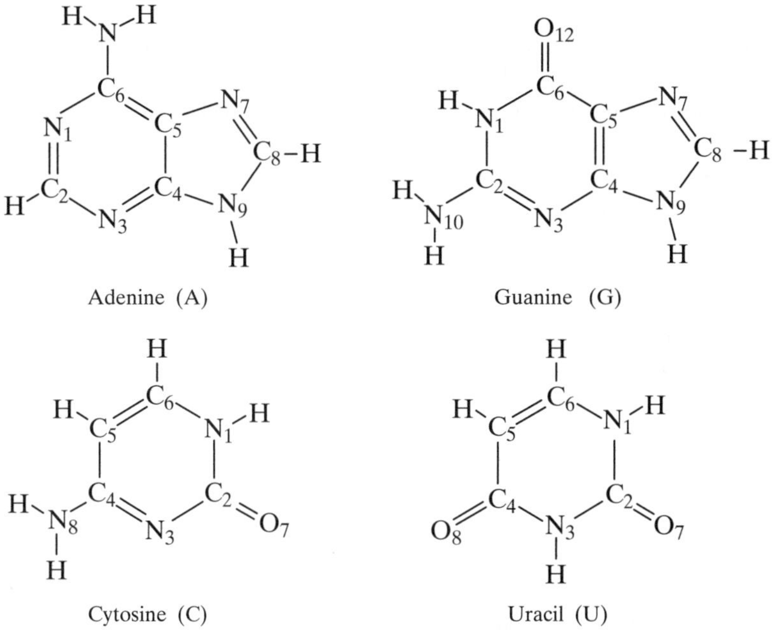

Adenine (A)

Guanine (G)

Cytosine (C)

Uracil (U)

12

We wished, of course, to be sure that we could calculate structures such as these adequately with molecular mechanics.

At this point (early 1980s) it became clear that we really needed to divide the treatment of aromatic heterocycles into two different, general cases. The first case is ordinary heterocyclic molecules. For those, the calculations are carried out as previously described. The second case would be when we wished to carry out a calculation on a macromolecule such as DNA, which has a great many molecules of nucleic acid bases present, and we do not want to do a quantum mechanical calculation on each of them one by one. We wished to carry out the quantum mechanical calculation for a given base once, and then develop a force field that will reproduce the structure of each of these bases adequately. These molecular force fields are then stored in a library. For future calculations involving those molecules then, one obtains from the library the appropriate force field for each of those molecules as needed, and they are then used in the ordinary molecular mechanical manner.

We had earlier done this for the particular case of a benzene ring, which can be treated by the general method most of the time, but when it occurs multiple times in a macromolecule (or other large system), there is a general force field for the benzene ring that can be transferred into the calculation, which avoids having to carry out the quantum mechanical part of the calculation on each individual benzene. By this approach one can expect to obtain more or less ordinary chemical accuracy for the nucleic acid bases, even though they are part of a truly large molecular system, and one can avoid the lengthy quantum mechanical part of the calculation. (And, of course, in macromolecules "chemical accuracy" is normally at least an order of magnitude less than usual.)

To apply this "library scheme" to the nucleic acid bases, one first had to know the structures of those molecules. There was, to be sure, quite a bit of crystallographic data on the structures of the nucleic acid bases available by 1980. However, as a group, these molecules and derivatives tend to form crystals that are not well ordered, the available X-ray structures had large R values, and the estimated standard deviations in the bond lengths and angles tend to be rather large. We did not feel that the crystallographic structures available were really adequate for our purposes. We noted that the Kollman group had previously determined force field formulations that they deemed adequate for their purposes from experimental or quantum mechnical data,[88] but only at a level that was less than ordinary chemical accuracy. It could have been that the discrepancy between what appeared to be the best structures from crystal data and those from quantum mechanical calculations (which was of the order of several hundreths of an angstrom, ten times larger than usual) was simply due to inaccurate results from the quantum calculations then possible, but there was also a serious question as to whether or not the crystal structures gave a good representation of the gas-phase structures. Molecules of this class tend to have rather high charges distributed about the aromatic system. When molecules with high charges are packed together in a crystal, not only must one worry about the usual problems of how well the electron density represents the nuclear structure (see Chapter 2 under X-Ray Crystallography and Neutron Diffraction), but also about whether or not the relatively large atomic charges might also change the structure relative to what would have existed in the isolated molecule, from the intermolecular interactions.

On the other hand, it was not possible to carry out accurate quantum mechanical (QH) calculations on these molecules with large basis sets including adequate electron correlation. When we first dealt with this problem a Hartree–Fock 6–31G* calculation was the best we could manage for molecules as large as adenine and guanine. And the weight of the available experimental evidence from the better crystal structures gave results that clearly differed substantially from the QM results at that level. The MM2 calculations that we carried out did not agree very well with the QM calculations, but the agreement with the experimental structures was even worse. And there was also the possibility of a breakdown of the transferability of parameter assumption in molecular mechanics.

When we became seriously concerned about this problem (about 1985[89]), it could not be convincingly decided which of these basic alternatives was correct (Hartree–Fock or X-ray), and it was not possible to proceed further. Subseqently, however, computers developed to the point where it became possible to carry out what are now considered modest quantum mechanical calculations directly on the nucleic acid base molecules. When this was done,[90] it was found that the main problem did not lie in the molecular mechanics approximations themselves, nor did it lie in the experimental comparison of a crystal structure with a gas-phase structure. Rather, the major problem was due to the neglect of electron correlation in the SCF calculation. If one carried out Gaussian HF6-31G* calculations on the nucleic acid bases, structures were obtained that were substantially different from the experimental ones. But when one added electron correlation into the quantum mechanical calculations (MP2/6-31G*), it was found that there were dramatic changes in those structures, and the calculated structures then corresponded rather well with the better experimental X-ray structures.[91] After further study, it was found that by adding electron correlation (MP2) to the Hartree–Fock calculations already in the MM3 program,[24] the bond lengths and overall geometries could be greatly improved.[91] So the errors causing the problems here were not in the general methodology, or the parameters, or in the σ–π separation approximations, or in the other multitude of possible problems that were considered, but were just due to the lack of electron correlation in the MM3 π-system calculations. Of course, we know from many quantum mechanical studies on molecules that the inclusion of electron correlation normally makes relatively small changes in structures of molecules. But in some cases, and we now know that in the nucleic acid bases in particular, the omission of an adequate correlation treatment leads to quite large errors in structure.

One final point might be added. Molecular mechanics is based on the assumption that simple bonding properties are transferable from one molecule to another. And generally speaking, this assumption works very well. Conjugated molecules are a somewhat exceptional case, however, because of the continuous range of bonds, and their corresponding properties, that must be dealt with. It is, however, noteworthy that the model developed, based on many semiempirical approximations, and in particular upon the σ–π separation, works very well, and as far as we know, can be extended in a very general way to all conjugated systems. It is possible for the most part to include the effects of correlation adequately in molecular mechanics implicitly (i.e., in the parameterization). But in some cases (e.g., the nucleic acid bases) the explicit inclusion of the results of correlation is essential if reasonably accurate structures are to be obtained.

Porphyrins

Porphin is the parent structure for a class of compounds called *porphyrins* having the rather large heterocyclic ring structure shown in Structure **13**:

Porphin   Porphin (Fe^{++} Complex)

13

Porphin itself is orange-red in color. The molecule contains two pyrrole rings and two similar rings from which the N–H hydrogens are missing. These latter two nitrogens each have a pair of electrons in an sp² orbital in the plane of the σ system. The two pyrrole hydrogens in the ring system are fairly acidic, and metallic derivatives (chelates) of these compounds are known. Numerous metals with a stable +2 oxidation state can form complexes with these compounds, where each of the hydrogens (protons) is removed and each nitrogen takes a −1 charge. The other two nitrogens are electron donors to the metal atom, so the four nitrogens become equivalent by resonance. A multitude of compounds having this general structure exist, and the iron complex of porphin shown is typical. One of the most important of these porphyrin derivatives is certainly the molecule *chlorophyll a*, Structure **14**. Note that it has a relatively complicated side chain and numerous other substituents. It is this porphyrin structure that gives the familiar green color to most plant leaves, grass, and the like. The large conjugated system leads the molecule to absorb light in the visible range, as well as in the ultraviolet. The red blood pigment, *hemoglobin*, is the other best known member of this class of compounds. Hemoglobin consists of a protein portion called *globin* and an iron porphyrin complex called *heme*, also shown (Structure **14**). It is of interest that this rather complicated ring structure exists and, in fact, plays extremely important, and quite different, roles in both the plant and animal kingdoms. This suggests that perhaps these structures were already present in living organisms very long ago, before the plant and animal kingdoms separated. Compounds such as this are modeled without difficulty by MM3 and MM4.

The porphyrin ring is best not thought of as a pyrrole derivative, but rather as a large aromatic ring system in its own right. There is a problem regarding the structure of the aromatic ring system in porphrin molecules, which is the same problem that arises with [18]annulene. This problem has not been previously discussed in the literature in connection with porphyrins to the knowledge of this author.

Chlorophyll a   Heme

14

Hartree–Fock calculations on porphin show alternating long and short bonds, as in polyenes. The structure from X-ray crystallography shows averaged bond lengths, as is found with [18]annulene. It was thought that the addition of electron correlation to the Hartree–Fock calculation here, as in [18]annulene, would probably lead to a stabilization of the porphin ring system in that averaged geometry. Accordingly, after the capability was added to MM3 so that an MP2 perturbation calculation could be carried out on the π system during the geometry optimization,[24] this procedure was applied to porphin. It was indeed found that similar to what happened with [18]annulene, the Hartree–Fock structure tended toward a structure with the bond lengths more nearly averaged. But, instead of having the bond lengths roughly 0.10 Å different, between the alternating long and short bonds as in the Hartree–Fock calculation, this value was only reduced to about 0.08 Å. It certainly did not go to the "equal" bond lengths that one might have anticipated. Thus, the porphin structure contains an open question regarding these bond lengths, in the view of the present author. It seems that the bond lengths are more likely alternating, but undergoing the long/short interchange over time, as suggested earlier for [18]annulene. See the discussion of [18]annulene (in Chapters 5, 10, and 11) for additional information regarding the bond lengths in that case, which would appear to be applicable here also.

REFERENCES

1. E. Hückel, *Z. Phys.*, **70**, 204 (1931); **76**, 628 (1932).
2. C. A. Coulson, *Valence*, Oxford University Press, Oxford, 1956.
3. A. Streitwieser, Jr., *Molecular Orbital Theory for Organic Chemists*, Wiley, New York, 1961.
4. K. Jug, P. C. Hiberty, and S. Shaik, σ-π *Energy Separation in Modern Electronic Theory for Ground States of Conjugated Systems, Chemical Reviews*, Vol. 101, P. v. R. Schleyer, Ed., American Chemical Society, Washington, D.C., 2001, p. 1477.
5. M. J. S. Dewar and H. N. Schmeising, *Tetrahedron*, **11**, 96 (1960).
6. (a) M. J. S. Dewar and G. J. Gleicher, *J. Am. Chem. Soc.*, **87**, 685 (1965). (b) M. J. S. Dewar, *The Molecular Orbital Theory of Organic Chemistry*, McGraw-Hill, New York, 1969, p. 179.
7. D. R. Hartree, W. Hartree, and B. Swirles, *Phil. Trans. R. Soc. London*, **A238**, 229 (1939).
8. V. A. Fock, *Z. Phys.*, **61**, 126 (1930).
9. G. G. Hall, *Proc. R. Soc. London*, **A205**, 541 (1951).
10. C. C. J. Roothaan, *Rev. Mod. Phys.*, **23**, 69 (1951).
11. R. Pariser and R. G. Parr, *J. Chem. Phys.*, **21**, 466, 767 (1953).
12. J. A. Pople, *Trans. Faraday Soc.*, **47**, 1375 (1953).
13. (a) R. D. Brown and N. L. Heffernan, *Australian J. Chem.*, **12**, 319 (1959). (b) N. L. Allinger, J. C. Tai, and T. W. Stuart, *Theor. Chim. Acta*, **8**, 101 (1967).
14. R. McWeeny, *Nature*, **166**, 21 (1950).
15. S. F. Boys, *Proc. R. Soc.*, **A200**, 542 (1950).

16. I. Shavitt, *Isr. J. Chem.*, **33**, 357 (1993).
17. (a) W. J. Hehre, W. A. Lathan, M. D. Newton, R. Ditchfield, and J. A. Pople, *Program 236, Quantum Chemistry Program Exchange*, Indiana University, Bloomington, Indiana: many subsequent versions. The latest available to us is: *Gaussian(03), Revision C.02*, M. J. Frisch, G. W. Trucks, H. B. Schlegel, G. E. Scuseria, M. A. Robb, J. R. Cheeseman, J. A. Montgomery, Jr., T. Vreven, K. N. Kudin, J. C. Burant, J. M. Millam, S. S. Iyengar, J. Tomasi, V. Barone, B. Mennucci, M. Cossi, G. Scalmani, N. Rega, G. A. Petersson, H. Nakatsuji, M. Hada, M. Ehara, K. Toyota, R. Fukuda, J. Hasegawa, M. Ishida, T. Nakajima, Y. Honda, O. Kitao, H. Nakai, M. Klene, X. Li, J. E. Knox, H. P. Hratchian, J. B. Cross, C. Adamo, J. Jaramillo, R. Gomperts, R. E. Stratmann, O. Yazyev, A. J. Austin, R. Cammi, C. Pomelli, J. W. Ochterski, P. Y. Ayala, K. Morokuma, G. A. Voth, P. Salvador, J. J. Dannenberg, V. G. Zakrzewski, S. Dapprich, A. D. Daniels, M. C. Strain, O. Farkas, D. K. Malick, A. D. Rabuck, K. Raghavachari, J. B. Foresman, J. V. Ortiz, Q. Cui, A. G. Baboul, S. Clifford, J. Cioslowski, B. B. Stefanov, G. Liu, A. Liashenko, P. Piskorz, I. Komaromi, R. L. Martin, D. J. Fox, T. Keith, M. A. Al-Laham, C. Y. Peng, A. Nanayakkara, M. Challacombe, P. M. W. Gill, B. Johnson, W. Chen, M. W. Wong, C. Gonzalez, and J. A. Pople, Gaussian(70), Gaussian, Inc., Wallingford, CT, 2004. (b) W. J. Hehre, L. Radom, P. v. R. Schleyer, and J. Pople, *Ab Initio Molecular Orbital Theory*, Wiley, New York, 1986.
18. (a) N. L. Allinger and J. T. Sprague, *J. Am. Chem. Soc.*, **95**, 3893 (1973).(b) J. T. Sprague, J. C. Tai, Y. Yuh, and N. L. Allinger, *J. Comput. Chem.*, **8**, 581 (1987). Note that the original way for calculating the *f*-factor (for correcting for nonplanarity) defined in this study was included in MM2(82) and earlier versions. It proved to be less accurate than desired, and in later versions [MM2(85), MM3, MM4] a better method was used. See Ref. 19.
19. T. Liljefors, J. C. Tai, S. Li, and N. L. Allinger, *J. Comput. Chem.*, **8**, 1051 (1987).
20. (a) J. T. Sprague, J. C. Tai, Y. Yuh, and N. L. Allinger, *J. Comput. Chem.*, **8**, 581 (1987). (b) N. L. Allinger and J. C. Graham, *J. Am. Chem. Soc.*, **95**, 2523 (1973).
21. N. L. Allinger, F. Li, L. Yan, and J. C. Tai, *J. Comput. Chem.*, **11**, 868 (1990).
22. N. Nevins, J.-H. Lii, and N. L. Allinger, *J. Comput. Chem.*, **17**, 695 (1996).
23. C. Møller and M. S. Plesset, *Phys. Rev.*, **46**, 618 (1934).
24. J. C. Tai and N. L. Allinger, *J. Comput. Chem.*, **19**, 475 (1998).
25. T. Liljefors and N. L. Allinger, *J. Comput. Chem.*, **6**, 478 (1985).
26. N. Nevins and N. L. Allinger, *J. Comput. Chem.*, **17**, 730 (1996).
27. N. L. Allinger, K. Chen, J. A. Katzenellenbogen, S. R. Wilson, and G. M. Anstead, *J. Comput. Chem.*, **17**, 747 (1996).
28. N. L. Allinger, K. Chen, and J.-H. Lii, *J. Comput. Chem.*, **17**, 642 (1996).
29. N. Nevins, K. Chen, and N. L. Allinger, *J. Comput. Chem.*, **17**, 669 (1996).
30. (a) M. Nishio, M. Hirota, and Y. Umezawa, *The CH/π Interaction Evidence, Nature, and Consequences*, Wiley-VCH, Canada, 1998. (b) M. Nishio, Y. Umezawa, K. Honda, S. Tsuboyama, and H. Suezawa, *Cryst. Eng. Comm.*, **11**, 1757 (2009).
31. R. A. Pascal, Jr., C. G. Winans, and D. Van Engen, *J. Am. Chem. Soc.*, **111**, 3007 (1989).
32. M. Faraday, *Phil. Trans. R. Soc. London*, **440** (1825).
33. P. v. R. Schleyer, Ed., Aromaticity, in *Chemical Reviews*, Vol. 101, American Chemical Society, Washington, D.C., 2001, p. 1115.
34. (a) J. A. N. F. Gomes and R. B. Mallion, Aromaticity and Ring Currents, in *Chemical Reviews*, Vol. 101, P. v. R. Schleyer, Ed., American Chemical Society, Washington, D.C., 2001, p. 1349. (b) A. R. Katritzky, K. Jug, and D. C. Oniciu, Quantitative Measures of

Aromaticity for Mono-, Bi-, and Tricyclic Penta- and Hexaatomic Heteroaromatic Ring Systems and Their Interrelationships, in *Chemical Reviews*, Vol. 101, P. v. R. Schleyer, Ed., American Chemical Society, Washington, D.C., 2001, p. 1421.
35. J. A. Elvidge and L. M. Jackman, *J. Chem. Soc.*, **856** (1961).
36. T. M. Krygowski and M. K. Cyraanski, Structural Aspects of Aromaticity, in *Chemical Reviews*, Vol. 101, P. v. R. Schleyer, Ed., American Chemical Society, Washington, D.C., 2001, p. 1385.
37. (a) D. L. Cooper, Valence Bond Theory, in *Theoretical and Computational Chemistry*, Vol. 10, D.L. Cooper, Ed., Elsevier, Amsterdam, 2002. (b) S. Shaik and P. C. Hiberty, *Valence Bond Theory, Its History, Fundamentals, and Applications: A Primer, Reviews in Computational Chemistry*, Vol. 20, K. B. Lipkowitz, R. Larter, and T. R. Cundari, Eds., Wiley, Hoboken, NJ, 2004, p. 1.
38. J. Trotter, *Acta Cryst.*, **16**, 605 (1963).
39. M. I. Kay, Y. Okaya, and D. E. Cox, *Acta Cryst.*, **B27**, 26 (1971).
40. (a) A. Borchardt, A. Fuchicello, K. V. Kilway, K. K. Baldridge, and J. S. Siegel, *J. Am. Chem. Soc.*, **114**, 1921 (1992). (b) A. H. Abdourazak, A. Sygula, and P. W. Rabideau, *J. Am. Chem. Soc.*, **115**, 3010 (1993). (c) A. Sygula and P. W. Rabideau, *J. Chem. Soc., Chem. Comm.*, 1497 (1994).
41. C. Lee, X. Wei, J. W. Kysar, and J. Hone, *Science*, **321**, 385 (2008).
42. W. Kratschmer, L. D. Lamb, K. Fostiropoulos, and D. R. Huffman, *Nature*, **347**, 354 (1990).
43. H. W. Kroto, J. R. Heath, S. C. O'Brien, R. F. Curl, and R. E. Smalley, *Nature*, **318**, 162 (1985).
44. (a) P. W. Stephens, Ed., *Physics and Chemistry of Fullerenes*, World Scientific, Singapore, 1993. (b) S. Iijima, *Nature*, **354**, No. 6348, 56 (1991).
45. J. Cioslowski, *Electronic Structure Calculations on Fullerenes and Their Derivatives*, Oxford University Press, Oxford, 1995.
46. E. Osawa, *Kagaku (Kyoto)* (in Japanese), **25**, 854 (1970); *Chem. Abstr.*, **74**, 75698v (1971).
47. D. A. Bochvar and E. G. Gal'pern, *Dokl. Akad. Nauk SSSR*, **209**, 610 (1973); *Proc. Acad. Sci. USSR*, (English translation), **209**, 239 (1973).
48. I. V. Stankevich, M. V. Nikerov, and D. A. Bochvar, *Russ. Chem. Rev.*, **53**(7), 640 (1984).
49. R. A. Davidson, *Theor. Chim. Acta*, **58**, 193 (1981).
50. H. W. Kroto, A. W. Allaf, and S. P. Balm, *Chem. Rev.*, **91**, 1213 (1991).
51. W. Krätschmer, L. D. Lamb, K. Fostiropoulos, and D. Huffman, *Nature*, **347**, 354 (1990).
52. J. M. Hawkins, A. Meyer, T. A. Lewis, S. Loren, and F. J. Hollander, *Science*, **252**, 312 (1991).
53. F. Li, D. Ramage, J. S. Lannin, and J. Conceicao, *Phys. Rev. B*, **44**, 13167 (1991).
54. C. J. Pope and J. B. Howard, *J. Phys. Chem.*, **99**, 4306 (1995).
55. M. Buhl and A. Hirsch, Spherical Aromaticity of Fullerenes, in *Chemical Reviews*, Vol. 101, P. v. R. Schleyer, Ed., American Chemical Society, Washington, D.C., 2001, p. 1153.
56. M. Saunders, H. A. Jimenez-Vazquez, R. J. Cross, and R. J. Poreda, *Science*, **259**, 1428 (1993).
57. (a) O. L. Chapman, C. L. McIntosh, and J. Pacansky, *J. Am. Chem. Soc.*, **95**, 614 (1973). (b) A. Krantz, C. Y. Lin, and M. D. Newton, *J. Am. Chem. Soc.*, **95**, 2744 (1973).
58. K. B. Wiberg, Antiaromaticity in Monocyclic Conjugated Carbon Rings, in *Chemical Reviews*, Vol. 101, P. v. R. Schleyer, Ed., American Chemical Society, Washington, D.C., 2001, p. 1317.

59. (a) R. Willstätter and E. Waser, *Chem. Ber.*, **44**, 3423 (1911). (b) R. Willstätter and M. Heidelberger, *Chem. Ber.*, **46**, 517 (1913).
60. L. J. Schaad and B. A. Hess, Jr., Dewar Resonance Energy, in *Chemical Reviews*, Vol. 101, P. v. R. Schleyer, Ed., American Chemical Society, Washington, D.C., 2001, p. 1465.
61. I. L. Karle, *J. Chem. Phys.*, **20**, 65 (1952); W. B. Person, G. C. Pimentel, and K. S. Pitzer, *J. Am. Chem. Soc.*, **74**, 3437 (1952).
62. F. A. L. Anet, *J. Am. Chem. Soc.*, **84**, 671 (1962).
63. Discussed in most textbooks of elementary organic chemistry.
64. K. Mislow, *J. Chem. Phys.*, **20**, 1489 (1952).
65. F. Sondheimer, R. Wolovsky, and Y. Amiel, *J. Am. Chem. Soc.*, **84**, 274 (1962); and references cited therein.
66. H. C. Longuet-Higgins, and L. Salem, *Proc. Roy. Soc.* (London): (a) **A251**, 172 (1959); (b) **A255**, 435 (1960); (c) **A257**, 445 (1960).
67. L. M. Jackman, F. Sondheimer, Y. Amiel, D. A. Ben-Efraim, Y. Gaoni, R. Wolovsky, and A. A. Bothner-By, *J. Am. Chem. Soc.*, **84**, 4307 (1962).
68. Y. Gaoni, A. Melera, F. Sondheimer, and R. Wolovsky, *Proc. Chem. Soc.*, 397 (1964).
69. (a) J. Bregman, F. L. Hirschfeld, D. Rabinovich, and G. M. J. Schmidt, *Acta Cryst.*, **19**, 227 (1965). (b) F. L. Hirschfeld and D. Rabinovich, *Acta Cryst.*, **19**, 235 (1965).
70. M. Gouterman and G. Wagniere, *J. Chem. Phys.*, **36**, 1188 (1962).
71. F. A. Van Catledge and N. L. Allinger, *J. Am. Chem. Soc.*, **91**, 2582 (1969).
72. (a) S. Winstein, *J. Am. Chem. Soc.*, **81**, 6524 (1959). (b) W. v E. Doering, G. Laber, R. Vonderwahl, N. F. Chamberlain, and R. B. Williams, *J. Am. Chem. Soc.*, **78**, 5448 (1956). (c) R. V. Williams, Homoaromaticity, in *Chemical Reviews*, Vol. 101, P. v. R. Schleyer, Ed., American Chemical Society, Washington, D.C., 2001, p. 1185.
73. R. B. Woodward, T. Fukunaga, and R. C. Kelly, *J. Am. Chem Soc.*, **86**, 3162 (1964).
74. (a) T. Clark, T. McO. Knox, M. A. McKervey, H. Mackle, and J. J. Rooney, *J. Am. Chem. Soc.*, **101**, 2404 (1979). (b) J. F. Liebman, L. A. Paquette, J. R. Peterson, and D. W. Rogers, *J. Am. Chem. Soc.*, **108**, 8267 (1986).
75. (a) J. W. Storer and K. N. Houk, *J. Am. Chem. Soc.*, **114**, 1165 (1992). (b) M. A. Miller, J. M. Schulman, and R. L. Disch, *J. Am. Chem. Soc.*, **110**, 7681 (1988). (c) J.-H. Lii and N. L. Allinger, unpublished. (d) S. P. Verevkin, H.-D. Beckhaus, C. Rüchardt, R. Haag, S. I. Kozhushkov, T. Zywietz, A. de Meijere, H. Jiao, and P. v. R. Schleyer, *J. Am. Chem. Soc.*, **120**, 11130 (1998).
76. F. Stahl, P. v. R. Schleyer, H. Jiao, H. F. Schaefer III, K.-H. Chen, and N. L. Allinger, *J. Org. Chem.*, **67**(19), 6599 (2002).
77. (a) J. B. Lambert, H. F. Shurvell, D. A. Lightner, and R. G. Cooks, *Organic Structural Spectroscopy*, Prentice Hall, Upper Saddle Riull, NJ, 1998. (b) J. M. Hollas, *Modern Spectroscopy*, 2nd ed., Wiley, Chichester, UK, 1987, 1991. (c) S. Grimme, *Calculation of the Electronic Spectra of Large Molecules, Reviews in Computational Chemistry*, Vol. 20, K. B. Lipkowitz, R. Larter, and T. R. Cundari, Eds., Wiley, Hoboken, NJ, 2004, p. 153.
78. (a) R. B. Woodward, *J. Am. Chem. Soc.*, **64**, 72 (1942). (b) L. F. Fieser and M. Fieser, *Steroids*, Reinhold, New York, 1959, p. 15. (c) D. L. Pavia, G. M. Lampman, and G. S. Kriz, *Introduction to Spectroscopy*, 3rd ed., Harcourt College, New York, 2001.
79. (a) N. L. Allinger and J. C. Tai, *J. Am. Chem. Soc.*, **87**, 2081 (1965). (b) N. L. Allinger, T. W. Stuart, and J. C. Tai, *J. Am. Chem. Soc.*, **90**, 2809 (1968).

REFERENCES

80. N. L. Allinger, J. C. Tai, and T. W. Stuart, *Theor. Chim. Acta*, **8**, 101 (1967).
81. J. C. Tai and N. L. Allinger, *Theor. Chim. Acta*, **15**, 133 (1969).
82. J. C. Tai and N. L. Allinger, *Quant Chem. Prog. Exchange Bull.*, **2**, 89 (1982); QCPE 443.
83. N. L. Allinger, *J. Am. Chem. Soc.*, **99**, 8127 (1977).
84. J. Tai and N. L. Allinger, *J. Am. Chem. Soc.*, **110**, 2050 (1988).
85. J. C. Tai, J.-H. Lii, and N. L. Allinger, *J. Comput. Chem.*, **10**, 635 (1989).
86. S. Leach, M. Vervloet, A. Despres, E. Breheret, J. P. Hare, T. J. Dennis, H. W. Kroto, R. Taylor, and D. R. M. Walton, *Chem. Phys.*, **160**, 451 (1992).
87. (a) C. Djerassi, *Optical Rotatory Dispersion. Applications to Organic Chemistry*, McGraw-Hill, New York, 1960. (b) P. Crabbe, *Optical Rotatory Dispersion and Circular Dichroism in Organic Chemistry*, Holden-Day, San Francisco, 1965. (c) K. Nakanishi, *Circular Dichroism: Principles and Applications*, N. Berova and R. Woody, Eds., Oxford University Press, New York, 1997. (d) L. Barron, *Molecular Light Scattering and Optical Activity*, 2nd ed., Cambridge University Press, Cambridge, UK, New York, 2004.
88. S. J. Weiner, P. A. Kollman, D. A. Case, U. C. Singh, C. Ghio, G. Alagona, S. Profeta, Jr., and P. Weiner, *J. Am. Chem. Soc.*, **106**, 765 (1984).
89. J. C. Tai, L. Yang, and N. L. Allinger, *J. Am. Chem. Soc.*, **115**, 11906 (1993).
90. (a) E. L. Stewart, C. K. Foley, N. L. Allinger, and J. P. Bowen, *J. Am. Chem. Soc.*, **116**, 7282 (1994). (b) J. Sponer and P. J. Hobza, *J. Phys. Chem.*, **98**, 3161 (1994).
91. J. C. Tai, L. K. Iwaki, Jr., and N. L. Allinger, unpublished.

6

"EFFECTS" IN ORGANIC CHEMISTRY

Covalent bonds hold the living world together and are of much interest in chemistry. We know a great deal about such bonds.[1] Solution of the Schrödinger equation at a high level of accuracy currently offers us fairly good numerical data regarding bond lengths (r_e) of covalent bonds.[2] However, these numbers themselves do not directly convey much understanding. One way of gaining such understanding is in terms of a *model* of molecular structure, based on molecular mechanics calculations. If molecular mechanics calculates numbers correctly, then we can understand those numbers in terms of our physical model. If molecular mechanics does not calculate correct numbers, our model is inaccurate, and there may be something fundamental that we do not understand.

The most important thing that determines the length of a covalent bond is the pair of atoms that are attached together.[1] Aside from this obvious fact, there are a number of environmental factors that contribute to the bond length of a given bond. Two factors that have received considerable scrutiny are the steric interactions between groups attached to the bond and electrostatic effects, when charges or dipoles on or near the bond interact with the bond or with each other. There are at least four other factors that can be important under different circumstances and that we will discuss here and in the

---

Molecular Structure: Understanding Steric and Electronic Effects from Molecular Mechanics, By Norman L. Allinger
Copyright © 2010 John Wiley & Sons, Inc.

"EFFECTS" IN ORGANIC CHEMISTRY

following chapter. They are the electronegativity effect, the anomeric effect, the hyperconjugative effect, and the Bohlmann or negative hyperconjugative effect.

In the early days of organic chemistry, chemists would isolate a compound and study it in great detail, trying to understand all of its physical and chemical properties. The subject seemed to be terribly complex because if you compare two or three different compounds chosen at random, their chemical properties are likely to be quite unrelated. But after accumulating data on many compounds, chemists recognized that compounds could be categorized by *functional groups*. Thus, two different alcohols are likely to have pretty similar chemical and physical properties, as would two different ketones, but the alcohols and ketones differed much more from one another. Hence, organic chemistry evolved as essentially functional group chemistry. Compounds that had related structures also had related chemical behavior. After that time, previous knowledge of chemistry could be put together in such a way that it became of predictive value. Fortunately, it worked out this way because otherwise dealing with the something like the 50 million now known organic compounds individually would certainly be a difficult chore.

Thus, many rules of organic chemistry developed, which were highly useful for predictive purposes. They were, however, far from infallible. For example, primary alkyl bromides are quite reactive toward S_N2 reactions, as in Reaction (1):

$$CH_3-CH_2-CH_2-CH_2-CH_2-Br \xrightarrow{CH_3O^-} \text{fast} \quad (1)$$

And that rule works well for many compounds. However, neopentyl bromide, a rather simple primary alkyl bromide, is quite inert to this reaction [Reaction (2)]. Why?

$$\underset{\underset{CH_3}{|}}{\overset{\overset{CH_3}{|}}{CH_3-C-CH_2-Br}} \xrightarrow{CH_3O^-} \underset{\underset{CH_3}{|}}{\overset{\overset{CH_3}{|}}{CH_3-C-CH_2-OCH_3}} \quad \text{very slow} \quad (2)$$

Well, the reason, which is now obvious, is due to the bulky *t*-butyl group attached to the primary carbon, and so this was called a *steric effect*. That is, the slow reaction was simply due to the bulky group attached to the reaction center that interfered with the two reacting species coming together. Actually, there was originally some question as to whether this might be an electronic effect of the *t*-butyl group, but that was shown not to be the case by a simple clever experiment. If there were an electronic effect from the *t*-butyl group to the bromomethylene group, then the molecule that has those two groups separated by an acetylenic linkage (as shown in Structure **1**),

$$\underset{\underset{CH_3}{|}}{\overset{\overset{CH_3}{|}}{CH_3-C-CH_2-Br}} \quad \underset{\underset{CH_3}{|}}{\overset{\overset{CH_3}{|}}{CH_3-C-C\equiv C-CH_2-Br}}$$

1

should still be pretty inert. But, if the effect were just steric, it should be pretty reactive. The compound is, of course, highly reactive, indicating that the *t*-butyl group, simply by virtue of its inert and bulky character, was blocking the two reacting species from getting together, and hence slowing the reaction.

Thus, steric effects are now relatively easy to predict and understand, and they are normally well calculated by molecular mechanics. They could not be calculated, however, by hand calculations. The reason was known for a long time and can be nicely illustrated by molecular mechanics. In the example at hand, for the reaction to occur, one needs to be able to calculate the geometry of the transition state and its energy. While it is not ordinarily useful to try to calculate the complete geometry of a transition state by molecular mechanics, one could assume that the carbon bromine bond in neopentyl bromide in the transition state is stretched by the same amount as it is in methyl bromide (which is known from quantum calculations or could have been estimated in earlier times). One could then calculate the interference of the incoming reagent with the remainder of the molecule, and the relative distortion and energy change that occurs as one goes from the ground state to the transition state. What made this problem hard for hand calculation is that the molecule can bend and stretch (or compress) in many degrees of freedom simultaneously, and to find the lowest energy arrangement with all of those distortions was just not practical by hand calculation. But it is trivial with the aid of a computer.

It might be worth a moment's digression here to make it clear why the distortions in a simple case like this would be so complex. It has long been known that when one distorts a molecule, say by stretching a bond, the energy for the distortion required is approximately proportional to the square of the deformation (harmonic approximation). What this means, in general, is that if you distort a molecule, it will lead to a lower energy system if you spread the distortion over as many degrees of freedom as you can. If you distort only one degree of freedom quite a bit, the energy increases as the square of a large number. If you distort several degrees of freedom to achieve the same geometry at the crucial point, the energy increases as the sum of the squares of many small degrees of freedom. The former number is always greater, usually very much greater, than the latter. So when molecules distort to lower their energy, they normally distort a large number of degrees of freedom. Finding the best way to distort all available of a large number of degrees of freedom simultaneously defeated attempts at hand calculations.

Electronic effects, on the other hand, tend to be more convoluted in their behavior than steric effects. It has been said that when one carries out an experiment, and gets a result that is widely different from what was expected (or what should have been expected at any rate), one probably has discovered a new "effect" in organic chemistry. Alas, as practicing organic chemists (and organic laboratory students) know, it is not that simple. There are endless reasons why an experiment does not come out as expected, and most of them are not very interesting or informative.

Nonetheless, the expression "effect" in organic chemistry has been used from time to time to indicate observations that were not in accord with reasonable expectations of the time. The steric effect has been mentioned as one that is now considered to be straightforward and easy to understand and to calculate. The division of these effects into steric and electronic is commonly made. Steric effects can usually be calculated rather easily by molecular mechanics. Here we will concentrate on *electronic effects*.

ELECTRONEGATIVITY EFFECT

Electronegativity is defined as the tendency for an atom (or a group) to take on electrons. Pauling electronegativities were established long ago[1] and consist of numerical values beginning with 1.0 for hydrogen, which is very lacking in electronegativity (one could say it is quite electropositive, but the latter term is not often used), and they go up to 4.0 for fluorine, which is the most electronegative element. The remaining halogens have numbers in the 2–3 range, as does oxygen. Metals tend to have quite low numbers.

Electronegativities are useful for a wide variety of purposes. After all, the tendency for electrons to be transferred is a fundamental phenomenon in chemistry. Accordingly, chemists have tried to refine these numbers to be as quantitative as possible, and to squeeze as much information as they could from such quantities. But it is not that easy. These numbers are not constants of nature, like π, or Planck's constant, and they vary somewhat depending on the details of how they are defined and measured.[3] This should not be surprising, as when electrons are given, or taken within molecules, they move between orbitals. These orbitals have various shapes, sizes, distances, and orientations, depending on the case at hand. Consequently, there is not just an electronegativity for an atom but a series of values, depending on the context. The *electronegativity effect* as we will discuss it here, is an observed distortion of molecular structures by the presence of an electronegative atom (or group). In particular, the presence of an electronegative atom tends to shorten the bond lengths of other bonds that are attached to the same atom (or a neighboring atom). And there are similarly angular distortions that go with these bond length changes (electropositive elements cause bond lengthening, and distortions of angles in the opposite direction, from those caused by electronegative elements). These distortions are all quantitative in nature and, hence, can be useful predictors of the details of molecular structures.

Most electronic effects in organic chemistry are caused by a pair of electrons (or the absence of a pair of electrons). Since these electrons normally occupy a specific orbital, such effects have stereochemical consequences, depending on the relationship between whatever occurs during the operation of that effect and the orbital in question. These stereospecific electronic (or stereoelectronic) effects are normally strongly dependent upon the three-dimensional geometry of the system in which they occur. Thus, for example, if we consider the solvolysis of a vicinal dihalide (such as 2,3-dibromobutane), there may be a strong stereochemical effect on the result (neighboring group participation), depending on the orientation between the two halogens in the reacting conformation. Most electronic effects, of which neighboring group participation is a typical example, are really stereoelectronic effects, for obvious reasons.

Electronegativity Effect on Bond Lengths

We will begin this discussion with an example of an electronic effect that does not have a strong stereoelectronic component. If we consider beginning with the ethane molecule, removing a hydrogen and replacing it by a halogen to give a molecule such as ethyl fluoride, we note that the overall structures of the product and reactant are pretty similar, that is, each of the carbon atoms is bound to its substituents in an approximately tetrahedral manner. However, there are distortions in the ethyl fluoride molecule, relative to the parent ethane molecule, well beyond those that would result from simple steric effects. Specifically, the C–C bond and most of the C–H bonds are noticeably shorter in ethyl fluoride than they were in ethane. This effect is sufficiently large and widespread that its existence was recognized quite long ago. It was sufficiently well understood by 1961 that a review article was published on the subject in that year.[4] The general rationale is straightforward. When a hydrogen in ethane is replaced by a fluorine, since the fluorine is very much more electronegative than the hydrogen was, the fluorine pulls electrons from the ethane molecule through bonds toward itself. This leads to a considerable distortion of the electronic system in the fluoroethane molecule, relative to what it was in ethane. Since p electrons are more polarizable than s electrons, more p character goes into the orbital from the carbon that is now bound to fluorine, and since the total s and p character at the carbon must be conserved, the amount of s character in the remaining bonds attached to that carbon increases. The geometric result of all of this is that adding s character into a sp^3 orbital tends to make the bond using that orbital shorter and stronger, while adding p character has the reverse effect. There is also an angular distortion, as adding p character tends to move a tetrahedral bond angle toward 90°, whereas adding s character tends to increase it toward 120°. And this simple interpretation is seen accurately fulfilled in the fluoroethane molecule. The C–C bond becomes shorter and the α C–H bonds both become shorter. The C–C–F and H–C–F angles all become relatively smaller than tetrahedral, whereas the angles on the "backside" of the carbon from the fluorination become larger.

Inductive effects, of course, propagate down a chain.[5] Thus, while α-chloroacetic acid is much stronger than acetic acid itself, β-chloropropionic acid is somewhat stronger than propionic acid, and γ-chlorobutyric acid is only marginally stronger than butyric acid.

Hence, we expect that in fluoroethane, not only are the bond angles and bond lengths of the C–H bonds in the α position changed as described, but the related geometric quantities of the hydrogens in the β position will similarly be changed, but by smaller amounts. Looking at this in a different way, attaching the fluorine to the ethyl radical is obviously adding a very electronegative group to the ethyl radical system. But, if we attach the –CH$_2$F group to a methyl radical, we're also adding a less electronegative group to the methyl radical system. And the results are in fact qualitatively as one would expect.

So, as is usual in organic chemistry, we discover an effect, and we work out a rationalization for that effect. And organic chemistry made immense advances over the last couple of hundred years in just this way. But now we are in a position where we can go a little bit further. If we have in fact a correct understanding of this effect, then

ELECTRONEGATIVITY EFFECT

we can develop a molecular mechanics model that can be applied to ethane, and to fluoroethane, and will allow us to calculate these differences in bond lengths and angles with chemical accuracy. And we can extend the formulation to other fluoro hydrocarbons and, more generally, to other hydrocarbons to which electronegative groups are attached. We can compare the quantitative predictions from our model with experiment. If the model is a good one, then it will have useful predictive value. At present, organic chemists can often make qualitative predictions of "what will happen if ..." But with the aid of our model, we should be able to now make quantitative predictions as well. And, of course, most organic substituents are electronegative, and most of those that aren't are electropositive (where all of the effects are exactly reversed, and hence also easily predictable), so here we have a proposal as to how we can obtain a fundamental understanding of a great many relationships in organic chemistry, which are in general usually referred to as the electronegativity effect.

So then how, exactly, are we going to formulate the electronegativity effect in molecular mechanics? And how are we going to show that our formulation is indeed correct and accurate in a predictive sense?

The way we usually define a structure, depending on the level of detail of interest, is with the aid of the set of internal coordinates and the **F** matrix. Most of the effects that we deal with in organic chemistry come into the **F** matrix, and thereby affect the structure. The electronegativity effect is unusual in that it comes most obviously into the coordinates rather than into the **F** matrix. What we know from early work in the area is that if we attach an electronegative atom to a bond, the length of that bond will shrink, and angles to that bond will distort. So how can we formulate this mathematically? Different substituents (or atoms) will have different electronegativities, and these will lead to different amounts of bond shortening. While in general electronegativities are understood qualitatively, these are not real constants that transfer throughout chemistry unchanged. So what we will need for this bond shortening will be a set of electronegativity constants that are appropriate for purposes at hand. Thus, we expect fluorine will have the largest and most negative (bond shortening) constant, whereas chlorine and oxygen will be second, and other elements of the periodic table will be decreasingly electronegative, until we come to sp^3 carbon (which will be our de facto standard since it is the basic element in most organic compounds). If we move further to the left in the periodic table, the elements become electropositive, and the values of these constants will become more positive, and increasingly so as we move still further to the left. These numbers can be gotten either from ab initio calculations or from experiments. Historically, they were obtained from experiment first, and we do want to make sure that calculated numbers give good agreement with experiment. But it is more rapid nowadays to determine them by quantum mechanical calculations.

So then we understand, in general, how these constants, which we will call electronegativity constants, could be, and were,[6] obtained, and how they can be used for calculating the bond shortening in organic molecules. The C–C bond is the most important one in organic molecules, so let us begin with that one. Of course, C–C bonds come in all different lengths, depending upon their hybridization and environment. But if we start with a saturated hydrocarbon and add, let's say, fluorines to that hydrocarbon, just how will this shortening occur? Consider starting with the ethane molecule and

TABLE 6.1. MM3[7] Electronegativity Bond Length Corrections (Å)[a]

Bond Type	Atom	Correction
C–C	O	−0.007
C–C	N	−0.005
C–C	F	−0.022
C–C	Cl	−0.008
C–C	Br	−0.012
C–C	I	−0.005
C–C	S	−0.001
C–C	Si	+0.009
C–H	O	−0.003
C–H	N	−0.001
C–H	F	−0.008
C–H	Cl	−0.008
C–H	Br	−0.006
C–H	I	−0.005
C–F	F	−0.030

[a]The corresponding MM4 table is still too incomplete for present purposes, but is similar.

adding fluorines one at a time to give in turn fluoroethane, 1,1-difluoroethane, 1,2-difluoroethane, 1,1,1-trifluoroethane, ... , hexafluoroethane. Qualitatively, what do we expect will happen? If one fluorine causes a certain amount of bond shortening, then a second one should cause more shortening, and a third one still more shortening, and so on. But how would we include this shortening in molecular mechanics? It would seem that what we are doing is shortening the bond length l_0, rather than l itself, since l is largely determined by steric effects. So electronegativity bond shortening is easily dealt with in molecular mechanics. One simply has a set of constants for various substituents and a specific number (Table 6.1) for fluorine. If there is a fluorine attached to a C–C bond, that particular bond in any molecule is shortened by that amount.

Generally speaking, the electronegativities of atoms increase as one goes to the right in the periodic table. They also decrease if one goes down the periodic table. If we look at the effects of different atoms attached to a carbon–carbon bond (first entries in Table 6.1 from the top), we note that oxygen, as in an ether or alcohol, leads to a shortening of the carbon–carbon bond to which it is attached by 0.007 Å. This is experimentally a substantial amount. If we look at the series of halogens, from fluorine down to iodine, we note that the fluorine is much more effective at bond length shortening (0.022 Å) than was the oxygen. The remainder of the halogens show less electronegative shortening as we go down to iodine, where the number is only 0.005 Å.

Continuing down the table, sulfur causes a negligible shortening, while silicon causes a substantial lengthening (+0.009 Å) of the carbon–carbon bond to which it is attached.

Perusing down Table 6.1 a little further, if we look at the effect of the different atoms previously discussed on shortening the C–H bond lengths, the shortenings tend to be smaller but in a similar order. It is also of interest that putting a second fluorine onto a molecule, attached to the same atom as the first fluorine, causes a mutual shortening of both fluorines by quite a large amount (0.030 Å).

The next question is: If we substitute two or more electronegative atoms onto the same bond, is the bond shortening effect additive? Based on what we know of the inductive effect, we would expect not. From studies going back more than 50 years,[5] it is known that if one adds a chlorine to the α position of say butyric acid, the acid becomes much stronger. This is an inductive effect that can be understood by examining Reaction (3):

$$CH_3-CH_2-\underset{Cl}{CH}-C\overset{O}{\underset{OH}{\diagdown}} \rightleftharpoons CH_3-CH_2-\underset{Cl}{CH}-C\overset{O}{\underset{O^{\ominus}}{\diagdown}} + H^{\oplus} \quad (3)$$

The chlorine withdraws electrons from the carbon to which it is attached, which leads to a stabilization of negative charge in the molecule, specifically on the oxygen atoms. Hence, it mainly stabilizes the anion on the right side of this equation and shifts the indicated equilibrium to the right, making the acid stronger. It was mentioned earlier that if the chlorine is on the β carbon, rather than the α, the same effect occurs but it is substantially weaker (about 0.4 as great). And if the chlorine is placed on the γ carbon, the effect is very weak indeed (perhaps about 0.4^2). This is down into the range where it is only marginally detectable experimentally, if at all. So we know about the effect of the location of the chlorine substituent on the system, but what about adding multiple chlorines? Suppose we add two chlorines on the α carbon? Well, in general what is found is that the second substituent has about 0.6 the effect that the first one had in the same direction, if it is in the same location. Thus, the effect of both α substituents would be about 1.6 times that of the first substituent alone.

Now suppose that we add to the parent structure two chlorines again, but one in the α position, and the second one β. The total effect would be $1.0 + 0.6 \times 0.4 = 1.24$ times that of the initial α substituent. And so on. Thus, we can make a prediction regarding the acidity of any kind of a chlorinated aliphatic carboxylic acid, and the predictions are reasonably good. So we might as a first approximation assume that similar numbers will apply in this case. The first fluorine added to a bond will shorten that bond by a certain amount. The additional shortening of the same bond brought about by the second fluorine attached to the same bond will be about 0.6 the amount brought about by the first fluorine. And the third fluorine will bring about 0.6^2 further shortening, and so on. We can add as many as six fluorines onto the same bond in ethane, but by the time we add the last fluorine or two, they are having a negligible electronegativity shortening effect (0.6^5 and 0.6^6 are small numbers). And, if we stop thinking at this point, and decide that the fluoroethane series should have continually shortened C–C bonds, according to the number of fluorines, we will come to the wrong answer. What actually happens is that if we add up to about three fluorines, the bond

TABLE 6.2. C–C and C–F Bond Lengths in Some Fluoroethanes $(r_g)^8$

Compound (Fluorine Positions)	C–C Bond Length		C–F Bond Length	
	Exper.[b]	MM4[a]	Exper.[b]	MM4[a]
1	1.504(5)	1.507	1.399(4)	1.400
1,2	1.505(3)	1.504	1.391(2)	1.392
1,1	1.498(4)	1.495	1.364(2)	1.369
1,1,1	1.496(3)	1.488	1.342(2)	1.350
1,1,2	1.502(5)	1.501(g)	1.355(4)CF_2H	1.361
			1.389(8)CH_2F	1.387
1,1,1,2	1.503(4)	1.504	1.391(6)CH_2F	1.378
			1.336(2)CF_3	1.335
1,1,2,2	1.520(5)	1.509(a)	1.352(2)	1.355(a)
Pentafluoro	1.527(4)	1.523	1.337(2)av.	1.336
Hexafluoro	1.545(6)	1.548	1.326(2)	1.328

[a]The MM4 values are for the conformer that is calculated to be of lower enthalpy.
[b]The experimental numbers are the Boltzmann averaged values for the conformers.

does indeed get shorter, and by decreasing amounts as the number of fluorines gets greater. But after we add three or four fluorines, the addition of still more fluorines does not cause further shortening; instead, it causes the bond to stretch out again, and by a large amount. So what is going on?

Actually, this was easily determined, simply by doing the molecular mechanics calculations. Molecular mechanics gives results quite similar to what is observed experimentally. The C–C bond length does reach a minimum at about three or four fluorines and then lengthens after that (Table 6.2). Why? The molecular mechanics calculations tell us in detail. Fluorine is much bigger than hydrogen, and a fluorine on one end of the molecule will exert a sizable van der Waals repulsion on a fluorine gauche to it on the other end of the molecule. There will also be a repulsion between the two C–F bond dipoles. So we can put three fluorines on one end of the molecule without any trouble, but if we start adding fluorines on the other end of the molecule, these repulsions come into play, causing the central bond to stretch. And by that time, the electronegativity shortening per fluorine being added is pretty small, so the repulsion wins out, and the bonds stretch instead of shrinking.

Pertinent information regarding bond lengths in fluoroethanes is given in Table 6.2. The C–C bond lengths are first given, according to the positions of substitution of fluorine into the molecule. The MM4 bond lengths are given in the next column. If we look down the two columns, we can see that the MM4 bond lengths duplicate the experimental ones, to approximately within the experimental errors. Accordingly, we will discuss only the MM4 bond lengths.

In the first compound, fluoroethane, the MM4 C–C bond length is 1.507 Å (compared with 1.532 for ethane itself). As we go down the table, this number decreases until we have added three fluorines (1,1,1-trifluoroethane, 1.488 Å). After that point, the addition of more fluorines causes the bond length of the C–C bond to increase, back up to 1.548 Å for the hexafluoride. So the idea of having the bond length decreased by electronegativity, and increased by steric effects, leads to a good fit of the MM4 values to experiment thoughout the series.

Next, we can look at the C–F bond lengths in the same series of molecules (right-hand columns in Table 6.2). If we have a single fluorine (fluoroethane), the C–F bond length is 1.400 Å from MM4. If we add a second fluoride to atom 2 in fluoroethane, the first C–F bond (as well as the second) shrinks down to 1.392 Å. But, if we add the second fluoride at the 1 position, we have an α effect rather than a β, and the bond shrinks considerably more to 1.369 Å. Note that the C–F bonds are more effective at shortening each other than they are at shortening C–C bonds, 0.030 vs. 0.022 Å (Table 6.1). And the 1,1,1-trifluoroethane has a still smaller C–F bond length (1.350 Å).

And, of course, if we have different numbers of fluorines on the different atoms, as in 1,1,2-trifluoroethane, the C–F bond lengths on the more highly fluorinated end are shorter than on the other. And so on down the table, until we come to the hexafluoroethane, where the C–F bond length is only 1.328 Å. Thus, our model of bond length shortening from electronegativity works very well for this class of compounds. The steric effects between the fluorines on opposite ends of the molecule can be greatly relieved by stretching the C–C bond, but stretching the C–F bond is of little help. Hence, the difference in behavior of the C–C and C–F bonds with increasing fluorination.

This electronegativity shortening (and the corresponding electropositivity lengthening) is well described by molecular mechanics, and it is an important determiner of molecular structure.

Electronegativity Effect on Bond Angles

As discussed earlier, the electronegativity effect causes not only a shortening of bonds, but also a deformation of bond angles, some getting smaller and some getting larger. The following discussion can best be understood with reference to Structure **2**:

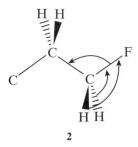

2

In the figure is shown a partial structure of propyl fluoride. As we increase the p character in the CF bond from carbon (relative to what it was in the corresponding hydrocarbon), the "front-side" angles indicated by the arrows, FC1H and FC1C, all tend to get smaller (toward 90°). The backside angles at C1 (those not mentioned in the previous sentence) tend to get correspondingly larger. This is the α-electronegativity effect of the fluorine on the bond angles at C1. Since the FCH and FCC angles each have a θ_0 of their own, those values can be adjusted to fit correctly the angle, and the "backside" angles will come out close to what is wanted. These can be further adjusted slightly with an angular electronegativity effect.

The effect of the electronegativity of the fluorine then continues on to the β carbon, so the front-side angles at C2 [which would include C1C2C3, and C1C2H (two of them)], all tend toward 90°, while the back-side angles at C2 tend to get larger. The result of these electronegativity bendings is that the θ_0 values at C1 are changed by highly electronegative elements on the order of several degrees, and by perhaps a third of that amount at C2. In other words, these are not negligible effects. If one wants to be able to correctly calculate molecular mechanics geometries with sufficient accuracy to reproduce moments of inertia of molecules, these effects must be taken into account.

Multiple substituents will presumably cause multiple effects, but they will also be diminished as successive ones are added, as was previously discussed for bond lengths. Some studies have been carried out along these lines, although not in very great detail.[6,8,9] But everything so far learned seems consistent with what one anticipates from the above discussion.

We can summarize the effect of electronegativity on the geometries of organic molecules in a quite straightforward way. The effect of an electronegative atom or group on a bond length in the molecular mechanics model depends upon a constant, the electronegativity, assigned to that atom or group. It causes a change in l_0, which carries over to bond length changes in the calculations. Electropositivity has the reverse effect. Multiple substitutions, or substitutions at the β position, are also accounted for as a fraction of the effect at the α position. For bond angle bendings, there are again constants to be determined, and they are similarly dependent on the substitution pattern. The test of all of this is whether we can assign constants in the model, and then carry out calculations and obtain molecular mechanical structures that agree, both with experimental structures and with those calculated by high-level quantum mechanical methods. That has been done for several different atoms and functional groups, and the model indeed works to the limit of accuracy that is consistent with the quality of the available data. It would seem reasonable to conclude that we understand the electronegativity effect fairly well.

C–H BOND LENGTH VERSUS VIBRATIONAL FREQUENCY

Over the next several pages we will discuss some properties of the C–H bond and how they are related, and what they can teach us about chemistry. But keep this in mind: The C–H bond is a typical covalent bond, which means that whatever we can learn about that particular bond is likely to be widely applicable to other covalent bonds.

It was mentioned earlier (Chapter 2) that the vibrational frequency of a C–H bond depends upon the bond length. Other things being equal, the shorter bond is the stronger bond. Since it is very easy to accurately measure these vibrational frequencies in most cases, and very difficult to measure the corresponding C–H bond lengths by neutron diffraction, or other methods, this relationship has potential uses.

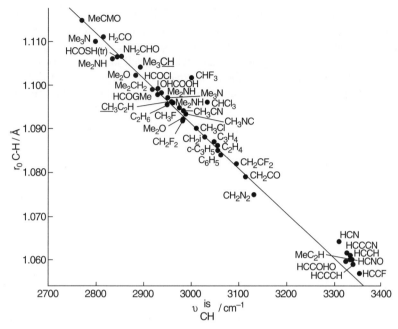

Figure 6.1. Correlation between bond lengths and vibrational frequencies for C–H bonds. (From McKean, Boggs, and Schäfer.[14] Copyright 1984 by Elsevier. Reprinted by permission of Elsevier.)

There have been several important attempts in the literature to find a correlation between bond length and bond stretching force constants. Badger[10] found a cubic relationship for a handful of diatomic structures in the ground state and in several excited vibrational states. Subsequently, Herschbach and Laurie[11] compiled a larger data set than was available to Badger and revised the constants in Badger's equation. They also used the same set of data to propose a different empirical equation that is exponential instead of cubic. Other workers have also used an exponential relationship.[12,13]

Figure 6.1 shows a plot of bond length versus vibrational frequencies for a number of C–H bonds in a variety of simple molecules.[14] The wide variety of simple structures used and the good correlation are evident upon inspection. Note that Figure 6.1 shows frequencies that are "isolated," in that there is only a single hydrogen attached to the carbon, or if there are two or more, they are mathematically isolated (decoupled). If the observed frequencies are simply used as they are observed (without isolating them), the relationship is less precise but still pretty good.

From the data in Figure 6.1, McKean[14] derived the relationship between bond length and vibrational frequency, which can be written as in Eq. (6.1):

$$v = \frac{1.3982 - r_0}{0.0001023} \tag{6.1}$$

The relationship for C–H bonds based on McKean's equation has an advantage over Badger's and Herschbach's equations in that McKean recognized the different ways in which the bond lengths can be described (Chapter 2) and always used the same physical quantity, r_0. Combining this equation with the well-known relationship between the stretching force constant (k), vibrational frequency (v), velocity of light (c), and C–H reduced mass (μ) in the harmonic oscillator approximation we find [Eq. (6.2)]:

$$k = v^2(2\pi c)^2 \mu \qquad (6.2)$$

and taking the partial derivative of the force constant with respect to bond length, we obtain Eq. (6.3)[15]:

$$\Delta k_s = \frac{2\mu(2\pi c)^2}{(-0.0001023)^2}(l_0 - 1.3982)\Delta l_0 \qquad (6.3)$$

McKean's Eq. (6.1) is expected to (and does) work well in the absence of serious differences in the steric and electrostatic effects exerted upon the bonds in question. By substituting the molecular mechanics parameter (l_0) for the actual bond length (r_0), Eq. (6.3) can be used in a general way with a molecular mechanics force field. To this end, one also needs to use Eq. (6.3) to determine the *force parameter* (k_s) for the bond under examination, and that replaces the ordinary force constant. The rest of the force field will take care of any unusual steric or electrostatic effects.

Since the C–H bond is a typical covalent bond, one would expect that this relationship between bond length and vibrational frequency should hold not only for the C–H bond but for covalent bonds in general. It was easier to see the existence of this relationship with the C–H bond because that particular bond shows vibrational frequencies that are well removed from the other bonds in the molecule (because of the small mass of the hydrogen), and the coupling of the particular bond in question with other bonds is normally negligible.

It was noted some time ago[8,9a] that if one began with ethane, and replaced the hydrogens successively by fluorines, to give a series of fluorinated ethanes ending with hexafluoroethane, a behavior similar to that in Eq. (6.3) for the C–C bond (with some different constants) could be observed. Since the fluorine is very electronegative, it tends to cause large bond shortenings. Hence, as more and more fluorines are attached to the same C–C bond, the bond tends to shrink very much (with respect to l_0, the actual bond length increases after three or four fluorines have been added from steric effects). The relationship between bond length and vibrational frequency is observed clearly, but only qualitatively. The main reason for the lack of a straight-line relationship here is that with the heavy atoms involved, the couplings between them (C–C and C–F stretchings) become much more serious, and the effect is less clear-cut. It is, however, still possible to "isolate" the frequencies in question in such heavy atom systems, and to show that related relationships between bond lengths and stretching frequencies do occur with heavy atoms as expected.[16]

HYPERCONJUGATION

It was noted in the early days of organic chemistry that compounds containing conjugated systems, such as butadiene, showed chemical reactivities that were quite different from those shown by the analogous unconjugated system (e.g., 1,4-pentadiene). A quantum mechanical interpretation of this reactivity is straightforward. In a conjugated system, the p orbital from one double bond or fragment overlaps with the p orbital from the second fragment, which is adjacent to it, leading to an interaction (π bond) between the two fragments. There result very significant chemical consequences since the π orbitals formed have energies that are higher than those of the other occupied orbitals in the molecule, while the π^* orbitals have the lowest energies of the unoccupied orbitals. Consequently, these orbitals and their electrons are more easily involved in chemical reactions (and hence more reactive) than similar molecules lacking conjugated orbitals. There are also physical consequences of conjugation in that the ultraviolet spectra, for example, are quite different for the conjugated and unconjugated systems.

Mulliken divided conjugation into two different kinds, depending on the valence bond wave functions, and the corresponding structures that one can write to represent those functions. The first, and best known, was called *isovalent conjugation*, and would be represented by benzene (Fig. 6.2). It is isovalent because there are the same number of double bonds and single bonds in one of the structures (*resonance forms*) as in the other one. The second type of conjugation is the type in butadiene, which he called *sacrificial conjugation*. It differs from isovalent conjugation in that the principle resonance form contains more double bonds than the others. And, of course, these two different types of conjugation confer somewhat different changes on molecules. Isovalent conjugation tends to have a strong stabilizing effect, which often leads to a lack of chemical reactivity. Sacrificial conjugation on the other hand, while leading to some small stabilization, moves the orbital energies around in such a way that it usually leads to the conjugated molecule being more reactive.

Isovalent Conjugation

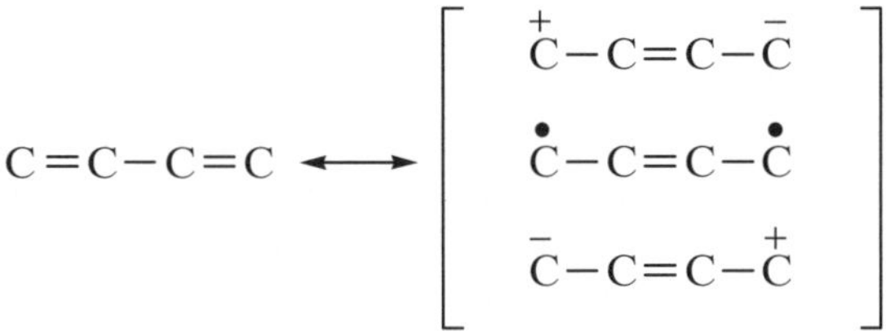

$$C=C-C=C \longleftrightarrow \begin{bmatrix} \overset{+}{C}-C=C-\overset{-}{C} \\ \overset{\bullet}{C}-C=C-\overset{\bullet}{C} \\ \overset{-}{C}-C=C-\overset{+}{C} \end{bmatrix}$$

Sacrificial Conjugation

Figure 6.2. ••.

In 1939, Mulliken proposed that there was also a similar kind of an effect between a saturated carbon atom and an adjacent unsaturated carbon atom, for example, in propene (Structures **3** and **4**).[17]

$$\begin{array}{c} \text{H} \\ | \\ \text{CH}_2-\text{C}=\text{CH}_2 \\ | \\ \text{H} \end{array} \longleftrightarrow \begin{array}{c} \text{H}^+ \\ \text{CH}_2=\text{C}-\bar{\text{C}}\text{H}_2 \\ | \\ \text{H} \end{array}$$

Propene

3

$$\text{H}_2\text{C}=\text{CH}-\text{CH}=\text{CH}_2 \longleftrightarrow \overset{+}{\text{C}}\text{H}_2-\text{CH}=\text{CH}-\bar{\text{C}}\text{H}_2$$

Butadiene

4

In this case, the p component of one or two of the CH σ orbitals of the methyl group (depending on the rotational orientation of the methyl group) could overlap strongly with the p orbital of the π system on the adjacent carbon, and this would lead to an effect similar to that of conjugation.

But here the effect would be less pronounced for at least two reasons. First, the sp³ orbital has a geometry such that it does not overlap with the adjacent p orbital as well as another p orbital would, and more importantly, the electrons in a π orbital are relatively high in energy, whereas the electrons in σ orbitals are generally much lower in energy (more stable, less easily delocalized). Hence, the strength of the interaction between the electrons in the sp³ orbital and those in the π orbital would be reduced because of this sizable energy difference. Nonetheless, there would be at least a qualitative similarity. Mulliken[17] called this weak conjugative effect *hyperconjugation*.

The effects of hyperconjugation are similar to those observed in conjugation (e.g., in butadiene), but they are smaller. Propene is more reactive in electrophilic addition reactions than is ethylene (but much less so than butadiene), and it shows absorption in the ultraviolet analogous to the π-π* transition observed in ethylene but at a longer wavelength. The corresponding transition in butadiene is at an even longer wavelength.

Earlier we discussed electronegativity and how it affected bond lengths and angles in organic molecules. Here we wish to discuss the first example of a stereoelectronic effect, which in our model is fundamentally different from the electronegativity effect. The electronegativity effect is not conformationally dependent. On the other hand, hyperconjugation is an effect that belongs in the class of stereoelectronic effects, and this effect is conformationally dependent. These two classes of effects divide mathematically in a way that we want to note at this point. The electronegativity effect changes the values for structural constants (l_0 and θ_0) that have to do with the atomic coordinates of a molecule. They do not appear in the **F** matrix. Hyperconjugative effects, on the other hand, do occur in the **F** matrix, where they appear as off-diagonal elements.

In order to have hyperconjugation of the Mulliken type, the C–H bond on the methyl has to align itself properly with the p orbital on the sp² carbon to which the methyl group is attached. The most effective interaction between the C–H bond and the p orbital on carbon will occur when the σ bond is oriented at 90° relative to the plane of the double bond. In the ground state of propene, one hydrogen is eclipsing the double bond (with a dihedral angle of 0° with respect to the plane) as shown in the left structure below. There would be no hyperconjugation between the double bond and the hydrogen in that position, as the two do not overlap for symmetry reasons (Structure **5**).

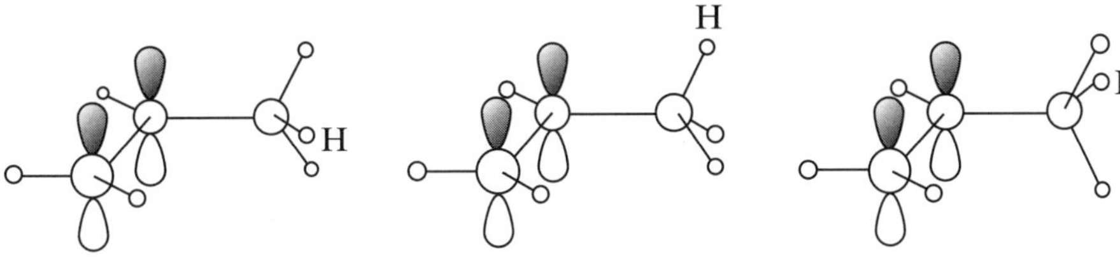

5

As the hydrogen shown at the left in Structure **5** is rotated around from 0° to 180° hyperconjugation would begin, and it would increase to a maximum value at 90° (center structure) and then decrease as the rotation continued on to 180° (right structure). Thus, a twofold cosine curve would describe this phenomenon. But since the geometries at 0° and 180° are different, we expect that environmental effects will also be present along with hyperconjugation. The phenomenon under examination will thus probably best be represented by a twofold cosine curve, superimposed upon a onefold cosine curve, so that the latter term allows for the cis–trans difference. And, of course, there are in fact three hydrogens with which to be concerned. Some properties of the system are a function of these hydrogens one at a time (e.g., the individual CH bond lengths), but other properties are a function of all three of them together (e.g., the C_{sp^3}–C_{sp^2} bond length). So some properties will change with torsional rotation according to the formulation mentioned, but others will probably stay constant, or very nearly so. That is the prediction that would come from examining Mulliken's proposal. What happens in the real world?

There are elementary predictions that can be made from the above model. These predictions involve, for example, the length of the C–H bond on the methyl group as a function of torsion angle for one thing, and the amount of positive charge on the hydrogen as a function of torsion angle, for another. The first of these can in principle be studied experimentally, but in practice this would be quite difficult. It is hard to accurately measure C–H bond lengths because the hydrogen has such small values for whatever property one chooses to measure. It has a small mass (if one wants to do microwave studies), and it has a small electron density (if one wants to do X-ray studies), and it has a small positive nuclear charge (if one wants to do electron diffraction). It also has a large and anharmonic vibrational amplitude because of its small mass, making it hard to specify the location (vibrational average). Hydrogens are often

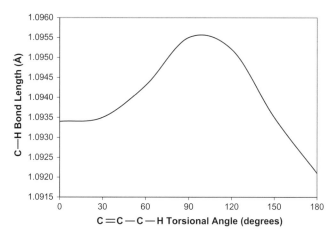

Figure 6.3. The C_{sp^3}–H bond length as a function of torsion angle in propene.

best located from neutron diffraction studies on crystals, but the accuracy attainable might additionally be limited by the fact that the molecule is in a crystal lattice. But one can easily carry out accurate quantum mechanical studies on a molecule as simple as propene.[18] And they show that as one would expect from the earlier discussion, the C–H bond length is at a minimum when the torsion angle with respect to the plane of the alkene is 0° since there is no hyperconjugation, and it is similarly at another (somewhat different) minimum at 180°. Additionally, it reaches a maximum in between, near 90°, as implied by the valence bond Structures **3** and **5**, showing that hyperconjugation leads to a reduced C–H bond order and hence a longer bond length. The kind of structure shown at the right of Structure **3** also has been said to represent *no-bond resonance*, for obvious reasons. The bond length variation here is sensitive to the size of the basis set used in the quantum mechanical calculations, and to the amount of electron correlation included, but only slightly so. The general result is really independent of those two things. So we may conclude that the Schrödinger equation tells us about this bond length variation of the C–H bond in propene with torsion angle, and what it tells us is in accord with Mulliken's predictions. Figure 6.3 shows the results of an MP2/6-31G* calculation. The bond length is a minimum at 180°. It lengthens by about 0.003 Å as the torsion angle reduces to about 90°, and then shrinks back down to another, somewhat higher, minimum at 0°. (The maxima and minima in these and most similar curves are artificially distorted slightly from their "ideal" values because the structures were generated using a driver routine; see Appendix.) So hyperconjugation is real, but as far as the C–H bond length, the effect is small, about 0.003 Å. The use of larger basis sets, or omission of electron correlation in the calculation, of course, leads to small changes in the numerical results, but the general conclusions remain unaffected.

The other thing that it is easy to calculate is the Mulliken charge (excess positive charge) on the hydrogen atom, as its torsion angle goes from 0° to 180°. That charge density is shown in Figure 6.4. The hydrogen atom loses electron density as it goes

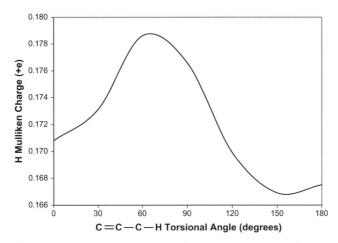

Figure 6.4. Mulliken charge density as a function of torsion angle in propene.

from 0° to 90° in torsion angle (becomes more positive), and then gains some of it back as the rotation continues to 180°, exactly as Mulliken suggested (Structures **3** and **5**).

The conclusion that one can draw from the above calculations is that hyperconjugation is a real effect, and it takes place in accordance with Mulliken's original suggestions. But it is a pretty small effect, and it would be difficult to detect it experimentally by measuring the hydrogen position using microwave or diffraction methods. One might, however, devise experiments that would measure this phenomenon in any of several different indirect ways. For example, one interesting and very general experiment suggested by Pross and co-workers[19] concerns measurement of the angle of tilt of the methyl group involved. This can be done with systems where the methyl is an electron donor (e.g., positive hyperconjugation to boron) or an electron acceptor (e.g., negative hyperconjugation from nitrogen; see following chapter), where the magnitude and direction of the tilt (often several degrees) indicates which way the electrons have moved, and to what extent. It is much easier and more reliable to calculate the magnitude of this effect than it would be to measure it experimentally.

Baker–Nathan Effect

Although today we might think that hyperconjugation is straightforward and obvious, there have been many arguments in the chemical literature regarding the existence of this effect. As recently as 1990 (more than 50 years after Mulliken's proposal), a standard textbook on physical organic chemistry concluded[20a] that "the weight of evidence for hyperconjugation in ordinary (closed-shell ground state) molecules did not support the existence of hyperconjugation." This relatively strong position has been somewhat relaxed now, however, and the later editions of the same text (2001 and 2007) state[20b,c] "there is a large body of evidence against hyperconjugation in the ground states of neutral molecules. A recent study[20d] of the one-bond coupling constants for the aromatic system 111 [Structure **6**], however, appears to provide the first structural evidence for hyperconjugation in a neutral ground state."

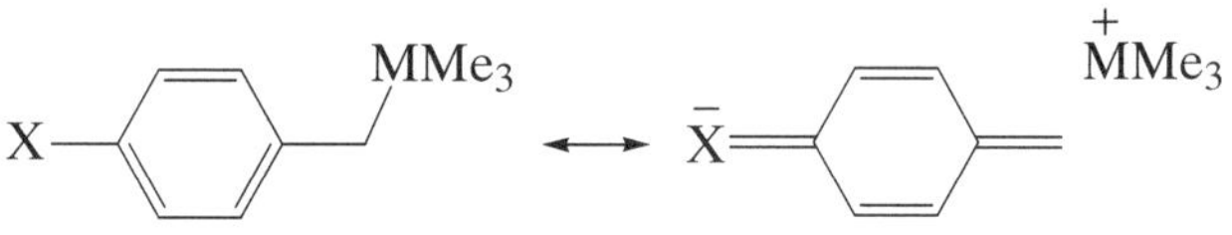

$$X-\text{C}_6\text{H}_4-\text{CH}_2-\text{MMe}_3 \longleftrightarrow \bar{X}=\text{C}_6\text{H}_4=\text{CH}_2 \cdots \overset{+}{\text{MMe}}_3$$

111

M = C, Si, Ge, Sn
X = NO$_2$, CN, H, Me, OMe

6

How may one establish experimentally the existence of hyperconjugation? Measuring the properties indicated (the charge and the C–H bond length) would seem to be a simple and direct approach. However, neither of these things is really susceptible to direct measurement in practice. There just are no available procedures of the required accuracy to make the determinations that we would like to have. So chemists tried to settle the problem by other kinds of experiments. But the kinds of experiments that were available, especially in the 1930s and for many years after that, were quite indirect. Here are the kind of reasoning and experiments that were employed.

First, it is known that, in general, the alkyl groups are electron-releasing groups into aromatic rings. The larger alkyl groups can delocalize electronic charge better since it can be more widely distributed. Hence, the measured values for the dipole moments in the gas phase for the series of compounds PhCH$_3$, PhC$_2$H$_5$, PhCH(CH$_3$)$_2$, and PhC(CH$_3$)$_3$ are, respectively, 0.37, 0.58, 0.65, and 0.70 D.[21] This electron donation is usually thought to predominantly occur by simple induction, just as would be the case when any sp^3 hybridized carbon is connected to any sp^2 hybridized carbon. (The sp^2 hybridized carbon has more s character than does the sp^3, and consequently electrons in the sp^2 orbital are of lower energy than those in the sp^3 orbital.) Electrons are simply withdrawn by induction from the sp^3 carbon, and then in turn, electrons are withdrawn from the attachments to that carbon. The negative charge builds up in the benzene ring.

If there is hyperconjugation, however, it can simultaneously lead to resonance structures wherein electrons are removed from the attached hydrogen (or other group) and pushed into the π electron system. One can write resonance forms for this hyperconjugation as in Structure **7**:

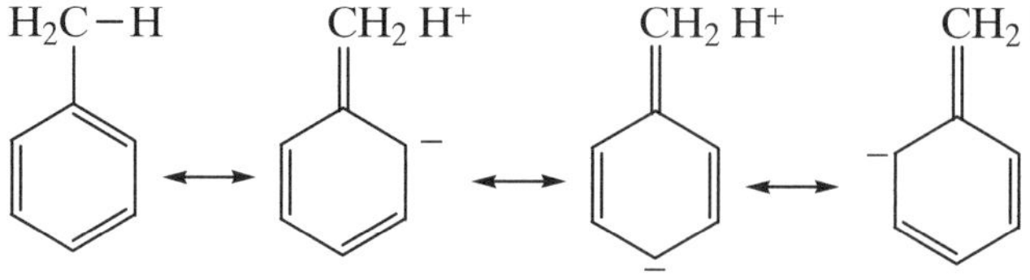

H$_2$C–H / Ph ⟷ CH$_2$ H$^+$ / C$_6$H$_5^-$ ⟷ CH$_2$ H$^+$ / C$_6$H$_5$ ⟷ CH$_2$ H$^+$ / C$_6$H$_5^-$

7

Evidently the *t*-butyl group donates electrons into the π system of the benzene ring more so than the methyl group does, hence the larger (gas-phase) dipole moment of the former.

Next, consider the benzyl bromide molecule, which has attached in the para position an R group, either a methyl or a *tertiary*-butyl. If we allow those bromides to react with pyridine by heating them in a sufficiently polar solvent (such as 90% acetone : 10% water), we can measure their rates of reaction. We know that the reaction goes through an intermediate that has strong carbocation character, so the rate is largely determined by the stability of that ion. The rate of the reaction then can be thought of as determined by the stability of the product in Structure **8**:

$$X-\text{C}_6\text{H}_4-CH_2-Br \longrightarrow X-\text{C}_6\text{H}_4-CH_2^+ + Br^-$$

8

To the extent that hyperconjugation leads to a stabilization of that carbocation, the pertinent resonance forms for the ions would be as shown in Structures **9** and **10**:

$$X=CH_3 \qquad CH_3-\text{C}_6\text{H}_4-CH_2^+ \longleftrightarrow \overset{H^+}{CH_2}=\text{C}_6\text{H}_4=CH_2$$

9

$$X=t\text{-Bu} \qquad CH_2-\underset{CH_3}{\overset{CH_3}{\underset{|}{C}}}-\text{C}_6\text{H}_4-CH_2^+ \longleftrightarrow CH_3-\underset{CH_3}{\overset{CH_3^+}{\underset{|}{C}}}\text{-C}_6\text{H}_4=CH_2$$

10

If there is hyperconjugation in the methyl derivative (Structure **9**), then that resonance should stabilize the transition state and lead to an increase in the rate of the reaction, relative to what it would be if there is no hyperconjugation. What about the *t*-butyl derivative? Here the resonance form involves a methyl cation, which is known to be an energetically unfavorable species (Structure **10**). So the assumption was made that the *t*-butyl derivative gave a base rate where there was negligible hyperconjugation, and if there were hyperconjugation in the methyl derivative, it would react at a substantially faster rate. And how does it come out? It comes out that the methyl derivative actually does react a little faster than the *t*-butyl derivative in polar solutions. This faster reaction of the compound expected to react more slowly on the basis of induction is called the *Baker–Nathan effect*.[22] It was originally taken to be evidence for hyperconjugation in the methyl derivative.

There are problems with this kind of experiment, however. For one thing, the reactions are carried out in a solvent, which then solvates the carbocations formed at the transition state of the reaction. The difference in the solvation energies between the methyl- and *t*-butyl ions may be substantial, and this would upset the comparison. But what seems to be even more important is this. This experiment only measures the *difference* in the amount of hyperconjugation in the two cases (methyl vs. *t*-butyl), not the absolute value. It is certainly true that CH_3^+ is a very unstable ion. But H^+ is not all that stable a species either. The H⁺ ion as we know it in daily life is really the relatively stable H_3O^+ molecule. And similarly the $CH_3OH_2^+$ ion is also relatively stable. Solvating the H⁺ ion, or the CH_3^+ ion, or reacting them with water or methanol if you prefer, will lead to strong stabilization. About all that one can really conclude from the Baker–Nathan experiments is that the difference between the hyperconjugative effects in the two cases appears to be small (a fraction of a kcal/mol in the activation energy). We can measure accurately the differences in the reaction rates, but we cannot relate these results in a reliable way to the difference in hyperconjugation between the two species. Even if we could, we still do not know if hyperconjugation is important in the *t*-butyl derivative, so it still doesn't really tell us what we want to know. Thus, if we are going to really decide on the importance of hyperconjugation experimentally, we need to have more definitive experiments than those that were used originally.

Subsequently[23,24] it was shown that the order of reactivity found by Baker and Nathan does in fact result from differential solvation because in the gas phase this order is reversed.[24a] Hence, it was concluded that the Baker–Nathan results were only a solvation effect and not evidence for hyperconjugation.

Are there special cases where hyperconjugation might be still more important (and easier to measure)? Yes, there are, and some of these have been studied. Again, the experiments would often be difficult to carry out, but one can easily do the calculation at a high enough level so that the conclusions are quite certain. In this case we can consider the molecule in which one of the methyl hydrogens of propene is replaced by a methyl group (to give 1-butene), and we rotate what is now the ethyl group relative to the double bond so that the dihedral angle that that ethyl makes with the alkene plane varies over the range of 0° to 180°. If we reexamine Structure **3**, we see that a proton is the detached fragment in the resonance form at the right. If we carry out the rotation shown in Structure **5** with butene instead of propene, instead of a proton being the detached fragment, a methyl cation becomes the detached fragment (Structure **11**).

$$CH_2-\underset{\underset{H}{|}}{\overset{\overset{CH_3}{|}}{C}}=CH_2 \quad \longleftrightarrow \quad CH_2=\underset{\underset{H}{|}}{\overset{\overset{CH_3^+}{}}{C}}-\bar{C}H_2$$

11

And we can calculate the methyl C–C bond length at various points throughout that rotation. Now instead of a hyperconjugative H⁺ fragment, we will have CH_3^+. This was done, and in line with what one would expect from hyperconjugation, the methyl C–C bond stretches as the bond goes from 0° to 90°, and then shrinks again back toward its original length as we continue from 90° to 180°. Thus, the methyl group C–C bond shows qualitatively the same hyperconjugative bond stretching that is shown by the C–H bond (Fig. 6.3). And when we look at the magnitude of the stretching, the C–C bond actually stretches by 0.0078 Å, more than twice the distance by which the C–H bond stretches. The calculation evidently shows that hyperconjugation is not only important with C–C bonds, but that it is more important with C–C than C–H bonds, at least with respect to the bond stretching phenomenon. But the stretching here is still not very large, 0.0078 Å. Can we imagine cases where it would be larger, and more accurately determined experimentally?

Well, suppose not just one double bond is hyperconjugating with the C–C bond that we are looking at, but suppose that we design a molecule such that two or three bonds hyperconjugate at the same time. The cumulative stretchings should then be greater and more readily determined.

Going back again to the propene molecule, we see that one thing happening due to hyperconjugation is that positive charge is being distributed onto one or more of the hydrogens of the methyl group, and negative charge is being distributed into the double bond. Is it possible to arrange a molecule such that the charges can be distributed (delocalized) still more widely, thus leading to a more stable molecule (more resonance forms)? Certainly, charges are distributed more widely in the benzyl anion than in the propene case, just because there are more places to put the negative charge (Structure **12**):

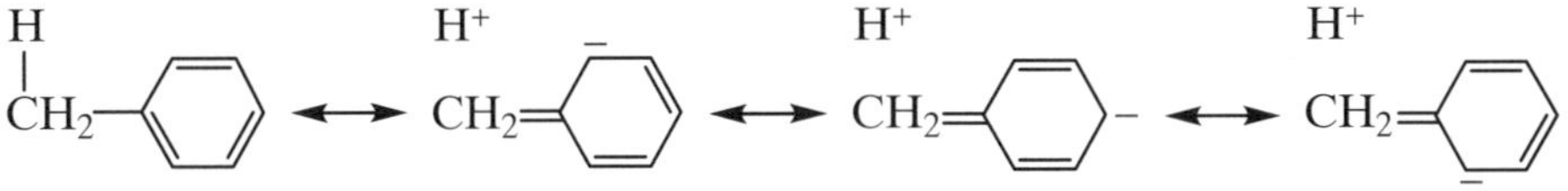

12

But we can go further than this. The trityl anion (Structure **13**) is much more stable than the benzyl anion, so that we can envision the possibility of extensive charge delocalization, which in part concentrates its effect into a small number of bonds.

A similar approach was taken here. Consider, for example, Structures **14a–c**:

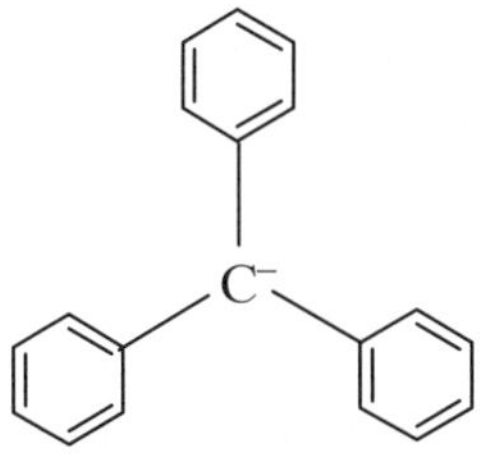

13

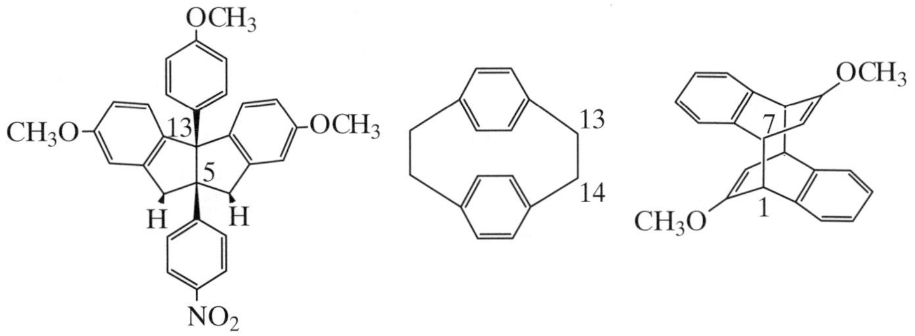

a b c

14

TABLE 6.3. Compounds Showing Long C–C Bonds (Å)[18]

Compound	Bond	Expt.[a,b]	MM4O[c]/Δ	MM4[d]/Δ	Hyperconj. Effect
a	C5–C13	1.621(2)	1.603/–0.018	1.618/–0.003	0.015
b	C13–C14	1.571	1.565/–0.006	1.574/0.003	0.009
c	C1–C7	1.617(5)	1.589/–0.028	1.623/0.006	0.034
		AVE[e]	–0.019	–0.003	
		rms[e]	0.025	0.012	

[a]X-ray values have been converted to r_g values.
[b]0.005 Å is added for the thermal libration correction.
[c]Hyperconjugative effect excluded. For further discussion of the MM4O program, see Appendix, under Jargon.
[d]Hyperconjugative effect included.
[e]Data are shown for three representative compounds (from Structure **14**) in Table 6.3. The actual study involved a total of 25 compounds, and the average and rms errors quoted in Table 6.3 are for the full 25.

The hyperconjugative resonance forms that can be written for these structures indicate that certain of the central bonds should be very long due to the hyperconjugation. These are the saturated C–C bonds that are aligned with the p orbitals of π system double bonds and are oriented in such a way as to hyperconjugate with them. These bonds are numbered, 5–13 in **a**, 13, 14 in **b**, and 1–7 in **c**. We can easily enough carry out the MM4 calculations and see what sort of bond length is predicted for each of these when no hyperconjugation is included in the MM4 force field. (The MM4 program that does not include hyperconjugation is here referred to as MM4O.) These structures are known experimentally from X-ray studies, to varying degrees of accuracy. The structural information is summarized in Table 6.3.[18] When a comparison is made with each of the bonds mentioned, the MM4O value (not including hyperconjugation), in each case gives quite a long bond length for steric reasons (a, b, and c are 1.603, 1.565, and 1.589 Å, respectively). However, these are all shorter than the corresponding experimental bond lengths (by –0.018, –0.006, and –0.028 Å, respectively). Because the error here is systematic, it certainly suggests that the MM4O bond lengths are calculated too short. As a calibration, the remaining bonds in the molecule not expected to hyperconjugate have MM4 values that agree on average with those found by experiment.

When hyperconjugation was included in MM4, using the parameters developed from the data given earlier regarding propene and 1-butene, and this version of the program was used to calculate the structures of the same molecules, for the most part the structures were the same or very similar, but the specific bonds mentioned in each case stretched out further to give the values shown in the column labeled MM4. In this case, the errors between MM4 and experiment were much reduced, and the hyperconjugative stretching effect in the three cases was calculated to be 0.015, 0.009, and 0.034 Å. This hyperconjugative stretching effect varies greatly from bond to bond and from one compound to another, because it varies with the number and kinds as well as the geometries of the σ-π overlaps of the particular bonds in question. A much longer table of similar related numbers was collected in the literature.[18] But the above shows the important facts. Namely, if hyperconjugation is not included in the MM4 calculation, the bond lengths calculated by molecular mechanics are systematically too short, but they stretch out to the experimental values when hyperconjugation, based upon the small molecule parameterization mentioned earlier, is included. Note that these stretchings can be very substantial, up to 0.034 Å. Hence, if our molecular mechanics model is to reproduce experiment (or ab initio calculations), this effect certainly needs to be included.

But, if we look at the hyperconjugative structure carefully in the case of propene, we note that the hydrogen becomes positive and the bond from that hydrogen to carbon weakens as that resonance occurs. But notice also that the other carbon at the end of the chain becomes negative. Suppose that instead of carbon, we substitute another atom there, one that will better accommodate negative charge? Should not the effect then be magnified?

Instead of propene, then, let us repeat the calculation of the C–H bond length for the rotating methyl group in acetaldehyde. The appropriate resonance forms are as shown in Structure **15**:

$$\overset{H}{\underset{H}{\overset{|}{CH_2-\underset{|}{C}=O}}} \longleftrightarrow \overset{H^+}{\underset{H}{\overset{}{CH_2=\underset{|}{C}-\bar{O}}}}$$

Acetaldehyde

15

Evidently, the C–H bond should hyperconjugate more strongly in acetaldehyde than it does in propene, and the C–H bond length difference as we go from 0° to 90° in torsion angle should increase more. And that's what we find. The calculations show that in this case the C–H bond length increase in acetaldehyde is 0.0047 Å, vs. 0.0030 Å in propene. This particular type of bond stretching has also been referred to in the literature as the *carbonyl effect*.[25]

The weight of the above evidence, in the author's estimation, convincingly favors not only the existence of hyperconjugation but also outlines part of its ubiquitous existence in organic chemistry. This is not a large effect in simple cases such as propene, but it is a clear and understandable effect, something that one expects should happen (Mulliken[17]), and it does happen. And it can become a large effect in some cases. This effect is not observed in molecular mechanics calculations when a simple diagonal force constant matrix is used. So how, then, do we allow for this effect in molecular mechanics?

Let us consider the part structure of propene involving the methyl group carbon and one of its attached hydrogens, plus the two sp^2 hybridized carbons, with a numbering system as shown in Structure **16**.

$$H_1-C_2-C_3=C_4$$

16

If we write a diagonal force constant matrix for this part structure, it would look exactly as shown in Eq. (4.6). The 1,2 bond is the one that should stretch when torsion occurs about the 2,3 bond, but that does not happen if one uses Eq. (4.6) in the usual way to describe the stretching constant of the C–H bond. But if we instead use Eq. (4.11) to describe the system, there is a term $k_{(s12)(\omega1234)}$ in the upper right corner of the matrix that connects the torsion angle and the bond length force constants. We want the 1–2 bond to stretch from a minimum length at 180° to a maximum length at 90° (twofold) and come back to a higher minimum at 0°. So the matrix element is calculated by Eq. (6.4):

$$k_{(s12)(\omega1234)} = k_a(l-l_0)\left[b(1-\cos\omega)+c(1-\cos 2\omega)\right] \tag{6.4}$$

and the constants k_a, b, and c can be chosen so as to cause the 1,2 bond to stretch with torsion exactly as found from the quantum mechanical calculations on propene.

It will be instructive to further examine Eq. (4.11) at this point. Note that the term just discussed interconnects a stretching and a torsion element, and thus makes one a function of the other. If the matrix were to be larger, as, for instance, if we include all the atoms of propene, then while, in principle, any element can be connected to any other element, normally if the atoms concerned are more distant than 1,4, those elements are set to zero. In other words, the mechanical quantities that we are dealing with involve only 1,2 (stretching), or 1,3 (bending), or 1,4 (torsion), or combinations of those, to a good approximation. (And, of course, we are talking here about the connectivity, and not the numbering system of the molecule.) But for large molecules, most of the explicit off-diagonal elements are in fact zero. By explicit, we mean that explicit equations such as the torsion–stretch Eq. (6.4) given above are used to calculate the values of those elements. In fact, nonbonded interactions (van der Waals and electrostatic) are usually calculated between all of the atoms in the molecule except for those few excluded by connectivity considerations. These van der Waals terms contribute something other than zero to most of those elements. However, explicit calculations for most of the matrix elements (e.g., stretching as a function of torsion for distant bonds) are not carried out.

It is instructive to think of hyperconjugation in terms of the valence bond structures, and then relate those structures to whatever motions (stretching, bending, torsion combinations) that are qualitatively predicted by those structures. And the forms of the equations that will describe those elements are usually evident from the valence bond structures themselves.

A simple molecular mechanics force field may be limited to just the diagonal elements in the force constant matrix. This is certainly a reasonable first approximation since these elements are generally an order of magnitude larger than the off-diagonal elements. But such a diagonal force constant matrix does not give very good molecular geometries in general, and it does not give very good representations of other properties for molecules either. For saturated acyclic hydrocarbons, a diagonal force constant matrix is in fact a pretty good approximation because most of the off-diagonal elements are small if not zero. But, if one adds perturbations into the molecule, such as the double bond of the alkene, or the nitrogen of an amine, then certain off-diagonal elements that have negligible values with saturated hydrocarbons become more important.

If we look at the valence bond structure of our propene molecule that yields a longer C–H bond upon rotation, and note that this is in the element previously discussed, we can also presume that there will be connections between the torsion element ($k_{\omega1234}$), and the other stretching elements (k_{s23} and k_{s34}), as suggested by the valence bond structures. But it turns out that with propene, these latter elements do not appear to contribute to the geometry. The reason is that although the elements do contribute, because of the symmetry of the methyl group, the contributions from all of the hydrogens summed together cancel out. But this is a special case because of symmetry. What about the more general case?

We know from studies on conjugated systems such as butadiene that changes in bond order lead to sizable changes in bond length. Quantum mechanical calculations (at any level) on the butadiene π system show that rotation about the single bond leads to a large change in the bond order of that bond. There is simultaneously a change in the bond orders of the terminal C–C bonds in the opposite direction, but this change is small. Thus, bond order changes, for example, due to rotation about the central bond in butadiene, lead to large changes in the length of that central bond, but the end "double" bonds remain nearly constant in length. So for molecular mechanics, it is convenient to consider two types of torsion–stretch interaction. If we use propene as the example, as torsion occurs about the central C–C bond, there are bond length changes of the methyl C–H, the middle bond (the C_{sp^3}–C_{sp^2}) bond about which torsion occurs, and the C=C bond. We divide these into type 1 (the middle bond) and type 2 (the end bonds). Both types are important in general.

It should be apparent to the reader at this point that, just as conjugation occurs in molecules that have π systems that belong to different bonds attached together, so hyperconjugation will occur when there is a π system attached to a σ system. It is a ubiquitous effect in organic molecules, which, if they contain π systems, are pretty sure to also contain hyperconjugation. Earlier arguments about whether hyperconjugation was important in one type of system or another were often misdirected. The comparisons made were ordinarily between C–H bonds and C–C bonds. And the question was really "assuming that hyperconjugation does not occur in the latter, is it important in the former?" The assumption that it does not occur in the latter is a poor one, and it is highly misleading. So what people were looking at and arguing about was the *difference* in hyperconjugation between the H⁺ and the C⁺ systems. That is a different matter, and one would expect that the difference between the interactions of these two similar systems would be a great deal smaller than the absolute interaction in either system. Additionally, the comparisons were usually made in polar solvents, where solvation tended to muddy up whatever else was happening.

To summarize: Baker and Nathan observed that hyperconjugation appeared to occur to a greater extent when H⁺ was the nonbonded fragment in the sacrificial valence bond form, compared with the case where CH_3^+ was the fragment. This difference between the two (called the Baker–Nathan effect) was not very great in any case and was subsequently shown to be due to a solvation effect. This has led to the statement that "hyperconjugation is unimportant in the ground states of molecules." This statement is simply wrong. There is ample evidence (as cited in this chapter) that indicates unequivocally that hyperconjugation is important and ubiquitous in organic molecules. It is the Baker–Nathan effect, which is the *difference between* the H^+ and CH_3^+ fragments in resonance forms, that is relatively unimportant in the ground states of molecules.

REFERENCES

1. L. Pauling, *The Nature of the Chemical Bond*, 3rd ed., Cornell University Press, Ithaca, NY, 1960.
2. W. J. Hehre, L. Radom, P. V. R. Schleyer, and J. A. Pople, *Ab Initio Molecular Orbital Theory*, Wiley-Interscience, New York, 1986.
3. L. C. Allen, *J. Am. Chem. Soc.*, **111**, 9003 (1989); L. C. Allen, *Acc. Chem. Res.*, **23**, 175 (1990).
4. (a) H. A. Bent, *Chem. Rev.*, **61**, 275 (1961). (b) W. Moffitt, *Proc. R. Soc. (London)*, **A196**, 524 (1949). (c) R. S. Mulliken, *J. Phys. Chem.*, **41**, 318 (1937). (d) A. D. Walsh, *Discuss. Faraday Soc.*, **2**, 18 (1947). (e) C. K. Ingold, *Structure and Mechanism in Organic Chemistry*, Cornell University Press, Ithaca, NY, 1953, p. 70.
5. G. E. K. Branch and M. Calvin, *The Theory of Organic Chemistry*, Prentice-Hall, New York, 1941.
6. N. L. Allinger, M. R. Imam, M. R. Frierson, Y. Yuh, and L. Schäfer, in *Mathematics and Computational Concepts in Chemistry*, N. Trinajstic, Ed., Ellis Horwood, London, 1986, p. 8.
7. (a) MM3 bond length electronegativity correction parameters are taken from the MM3 (2000) program. *QCPE Bull.*, Indiana University, Vol. 15, No. 1, Feb. 1995, p. 15. (b) N. L. Allinger, Y. H. Yuh, and J.-H. Lii, *J. Am. Chem. Soc.*, **111**, 8551 (1989).
8. K.-H. Chen, G. A. Walker, and N. L. Allinger, *J. Mol. Struct. (THEOCHEM)*, **490**, 87 (1999).
9. (a) K.-H. Chen, J.-H. Lii, G. A. Walker, Y. Xie, H. F. Schaefer, III, and N. L. Allinger, *J. Phys. Chem.*, **110**, 7202 (2006). (b) N. L. Allinger, K.-H. Chen, J.-H. Lii, and K. A. Durkin, *J. Comput. Chem.*, **24**, 1447 (2003). (c) J.-H. Lii, K.-H. Chen, K. Durkin, and N. L. Allinger, *J. Comput. Chem.*, **24**, 1473 (2003). (d) J.-H. Lii, K.-H. Chen, T. B. Grindley, and N. L. Allinger, *J. Comput. Chem.*, **24**, 1490 (2003). (e) J.-H. Lii, K.-H. Chen, and N. L. Allinger, *J. Comput. Chem.*, **24**, 1504 (2003). (f) J.-H. Lii, K.-H. Chen, and N. L. Allinger, *J. Phys. Chem.*, **108**, 3006 (2004).
10. R. M. Badger, *J. Chem. Phys.*, **2**, 128 (1934); *J. Chem. Phys.*, **3**, 710 (1935).
11. D. R. Herschbach and V. W. Laurie, *J. Chem. Phys.*, **35**, 458 (1961).
12. J. A. Ladd, W. J. Orville-Thomas, and B. C. Cox, *Spectrochim. Acta*, **20**, 1771 (1964).
13. D. M. Byler, H. Susi, and W. C. Damert, *Spectrochim. Acta*, **43A**, 861 (1987).
14. D. C. McKean, J. E. Boggs, and L. J. Schäfer, *J. Mol. Struct.*, **116**, 313 (1984).
15. H. D. Thomas, K.-H. Chen, and N. L. Allinger, *J. Am. Chem. Soc.*, **116**, 5887 (1994).
16. J. A. Larsson and D. Cremer, *J. Mol. Struct.*, **485–486**, 385 (1999).
17. (a) R. S. Mulliken, *J. Chem. Phys.*, **7**, 339 (1939). (b) R. S. Mulliken, C. A. Rieke, and W. G. Brown, *J. Am. Chem. Soc.*, **63**, 41 (1941).

18. N. L. Allinger, K. Chen, J. A. Katzenellenbogen, S. R. Wilson, and G. M. Anstead, *J. Comput. Chem.*, **17**, 747 (1996).
19. A. Pross, L. Radom, and N. V. Riggs, *J. Am. Chem. Soc.*, **102**, 2553 (1980).
20. (a) J. March, *Advanced Organic Chemistry*, 4th ed., Wiley, New York, 1992, pp. 47, 69. (b) M. B. Smith and J. March, *March's Advanced Organic Chemistry*, 5th ed., 2001, and 6th ed., 2007, Wiley, New York. (c) For evidence in favor of hyperconjugation, see T. Laube and T. K. Ha, *J. Am. Chem. Soc.*, **110**, 5511 (1988). (d) J. B. Lambert and R. A. Singer, *J. Am. Chem. Soc.*, **114**, 10246 (1992).
21. J. W. Baker and L. G. Groves, *J. Chem. Soc.*, 1144 (1939).
22. J. W. Baker and W. S. Nathan, *J. Chem. Soc.*, 1840, 1844 (1935).
23. This idea was first suggested by W. M. Schubert and W. A. Sweeney, *J. Org. Chem.*, **21**, 119 (1956).
24. (a) W. J. Hehre, R. T. McIver, Jr., J. A. Pople, and P. v. R. Schleyer, *J. Am. Chem. Soc.*, **96**, 7162 (1974). (b) E. M. Arnett and J. M. Abboud, *J. Am. Chem. Soc.*, **97**, 3865 (1975). (c) R. Glyde and R. Taylor, *J. Chem. Soc., Perkin Trans.*, **2**, 678 (1977). (d) R. Taylor, *J. Chem. Res. (S)*, 318 (1985).
25. N. L. Allinger, L. Schäfer, K. Siam, V. J. Klimkowski, and C. Van Alsenoy, *J. Comput. Chem.*, **6**, 331 (1985).

7

MORE "EFFECTS"—NEGATIVE HYPERCONJUGATION

A popular saying from the time of this author's youth was something like this: "If a reaction does not come out the way you thought it should have, you have probably discovered a new *effect* in organic chemistry." The fact is, elementary organic chemistry is pretty straightforward. (Many students in the elementary course would probably disagree.) But when you look at things more carefully, it tends to get more complicated. These complications are everywhere in the experimental world. Infrared frequencies are displaced from where you thought they should be, chemical shifts in the NMR are in unexpected places, bond lengths are too long, or too short, a compound shows a stability, or instability, that was not expected, and so on. Many of the reasons for these anomalies are conveniently assigned to what organic chemists refer to as "effects." In the previous chapter we discussed the effects assigned to electronegativity and to hyperconjugation. Steric effects are qualitatively pretty obvious, and they have been mentioned in passing previously in this volume. There are a few other effects that are well known but perhaps not too widely known. The more important of these will be discussed in this chapter.

If molecular mechanics is to be an accurate model of molecular structure, then these various effects need to, first, be understood and, second, be included in our

Molecular Structure: Understanding Steric and Electronic Effects from Molecular Mechanics,
By Norman L. Allinger
Copyright © 2010 John Wiley & Sons, Inc.

BOHLMANN EFFECT

molecular mechanics calculations in an accurate and well-defined way, and they must be transferable like everything else.

BOHLMANN EFFECT

It was recognized in the 1950s by F. Bohlmann that under some circumstances, alkaloids showed in the infrared an intense C–H stretching vibration of unusually low frequency.[1] After some study, it was determined that these low-frequency C–H vibrations (which were by then called *Bohlmann bands*) came about when the lone pair on the nitrogen of the alkaloid was anti co-planar to a C–H bond attached to a carbon that was in turn attached to the nitrogen. In the valence bond representation of ordinary hyperconjugation it is common to have H⁺ and no bond between that hydrogen and the attached atom. In so-called *negative hyperconjugation* (of which the Bohlmann effect is an example), there is similarly an H⁻ and no bond from that hydrogen to the adjacent atom. If we take methylamine as an example, in the ground state the hydrogens take up staggered positions relative to one another (as in ethane), and the stretching frequency assignable to the hydrogen that is attached to carbon anti co-planar to the lone pair is about $100\,\text{cm}^{-1}$ lower in frequency than the other two C–H stretching frequencies from the methyl group. This is a simple clear-cut and easily recognizable example of negative hyperconjugation, which is called the *Bohlmann effect*.[2,3] Compare hyperconjugation with negative hyperconjugation (Structure 1).

$$\underset{\text{Hyperconjugation}}{\overset{H}{\underset{|}{C}}-C=C \longleftrightarrow C=C-C^-} \qquad \underset{\text{Negative Hyperconjugation}}{\overset{\ddot{}}{N}-C-H \longleftrightarrow N^+=C\ \ H^-}$$

1

With (positive) hyperconjugation, the usual typical example of the effect involves specifically a hydrogen (proton) as the detached fragment, but as discussed in Chapter 6, it occurs with other elements too. It may be presumed that the same is true for negative hyperconjugation. That is, the example shown involves a hydrogen with a negative charge, but one would anticipate that other elements could take the place of the hydrogen, especially electronegative elements that would better support negative charge. The Bohlmann effect is of interest because the shift of the C–H frequency is so large and because its intensity is enhanced. Hence, it is easy to see by inspection when one examines an infrared spectrum, and it has been useful in assigning stereochemistry in the alkaloid field.[1]

The reason for the longer bond length seems clear enough (Chapter 6). As the bond lengthens, the potential curve that describes that bond stretches out. And the result is that the vibrational levels sink down into the potential well, and hence the frequency observed in the infrared becomes lower.[2,3] Are there other consequences, perhaps more generally interesting consequences, related to the Bohlmann effect, or more generally speaking, to negative hyperconjugation?

It turns out that in fact the Bohlmann effect is the tip of an iceberg. It, and other forms of negative hyperconjugation, have a large impact on the structures of organic molecules. Look again at the example in Structure **1**. All that is really required here is a lone pair, which can donate electron density to an atom (hydrogen in the case shown), and we can have negative hyperconjugation. Structure **2** shows the more general formulation of negative hyperconjugation.

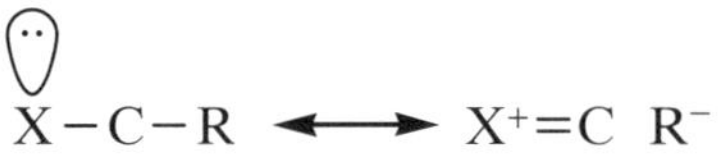

$$X-C-R \longleftrightarrow X^+=C \; R^-$$

Negative Hyperconjugation (General)

2

What this means is that any atom (X) bearing a lone pair (and in particular oxygen, nitrogen, or sulfur) will exhibit this effect. And the R group can be anything that will accept electrons (hydrogen and alkyl are electronegative enough). And even the central C could be a different atom. So the Bohlmann effect is just a simple example of something that is widespread throughout organic molecules. This was recognized by Laube and Ha and discussed in some detail from a quantum mechanical viewpoint in 1988.[4]

Let us consider as an example the specific case of the gauche and anti conformations of ethylamine, with respect to the methyl group and the lone pair (Structure **3**). This Bohlmann effect is stereospecific. If a hydrogen on the methylene carbon is anti to the lone pair (hydrogen 3 as shown in the gauche conformation), the p components of the C–H bond and of the lone-pair orbital overlap very well, and the effect is at a maximum. In this case the C–H bond is unusually long. On the other hand, if the hydrogen is gauche to the lone pair (as for hydrogen 3 in the anti conformation), then there is a 60° dihedral angle between the hydrogen and the lone pair, and the effect is small. Hence, the hydrogens attached to the methylene carbon both have ordinary bond lengths in the anti conformation, but one of them has a very long bond length in the gauche conformation. This information is summarized in Table 7.1.[5,6]

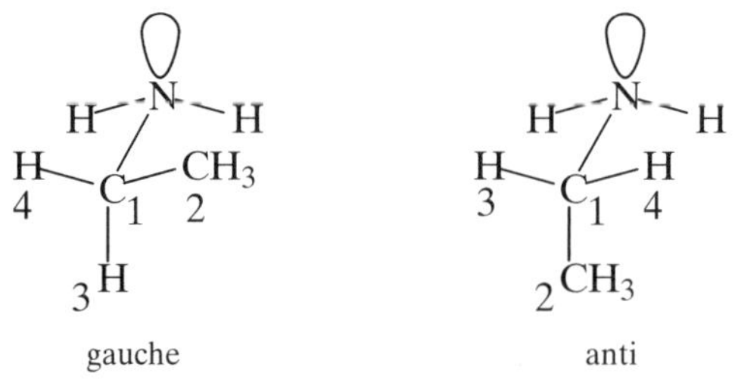

gauche   anti

3

Specifically, the C_1–H_3 bond has a length of 1.094 Å in the anti conformation from the MP2/BC calculation, and it lengthens out to 1.099 Å in the gauche conformation.

But, if the effect is general, we expect that the C_1–C_2 bond will similarly lengthen when anti to the lone pair (the anti conformation), and indeed the MP2 bond length is

TABLE 7.1. Geometries (r_e) of Ethylamine Conformers[a]

	MP2/BC		MM4	
Conformer	Anti	Gauche	Anti	Gauche
C_1–C_2	1.526	1.519	1.527 (849)	1.520 (887)
C_1–H_3	1.094	1.099	1.095 (2924)	1.104 (2796)
C_1–H_4	1.094	1.093	1.095	1.096
C_2–C_1–N	115.4	109.7	114.2	109.5

[a]The appropriate corrections for the MP2/BC calculations (for definition of MP2/BC, see Jargon in the Appendix) have been made so as to yield r_e geometries.[5,6] The geometric units are angstroms and degrees. The corresponding MM4 vibrational frequencies are given in reciprocal centimeters (cm^{-1}) in parentheses.

1.519 Å in the gauche conformation and 1.526 Å in the anti. In accordance with the Bohlmann effect, the C–H_3 stretching frequency is much lower in the gauche conformation than in the anti. And indeed, this is seen (Table 7.1). One sees a large frequency shift when the hydrogen undergoes the Bohlmann effect, but there is a similar shift (although smaller) for the carbon–carbon bond. The C–C stretching frequencies for these conformations were calculated (quantum mechanically using Baker and Pulay's method[7]) to be 883 cm^{-1} for the gauche conformation and the 842 cm^{-1} for the anti conformation. So the facts are clear, and how do we formulate this effect in MM4?

The effect is represented in MM4 as a torsion–stretch interaction. It has a maximum value in the anti conformation, which is smaller but significant at the cis conformation, and quite small when the bond and the lone pair have a torsion angle of 90°.[5,6] The observed stretching frequency for the C–H bond that is gauche or anti to the lone pair is automatically calculated correctly by MM4 because it only depends upon having a correct C–H bond length. And this will occur not only for C–H bonds but also for C–X bonds, for example. Specifically, in the case of ethylamine, where the ab initio values for the C–C stretching are 883 cm^{-1} for the gauche conformation and 842 cm^{-1} for the anti conformation, the corresponding MM4 values are 887 and 849 cm^{-1}. So the large and conspicuous shift in the Bohlmann bands for C–H to lower frequencies by 100–150 cm^{-1} is indeed found here (and in general) for the C–C stretching frequencies, but here it is much smaller (about 40 cm^{-1}). The Bohlmann effect for C–H bonds is well known and practically useful. The shift is large enough that this particular vibrational frequency is well removed from the general band envelope of C–H bands, and is an otherwise vacant part of the spectrum. The C–C band, on the other hand, is just another vibration buried in the fingerprint region and is not ordinarily experimentally identifiable without considerable effort. So in this case the shift is of limited practical importance, but it is there nonetheless.

A similar but somewhat smaller Bohlmann effect occurs with ethers and alcohols, where the lone pairs on the oxygen can delocalize into C–H and C–C bonds. And, of course, for oxygen there are two lone pairs, each of which must be separately considered. The spectroscopic shift in this case is of limited practical use, however, since although the bond lengthening of the C–H bond from the Bohlmann effect does occur,

TABLE 7.2. Bohlmann Distortions for Bonds (Å) and Bond Angles (deg) in Ethylamine and Ethanol (r_e)

Bohlmann Effect	Ethylamine[a]		Ethanol[b]	
	MP2/BC	MM4	MP2/BC	MM4
ΔCC	+0.007	+0.006	+0.006	+0.005
ΔCH	−0.005	−0.009	−0.005	−0.005
ΔCCX	+5.7°	+4.7°	+5.1°	+4.4°

[a]Anti conformation minus gauche. The anti form has the lone pair anti to the methyl (Structure **3**).
[b]Gauche conformation minus anti. The gauche form has a lone pair anti to the methyl.

the bond shortening from the electronegativity effect pretty much cancels it out, and the frequency of a "Bohlmann hydrogen" is not cleanly shifted out and away from the general C–H bond envelope of the molecule.[3]

Some geometric comparsons between ethylamine and ethanol are given in Table 7.2. There is an even smaller shift that is brought about by fluorine. This relationship between the magnitudes of the Bohlmann effects from N, O, and F is expected from the electronegativities of the atoms. The atom with the greater electronegativity holds its electrons more tightly and allows less delocalization, and hence shows a smaller Bohlmann effect. (The fluorine has lone pairs all around, of course, and thus shows no stereochemical effect, but it is nonetheless possible to separate the electronegativity bond shortening from the Bohlmann bond lengthening, and the preceding statement is found to be true, as expected.[3])

If we refer to Structure **2** again, the right-hand resonance structure indicates clearly enough that the C–R bond stretches from the donation of the electron pair from the nitrogen into the anti bonding orbital for the C–R bond. This means that the nature of the bonding at the central carbon moves from something that might be described as approximately sp^3 in character toward something that is more nearly sp^2. Thus, the Bohlmann effect not only leads to abnormal bond lengths, but it also leads to abnormal bond angles, corresponding to that type of change. In MM4 this effect is included as a torsion–bend interaction. As with the torsion–stretch, and for the same reason, we anticipate that V_1 and V_2 terms will be required, as they describe the results of the delocalization of the lone pair analogous to the stretching case. The C_2–C_1–N bond angle is recorded in Table 7.1 for ethylamine from MP2/BC calculations. The angle is "normal" for the gauche conformation, 109.7°, but opens out toward sp^2 considerably in the anti conformation, where it has the value 115.4°, for an opening of 5.7°. The opening from ethanol (from both lone pairs) is 5.1°. The MM4 values are quite similar.

If we make the assumption that this negative hyperconjugation is applicable to atoms in general, and not a function of the nitrogen or hydrogen explicitly, that is, a *generalized Bohlmann effect*, then clearly there will be countless possible compounds in which this effect might be expected to occur, and it is not limited to first-row atoms. The effect is quite large with sulfides[8a] and phosphines.[8b] Since the overlap of a 2p orbital with a 3p orbital is relatively small because of the long bond distance and size disparity, it might have been supposed that the effect would be small in such molecules. However, the valence electrons in second-row atoms are much more polarizable than those in first-row atoms, and they tend to delocalize more readily. For methyl ethyl sulfide, for example, the C–C–S angle in the gauche form (where the ethyl is anti to a lone pair) is opened out to 114.5° relative to the trans value of 108.7°,[8a] an opening of 5.8°.

It will be evident at this point that this generalized Bohlmann effect will be ubiquitous in organic molecules. If there is an atom carrying a lone pair, there will be this type of effect on each of the adjacent bonded atoms, and also on all of the atoms geminal to the atom with the lone pair. In molecular mechanics, if the structures of molecules are to be calculated with chemical accuracy, torsion–stretch and torsion–bend terms will be required for all of the commonly found heteroatoms. Thus, while the molecular mechanics of hydrocarbons is relatively simple and straightforward, the molecular mechanics of organic molecules in general is a whole different ballgame because of the generalized Bohlmann effect. Good accuracy can still be obtained for organic molecules in general as far as we know, but only with considerably greater effort than that required for hydrocarbons. And the following should be noted. A good (large) quantum mechanical calculation will give us a good geometry for a molecule. But bonds and angles may be stretched or compressed for a whole variety of reasons, for example, steric effects and an assortment of electronic effects (Chapters 6 and 7). The quantum mechanical result gives us the sum of all of these effects. The molecular mechanical results give us in addition a breakdown of just how much of the distortion can be attributed to each of those numerous effects, as defined by the molecular mechanics model.

ANOMERIC EFFECT*

Conformational analysis is usually thought of as having become important in organic chemistry after the appearance of the study by D. H. R. Barton that spelled out several important consequences of the subject.[9] A fundamental principle of the subject is that anti conformations are ordinarily more stable than their corresponding gauche counterparts in open-chain carbon compounds, and in six-membered rings this translates to an equatorial conformation being more stable than an axial. It was, however, found that in carbohydrates, and specifically in hexopyranoses, a hydroxyl, methoxyl, or acetoxyl at the anomeric carbon is usually more stable in the axial than the equatorial position, contrary to the usual situation. As has long been common in organic chemistry, this curiosity (the unexpected stability of the axial hydroxyl) was identified by giving it a formal name, the *anomeric effect*.

---

*In simple carbohydrates there is one hydroxyl group (which might be methylated, acetylated, or otherwise modified) attached to each carbon atom except one. On that one carbon there are two such groups. That particular carbon is called the *anomeric carbon*. It is of special interest because it is really a masked carbonyl group, and it has quite different properties from the singly oxygenated carbons.

This effect is the best known example of negative hyperconjugation or the generalized Bohlmann effect. It was discovered in the carbohydrate field and is still best known in that area. Carbohydrates tend to be rather formidable molecules, so before we examine the importance of the anomeric effect in such systems, it will be advantageous to study it in simple molecules. There are sizable changes in the bond lengths and bond angles, as well as energies, as a result of the anomeric effect.[10] We will examine all of this by looking first at the simple molecule dimethoxymethane, and take up in detail and in order the energetics and the structural consequences of the anomeric effect. Later (in Chapter 8) we will come back to the importance of the anomeric effect in the carbohydrate field.

To see the relationship between the Bohlmann effect and the anomeric effect we can look at the valence bond structures for methylamine and dimethoxymethane, Structure **4**:

Methylamine

Dimethoxymethane

4

In the former molecule, the nitrogen lone pair can delocalize into the C–H bond in the anti position. In the latter, electrons from an oxygen lone pair delocalize into a C–O bond in the anti position. The nitrogen in methylamine does not hold its lone pair very tightly, hence, the greater importance of the Bohlmann effect in amines than in alcohols. On the other hand, the dimethoxymethane is delocalizing the electron pair into a C–O bond, and the electronegativity of the oxygen makes this a relatively favorable process. Thus, the anomeric effect leads to similar, but usually greater, structural consequences than does the Bohlmann effect. It should also be noted that the anomeric effect can occur with a wide variety of heteroatoms. The case where there are two geminal oxygens (O–C–O) was the first noted, but two nitrogens (N–C–N) behave similarly, or one of each (O–C–N) is even better. Other atoms (such as sulfur, phosphorus, and halogen) in various combinations also show anomeric effects. Note also that the anomeric effect goes in both directions at the same time, so the importance of the effect in a given molecule is dependent on the conformations about both of the C–O (or C–X) bonds.

DIMETHOXYMETHANE

The most simple compound that shows the classical anomeric effect is dihydroxymethane. This compound exerts hydrogen bonding characteristics that may complicate the situation, so we will here examine the corresponding ether, dimethoxymethane, as our parent model compound. We will take up the energetic effects first, as historically these were the first indication of the anomeric effect.

Energetic Effects

The entire potential surface for dimethoxymethane was mapped in 1996,[11] at what was then a substantial level of calculation, MP2/6-31G**. In reasonable agreement with earlier studies (1978) using smaller basis sets at the Hartree–Fock level,[12] it was found that the surface contained structures with energy minima at all of the possible anti and gauche conformations for the CO bonds, and no others. Thus, the stable conformations are anti–anti (aa), gauche–anti (ga), gauche–gauche (gg), and gauche–gauche' (gg'). The energies of the structures that resulted from energy minimization at the MP2/6-311++G(2d,2p) level, with bond length corrections to the equilibrium (r_e) geometry (which we call MP2/BC, or often just BC), were further improved[13] relative to earlier studies, and are summarized in Table 7.3.

It was found, as in all previous studies,[10] that the gg conformation is the most stable.[11] (This conformation is the only one observed by gas-phase electron diffraction.[14]) The ga conformation is next in stability, followed in order by the gg' and the aa conformations.

The MM4 calculations were then fit to a torsional potential from the BC calculation shown in Figure 7.1. It was also necessary to fit torsion–stretch and torsion–bend interactions simultaneously in order to fit the structural features that will be described below.

The conformational energies obtained from the BC calculation, relative to the gg, were 2.89, 3.85, and 6.11 kcal/mol, respectively (Table 7.3 and Fig. 7.1). The corresponding MM4 conformational energies are 2.14, 3.22, and 4.95 kcal/mol, so that the errors between the best ab initio values and the MM4 values have now been considerably reduced, relative to what was found earlier comparing earlier force fields and small basis set values. The largest remaining error is the aa conformational energy, where the BC and MM4 calculations still differ by 1.2 kcal/mol. It is not known how much of the error lies in the ab initio value and how much in the MM4 value.

TABLE 7.3. Conformational Energies of Dimethoxymethane (kcal/mol)

Method	gg	ga	gg'	aa
MP2/6-31G**[11]	0.00	3.47	4.84	7.31
BC[13]	0.00	2.89	3.85	6.11
MM4[13]	0.00	2.14	3.22	4.95

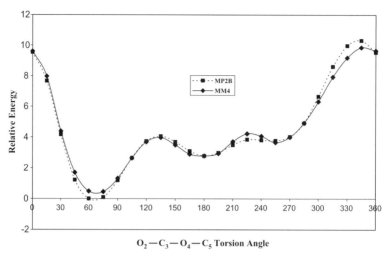

Figure 7.1. O_2–C_3–O_4–C_5 torsion potential of $(MeO)_2CH_2$. The C_1–O_2–C_3–O_4 torsion angle is fixed at 67°.

This molecule is a little bit unusual in that the most stable conformation is gauche–gauche. If we energy minimize the structure, and then hold one of the C–O–C–O torsional angles in the stable gauche conformation (67°), and rotate the other end of the molecule from 0° to 360° in 15° increments (using the Driver routine; see Appendix), we obtain the potential function shown in Figure 7.1. Note that the stationary points do not correspond exactly to the energy minima shown here, partly because the one torsion angle in the MM4 calculation in Figure 7.1 was fixed at a value of 67°. There are three minima found, as expected. If we proceed from left to right across Figure 7.1, these three minima in order are gg, ga, and gg′, where the first g corresponds to the end that is fixed at 67°. The gg conformation is a much deeper well than the other two, and the values for these energies are shown in Table 7.3. The aa conformation does not appear in Figure 7.1, of course, but is calculated separately. Note that the gg conformation is more stable than any of the others by over 2 kcal/mol. This means that in ordinary unconstrained anomeric systems, we will seldom see anything other than the gg conformation. In cyclic molecules (sugars), however, the presence of rings and other constraints means that we will in fact frequently encounter some of these other conformations.

Structural Effects

We have also made detailed geometric comparisons between the structures of dimethoxymethane from the MM4 and BC calculations. Ab initio calculations in general give an approximation to the equilibrium (r_e) structures for molecules. To compare these with experimental structures, the effects of vibrational motion have to be taken into account (Chapter 2). From a study of the basis set/correlation truncation errors from BC calculations, the CO bond needs a correction[15] from the raw value (an

TABLE 7.4. Geometry (r_e) and Energy (kcal/mol) Data for Dimethoxymethane[13]

	aa	ga	gg'	gg	gg − aa	gg − ga
Bond Lengths (Å)						
CH_3–O						
BC	1.411	1.419/1.411	1.416/1.414	1.418	+007	
MM4	1.412	1.417/1.411	1.419/1.417	1.416	+004	
O–CH_2						
BC	1.389	1.377/1.408	1.396/1.391	1.396	+007	−019/+008
MM4	1.389	1.377/1.405	1.400/1.395	1.392	+003	−015/+013
Bond Angles (°)						
C–O–C						
BC	111.5	112.3/110.6	114.2/112.7	111.8	+0.3	
MM4	111.7	112.2/111.2	114.2/115.1	112.4	+0.7	
O–C–O						
BC	106.2	109.9	113.1	113.7	+7.5	+3.8
MM4	106.8	109.8	115.4	113.0	+6.2	+3.2
Torsion Angle (°)						
C–O–C–O						
BC	180	67/178	111/−66	66		
MM4	180	68/179	98/−74	69		

approximation to the r_e value) to the r_e value of −0.0077 Å, so these corrections have been added to the bond lengths as are listed in Table 7.4.

The MM4 calculation gives r_g structures as the normal result, but the program also makes the vibrational corrections and calculates the r_e (and other) bond lengths.[16] These r_e bond lengths are also included in Table 7.4, so that the accuracy of the MM4 calculations with respect to the BC calculations as a standard are as shown. The energy value used is that obtained before the bond length correction is added. (The differences between r_e and r_g for bond angles are usually only fractions of one degree and can be ignored.) The discrepancies between the BC and MM4 values are relatively small, and, again, it is not known how much error there is in each calculation. But the agreement suggests that there is not very much error in either.

The bond length changes for the methoxyl group CO bonds (i.e., external to the anomeric system) from gg to aa that result from the anomeric effect are a shortening of the order of 0.007 Å by BC and 0.004 by MM4 calculations (Table 7.4). (These will be discussed in Chapter 8.) Note that the corresponding changes in the internal C–O bonds are more dramatic. In the gg → ga change, one bond shrinks and the other stretches, −0.019/+0.008 BC and −0.015/+0.013 MM4, respectively. The other large conformational structural change is the O–C–O angle, which is (BC) 106.2° for aa, 109.9° for ga, and 113.7° for gg. The corresponding MM4 values are 106.8°, 109.8°, and 113.0°. These MM4 results all seem satisfactory, and the model of the anomeric effect as a generalized Bohlmann effect is a good one.

There was an explicit, and somewhat involved, set of equations that was used to describe the anomeric effect on bond lengths in MM2[10,17] and MM3.[18,19] These equations were used to derive the bond shortening/lengthening, which affected l_0 values for the different bonds. The inconvenience of this method of approach was that one had to do the calculation in an interative manner. That is, one had to calculate the change in l_0, and then calculate the structure, which led to further change in l_0, so the structure had to be recalculated, analogous to what is done in a self-consistent field calculation. A more simple, direct way of accomplishing the same result here is to use a torsion–stretch interaction element in the **F** matrix. The latter approach is used in MM4. The disadvantage of this method is that one cannot directly see what the bond shortening and lengthening effects from the anomeric effect are, one can only see the overall result including the changes in effects of van der Waals and electrostatic interactions on the bond lengths with torsion. (It would, of course, be possible to obtain this information from the calculations, but it would require dissection of the matrix element. This has not been done as of this writing.) One of the nice things about molecular mechanics is that one can see exactly what kind of assumption, approximation, or equation leads to exactly what difference in the result. In this case the results are muddied by the problem mentioned. (One can, however, alter the interaction elements in the **F** matrix and note the effect.) However, what is going on has been clearly established with MM3,[18] and that work and type of calculation can be referred to for additional information if desired.

As suggested by the valence bond structure on the right in Structure **4**, relative to the anti–anti conformation (where there would be little resonance), the gauche–anti conformation has the resonance as shown in the structure at the far right. This shortens the 2–3 bond and lengthens the 3–4 bond. In Structure **5** are shown the calculated C–O bond lengths (in angstroms) for the different conformations. The bond lengths in parenthesis are from BC, and the others are from MM4 calculations. If we look at the gauche–gauche conformation, the bond length changes induced by the different oxygens are in opposite directions and cancel out.

The gauche–anti conformation is stabilized relative to the anti–anti by about 3.0 kcal/mol (Table 7.3). If we go from the gauche–anti to the gauche–gauche conformation, there is another increment of energy lowering, about 2.5 kcal/mol this time. The gg′ conformation is also stabilized by this kind of resonance, but it is destabilized by steric repulsions and unfavorable dipole alignments. The gg′ conformation turns out to be more stable than the aa but less so than the ga.

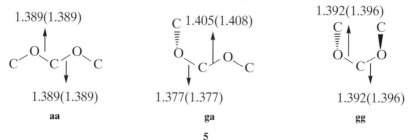

5

Angle Effects

Except for the gg' conformation where there are rather large steric effects, the C–O–C angles show only minor conformational dependence, a degree or so, largely from steric effects (Table 7.4). The O–C–O angle, on the other hand, has a rather small BC value of only 106.2° (MM4 106.8°) for the aa conformation, and such a shrinkage is well accounted for with MM4 by the electronegativity effects of the two oxygens. The anomeric effect predicts that this angle will open as a conformation goes a→g. Delocalization of the lone pair causes expansion of the O–C–O angle as the carbon tends to go from sp^3 toward sp^2 hybridization, as should be clear from the valence bond pictures shown in Structure **4**. Contrary to the bond lengths (where the effects in different directions tend to cancel out), here the angle effects are additive. The calculated changes in the values for the O–C–O angle with conformational changes as we go aa→ag→gg are quite large: MM4(BC) 106.8 (106.2), 109.8 (109.9), and 113.0 (113.7°).

2-METHOXYTETRAHYDROPYRAN

The anomeric effect is of particular importance in carbohydrates. To that end, we will examine the title compound as a prelude to the study of glucose, and its diastereomeric aldohexapyranose isomers. The pyran contains the anomeric part of the carbohydrate systems, while avoiding the complications of the hydroxyl groups and C-6 substituent present in glucose itself. So for present purposes we will first consider 2-methoxytetrahydropyran, as a glucose analog, Structure **6**:

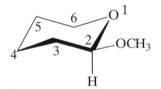
Eq-2-methoxytetrahydropyran
β-Glucose analog

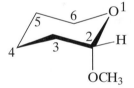
Ax-2-methoxytetrahydropyran
α-Glucose analog

6

The conformational equilibrium in 2-methoxytetrahydropyran has been studied experimentally, with ab initio calculations, and with MM2,[17] MM3,[18] and MM4.[13] It is important to fit this equilibrium well with molecular mechanics, if one is to have a good force field for carbohydrates. For a study of the anomeric effect what we really need is information regarding the axial/equatorial equilibrium. This is a little more complicated than it sounds because the methoxyl group can rotate to give three equatorial conformations and three axial conformations. Their relative MM4 energies are shown in Table 7.5.

We also are interested in this equilibrium as a function of dielectric constant (D), since for simple compounds, gas phase and solution data are often available, and for more complex compounds, particularly carbohydrates, the measurements are usually

TABLE 7.5. Variation of MM4 Conformational Energies of 2-Methoxytetrahydropyran (kcal/mol) with Dielectric Constant[13]

D			
1.5	1.42	3.09	4.64
4.0	1.01	2.43	3.58
20.0	0.76	2.14	3.02
D			
1.5	0.00	6.44	2.86
4.0	0.00	5.45	2.34
20.0	0.00	5.03	2.10

made in water at high dielectric constant. So in Table 7.5 are given calculated energies at dielectric constants of 1.5 (the gas-phase value for a molecule with MM4), 4.0, which is the general value used for relatively concentrated solutions of arbitrary molecules in most organic solvents, and 20.0, which is a general large number that is suitable for polar solvents such as dimethyl sulfoxide, acetonitrile, and water.

The numbers in Table 7.5 at a dielectric constant of 1.5 show that one of the axial conformations is the most stable, and it is given a relative energy of 0.00. The other axial conformations are much higher in energy under all circumstances. The lowest energy equatorial conformation has an energy of 1.42 kcal/mol, and this is reduced with increasing dielectric constant to a value of only 0.76 kcal/mol at a dielectric constant of 20. The other equatorial conformations are about 1.5 kcal or more higher than this under all conditions. Therefore, to a good approximation, we need only look at the single conformation of lowest energy for each the axial and the equatorial conformations.

It will be of interest to note at this point some facts regarding the cause or basis of the anomeric effect. There has been considerable discussion regarding this point.[10] It is evident from Table 7.5 (and see also Table 7.6) that the value of the anomeric effect in the gas phase is substantial, with an energy difference between the most favorable axial and equatorial conformations of 1.42 kcal/mol. But also note that the value is reduced to about half (0.76 kcal/mol) at a dielectric constant of 20. This indicates that the anomeric effect contains a substantial contribution from electrostatics (i.e., dipole–dipole repulsions between the C–O bonds, and preferential solvation of the equatorial conformation). But they also show that a sizable portion of the effect is from something other than simple electrostatics. If the cause were simply electrostatics, the 1.42 kcal/mol number would be reduced to a sizable negative number, and while there is substantial energy reduction, it does not go nearly that far.

TABLE 7.6. Solvent Effects on the Equilibrium in 2-Methoxytetrahydropyran (kcal/mol)[13]

	Experiment		MM4	
Solvent	Est. D^a	$\Delta G°$	$\Delta G°_{298}$	ΔE
Gas	1.5	—	1.15	1.42
CCl_4	2.2	0.89	0.99	1.23
$CDCl_3$	10	0.76	0.63	0.83
CH_3CN	39	0.37	0.54	0.73

^{a}The solvents used have well-known dielectric constants (D), but these bulk values are not expected to be very good estimates of the effective dielectric constants of the solutions at the modest concentrations employed.

The axial conformation in MM4 is stabilized by the anomeric effect, which is represented largely by substantial V_1 (−1.85) and V_2 (−1.30) torsion potentials (Fig. 7.1). The bond dipoles in the equatorial conformation add up to give a greater dipole moment, and a higher dipole–dipole repulsion energy than in the axial conformation. Thus, increasing the dielectric constant tends to stabilize the equatorial conformation more than it stabilizes the axial, as shown in Table 7.5. The anomeric effect in MM4 is the summation of the electronic delocalization effect included in the torsion potential, plus contributions from torsion–stretch and torsion–bend interactions, plus the electrostatic effect from the interaction of the bond dipoles.

We wish to compare these MM4 results with values obtained from ab initio calculations, and also with values obtained from experiment. The ab initio values are easier to discuss. The earlier calculations at the MP2/6-31G** level indicated that the axial conformation was more stable by 2.66 kcal/mol. That turns out to be a substantial overestimation. The BC calculation reduces this value to 1.66 kcal/mol.[13] On the other hand, the MM3 value was quite a bit too small (0.76 kcal/mol) and is significantly increased in MM4 (1.42 kcal/mol).

The experimental information that we wish to examine is given in Table 7.6. The experimental quantities measured are $\Delta G°$, and in the table they are given, together with the corresponding MM4 values for $\Delta G°$. The entropy effects are small but not completely negligible, and the $\Delta G°$ values are systematically about 0.2–0.3 kcal/mol lower than the $\Delta E°$ values. We have experimental measurements of the equilibrium constants between the conformations from dipole moment measurements in various solvents, as shown in the experiment column. The MM4 (gas-phase) conformational energy difference is a compromise and is about 0.2 kcal/mol too small relative to the best ab initio value, and about 0.3 kcal/mol too large relative to the experimental values.

There have been extensive discussions in the literature regarding the nature of this effect,[20] and this work has been well reviewed by Juaristi and Cuevas.[10a] The conclusions now seem clear. This effect can best be understood by looking at the valence bond pictures as shown (Structure **7**):

[Structure 7: two resonance structures of a tetrahydropyran derivative with axial OCH₃ group, showing negative hyperconjugation between ring oxygen lone pair and C–OCH₃ bond]

7

When one lone pair on the ring oxygen is thought of as trans to the methoxyl group, the resonance shown is important. This means that the C–O bond within the ring becomes more double in character and is shortened, while the exocyclic C–O bond becomes less bonding and is lengthened by this resonance. The energy of the molecule is also lowered by this interaction. For the axial conformation, the axial lone pair–oxygen–carbon–OMe dihedral angle is about 180°. In the equatorial form the corresponding angle is about 60°. The overlap of the p component of the axial lone pair orbital with the p component of the C–O bond is large because they have their axes parallel. The overlap of the axial lone pair with the equatorial C–O bond is much smaller because they are close to perpendicular. Thus, the resonance is much more effective for the α (axial) methoxyl in Structure **7** than for the β (equatorial) conformer. This interaction (negative hyperconjugation) lowers the energy for the axial form and accounts for the main part of the energetics of the observed anomeric effect. (Most of the remainder of the energetics is a result of the intramolecular dipole–dipole interactions.) The other lone pair on the ring oxygen is at about 60° relative to the methoxyl in both conformations and thus plays no significant role here.

The anomeric effect has sizable energetic consequences, and these are accompanied by significant geometric changes, as might be gathered by examining Structure **7**. For example, the C–O bond, which becomes double in the resonance form on the right in Structure **7**, has a higher than normal bond order, and it is therefore shorter than the corresponding bond in the equatorial conformer, where this resonance is minimal. The bond that has disappeared in the right-hand structure has a lower bond order and is therefore longer. There are also the expected small angular changes associated with the hybridization change at carbon-1 as implied by this resonance structure.

MM3 included a special torsion–stretch equation with an electronegativity-based bond length correction designed to deal with the anomeric effect,[18,19] as mentioned previously. Since MM4 contains a torsion–stretch term for general use, a separate, explicit anomeric effect equation is no longer required. The new cross term, together with an electronegativity-based bond length adjustment for the C–O bond from the second oxygen on the carbon, implicitly model the anomeric effect in MM4. The MM3 and MM4 treatments, although formally different, in practice yield very similar geometries.

The MM3 treatment of the anomeric effect was fairly good based on the data available at the time. But better quantum mechanical calculations are now possible, and they indicate that the anomeric energy effect was somewhat larger than was incorporated in MM3. Improvements seen in the MM4 structures discussed here are partly due to improvements in the hydrocarbon section of the force field.[21] Furthermore, torsion–stretch and torsion–bend effects are important in the alcohols and ethers in general, but they were not included in MM3. Since these terms are included[15] in MM4, alcohols and ethers are more accurately treated, and all of this translates into better structures and energies for anomeric compounds as well. Some results of the anomeric effect in sugars will be discussed in Chapter 8.

α-HALO KETONE EFFECT

In 1953 E. J. Corey began publishing what was to become an important series of articles describing detailed studies on the conformational analysis of α-bromocyclohexanones.[22] At that time the importance of steric effects in organic chemistry was qualitatively well understood but electrostatic effects much less so. The α-bromocyclohexanones represented compounds where electrostatic effects were expected to be important, which could be studied with the then available techniques, particularly infrared spectra. The parent system would be 2-bromocyclohexanone, for which the axial and equatorial conformations are in equilibrium as in Structure **8**:

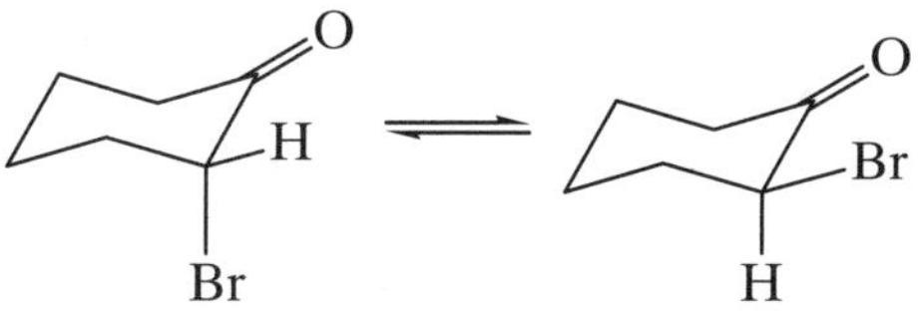

8

By examining model compounds with fixed conformations and from earlier work, Corey determined that there were distinct differences in the carbonyl frequencies of cyclohexanones that had a bromine in the α-position, and that depended upon whether the bromine was in an axial or in an equatorial position. Model compound studies showed that the parent cyclohexanone itself had a carbonyl stretching frequency at $1712\,cm^{-1}$ in carbon tetrachloride, and cyclohexanones that contained an axial bromine in the α position showed their carbonyl absorption at very nearly the same place, about $1712–1716\,cm^{-1}$. On the other hand, when the bromine was in the equatorial position, the absorption was found about $1728–1730\,cm^{-1}$. If one had a mixture of conformations with the bromine partly equatorial and partly axial, one could usually more or less resolve these two frequencies. This information could be used to tell in an equilibrium situation whether the bromine was mostly axial or mostly equatorial, and by approximately how much. (The assumption is made here that the inherent intensity of the vibrational band is the same in both the axial and equatorial arrangements. It was later shown that this is not quite true, and the equatorial stretching band has a somewhat lower intensity, in the range of 70–90%, depending on the particular compound, and whether the integrated or maximum intensity is referred to. But it is a reasonable approximation.) Corey measured the infrared spectrum in the carbonyl region for 2-bromocyclohexanone itself and found there was only a small shoulder at 1730, and a strong band at 1716 cm^{-1}, which was taken to indicate that the axial conformation strongly predominated. Corey concluded from these data that for 2-bromocyclohexanone itself, the mixture was at least 97% axial.

Now why would it be that the stretching frequency of a cyclohexanone (1712 cm^{-1}) is essentially unaffected by an axial bromine, but be raised to about 1730 cm^{-1} by the presence of an equatorial bromine? The explanation offered was as follows. We can think of the carbonyl group as having two resonance forms, as shown in Structure **9**:

$$\diagup\!\!\!\!\diagdown\!\!\text{C}=\text{O} \longleftrightarrow \diagup\!\!\!\!\diagdown\!\!\text{C}^+ - \text{O}^-$$

9

The presence of the dipole of the equatorial bromine, since it is nearly parallel and co-planar with the carbonyl dipole, would tend to cause an induced dipole in the carbonyl group, reducing the importance C$^+$–O$^-$ structure and increasing the importance of the doubly bonded structure. The more double the C=O bond, the stronger and shorter it will be and, hence, the higher the stretching frequency. This is a reasonable interpretation that fits the known facts, but it proves not to be correct, as will be discussed later.

Corey also carried out calculations based on the classical interaction between the C–Br and C=O dipoles to determine exactly what difference in electrostatic energy would be expected between the two conformations. This calculation would be for an interaction in vacuum (which could be used to approximate the gas phase). However, the experimental measurements were in a carbon tetrachloride solution. That this was of significance was not recognized at the time. Those calculations were quite formidable at the time, involving three-dimensional objects containing quite a few atoms, and with no aid from computers, which were not yet available. But Corey determined that the axial conformation was favored by about 2.7 kcal/mol, which he considered to be consistent with what was observed experimentally.

Slightly later, Kumler and Huitric[23] carried out dipole moment measurements on 2-bromocyclohexanone and found that from calculation of the dipole moments based on the experimentally known moments of model compounds, the equilibrium in 2-bromocyclohexanone did have the axial form predominating slightly, from 50 to 70%, depending upon the solvent. But Corey's results were different from these by something like 2–3 kcal/mol.

It was long known from classical electrostatics that if one places a material between two condenser plates, the electrostatic energy between the plates varies inversely as the dielectric constant of the material. It was therefore suggested[24] that the equilibrium between the two conformational isomers of 2-bromocyclohexanone would vary strongly with the effective dielectric constant of the medium, that is, the nature of the solvent, and the concentration, used for the experiment, as in Eq. (7.1):

$$E = \frac{E_{\text{VAC}}}{D} \qquad (7.1)$$

This equation then could be used as an approximation to relate the conformational energies (E) in various solvents, where E_{VAC} is the electrostatic energy in vacuum, and D is the effective dielectric constant of the solution. These D numbers range from 2 up to 10 or so for most solvents. When D is 10 or higher, the electrostatic effect is essentially wiped out. The effective dielectric constant of a solvent is usually smaller than the bulk value, since the interior of the molecule is usually hydrocarbon with a D value of about 2. The electric field interacts partly through the molecule, and partly empty space (vacuum, $D = 1$) or solvent. This equation is, of course, an approximation, but a good enough approximation to be useful. In more polar systems, better approximations are usually desirable.[25] In the absence of other information, the value of D in solution in polar solvents has been widely taken to be 4.

A study of the infrared spectra in a variety of solvents showed[24] that this interpretation brought into reasonable agreement the work by Corey[22] and the work by Kumler and Huitric,[23] and suggested that it would be necessary to pay much more attention to the solvent when discussing equilibria between polar compounds than when one was studying steric effects where the solvation effect is typically small.

It was also shown that the pair of infrared frequencies attributed to the axial and equatorial conformations changed their relative intensities markedly with changing solvent, in accord with the axial–equatorial equilibrium changing with solvent.[24] In infrared spectroscopy it is common to use very concentrated solutions, and with 2-bromocyclohexanone, this meant effectively using a highly polar solvent because the compound itself was partly acting as its own solvent. Additionally, at these high concentrations, the compound tended to dimerize (as indicated by freezing point measurements), which further complicated the situation.

An alternative and more decisive method of establishing this equilibrium involved the synthesis of the two diastereomerically related compounds, trans- and cis-2-bromo-4-t-butylcyclohexanone, which exist in the axial and equatorial conformations, respectively.[26] Here we're not dealing with a mixture of rapidly interconverting conformations, but we have two separate isolable compounds available for study, one with a pure axial conformation for the bromine and the other pure equatorial. And they could be equilibrated in the presence of HBr catalyst. The presence of the t-butyl group does not significantly affect the dipole moments, or various other properties of interest, and the use of the t-butyl group as a "holding group" had previously been well established.[9b,27]

From the determination of the dipole moments of the 2-bromo-4-t-butylcyclohexanone isomers, and the dipole moments of 2-bromocyclohexanone itself in various solvents (at infinite dilution), it was possible to show that the equilibrium between the conformations in the latter was a sensitive function of the dielectric constant of the medium in which the measurements were made.[24,26,28] The more polar the solvent, the more equatorial conformation was present at equilibrium.

In 1955, Corey and Burke published an article discussing conformational studies with 2-chlorocyclohexanone, similar to those reported earlier for the corresponding bromoketone.[29] It had been anticipated that the amount of axial isomer would be greater in the chloro compound than in the bromo. The C–Cl and C–Br dipoles are similar, so the electrostatics in the two cases should be similar. But because the axial position is crowded in a cyclohexane ring, the smaller chlorine atom was expected to fit there more easily than the bromine. However, the reverse was found to be true. Subsequent experimental studies showed that this is part of a steady progression from fluorine[30] through bromine, where the percentage axial conformer in the equilibrium increases steadily and appreciably as we go down the periodic table.

Corey and Burke[29] wondered about the reason behind this stability order and proposed two possibilities. First, they mentioned that the equatorial ketone oxygen might exert a greater steric repulsion with the halogen than they had previously considered, and hence the larger halogen was pushed harder into the axial position by the van der Waals repulsions from the oxygen. They also suggested a second possibility, which might be called "Corey resonance." This is a hyperconjugative effect (see Chapter 6 under Hyperconjugation), where the halogen–carbon σ bond overlaps with the carbonyl π bond when the halogen is in the axial position. This overlap generates a hyperconjugative resonance form, of which both the valence bond and molecular orbital formulations are shown in Structure **10**:

Structure 10. Corey resonance.

This resonance would preferentially stabilize the axial conformation. The bromine is less electronegative than the chlorine, and its σ bond is weaker and so lies at a higher energy. Consequently, one might expect the importance of this resonance to be greater with bromine than with chlorine, and this was borne out by quantum mechanical calculations in due course.[4,31]

The axial halogen C–X σ orbital does have a sizable overlap with the carbon p orbital of the C=O π bond, whereas the equatorial C–X bond is almost perpendicular to that p orbital, and the overlap is small. The presence of the axial halogen thus leads to an interaction, and further splitting apart of the carbon halogen σ* orbital, and the π* orbital, with the latter moving to lower energy. The axial halogen causes the $n \rightarrow \pi^*$ transition to be shifted to the red by a considerable amount (10–30 cm^{-1}), calculated, and observed, compared to the parent cyclohexanone. The corresponding transition is negligibly shifted (a few cm^{-1}) in the case of an equatorial halogen.

The amount of energy corresponding to 10 cm^{-1} is quite small in more conventional units, approximately 0.03 kcal/mol. The question might be asked; Do we really care about this small amount of energy? The answer would be: Not usually but in some cases we might. For example, if we are concerned with an equilibrium constant, or a heat of formation, that amount of energy normally would be considered as negligible. On the other hand, this rather small amount of energy is a kind of a peephole, that lets us look into some details of chemistry that might otherwise be accessible only with great difficulty. These small energy differences allowed Corey and his contemporaries to identify different conformations in equilibrium, and to show how that equilibrium varied with temperature, solvent, molecular structure, and the like. Ordinarily, "chemical accuracy" means a few parts per thousand. And measuring something much more accurately than that will, for the most part, not lead to much that is useful. But there are exceptions. Perhaps a reminder from physics will be of interest to those not familiar with the following matter.

Around 1900, there was a general feeling among the physics community that the most important thing left to do in physics was to measure the next decimal place, for everything that had already been measured. There were many that thought there really was nothing new to learn—it's just that it could be learned a little better.

Of course, shortly after that topics like relativity and quantum chemistry became recognized and gradually understood. It is still necessary and sometimes important to measure the next decimal place, but twentieth century physics and nineteenth century physics are very different subjects, regardless of the decimal places. Or, putting it another way, "if you do not look, you may not find."

A competing proposal at the time regarding Corey's work was that the observed axial/equatorial splitting involved not the C–X σ^* orbital but rather the lowest empty s orbital (Rydberg orbital) on the halogen. This orbital could also interact with the π^* orbital and cause the latter to be lowered in energy. Quantum mechanical calculations were made at the time[31] and showed that sizable shifts of the $n \rightarrow \pi^*$ transitions were expected from each of the halogens in the axial position, but the calculated shifts from the Rydberg orbitals were negligibly small. The latter orbitals are just too high in energy to cause an observable shift of the $n \rightarrow \pi^*$ transitions.

So either, or some combination of both, the oxygen–halogen van der Waals repulsion and the Corey resonance appeared to be responsible for the observed progression in the axial–equatorial equilibria as the halogen was changed from bromine to chlorine to fluorine.

Since the time of Corey's work both experimental and computational methods have advanced very much. The use of modern NMR methods[32,33] makes it much easier to unravel conformational details experimentally than was possible in the 1950s. The advent of quantum mechanics in computational chemistry, and the availability of computers to solve real problems by that method, has made it easier for us to understand in considerably more detail just what is involved in problems of this nature. Further, by modeling this type of problem with molecular mechanics, we can also quantitatively understand the problem in the familiar terms of physical organic chemistry.

Let us first consider the application of quantum mechanics to the conformational problem of the 2-bromocyclohexanones and see what we can learn here. We utilized MP2B calculations,[34] which we believe give reasonably good results for the type of information we wish to obtain. (For definition of MP2/B, see under Jargon in the

Appendix.) We can readily obtain the energy difference between the conformations, and this shows us that the axial form is the more stable as long ago deduced by Corey. The difference in the quantum mechanical energies is 1.3 kcal/mol. This is, of course, the energy difference on the potential surface and refers to the isolated (gas-phase) molecule. So, we determine the geometries of the conformations in the course of the energy minimization. There is a great deal of interesting information to be found in these structures. We submit that understanding these data is easier and more straightforward if we convert the quantum mechanical information into a corresponding molecular mechanics model. The axial conformation is more stable than the equatorial. But why, exactly, is this true?

Molecular Mechanics Model

To develop the model, we used the MM4 program starting with cyclohexyl bromide and cyclohexanone, which were previously studied and well-understood systems. In addition to the parameter set from these compounds, we needed a little further information. Specifically, we examined the unit Br–C–C'=O. We needed for our model the torsional function that is required to describe that unit, and we also needed any torsion–bend and torsion–stretch interactions that correspond to that torsion. All of these were obtained by studying bromoacetone in the usual way.[34]

Having then a complete molecular mechanics model for the 2-bromocyclohexanone system, it was straightforward to determine the complete structures of the conformations and their energies. If we have a good molecular mechanical model, then we will reproduce the quantum mechanical structure with chemical accuracy. Some pertinent results are shown in Figure 7.2.

In the left column of Figure 7.2 are shown partial structures. The MM4 numbers are given, followed by the MP2/BC numbers in parentheses. Bond lengths are all given in terms of r_e. The standard bond length corrections were used to obtain the r_e values from the MP2/BC values (C–H, +0.0056 Å, C–Br, −0.0124, C–C, +0.002, and C=O, −0.0116 Å).

If we begin with cyclohexanone itself, the MM4 and MP2/B values for each bond agree to within about 0.003 Å. Adding the bromine causes shortening of the C–C bonds to which it is attached, and also of the α C–H bond on the carbon with the bromine. The C–H shortening is similar in the two isomers. The bromine also causes a shortening of the C=O bond, as a β-electronegativity effect. This quantity is also similar in both isomers. So for the equatorial bromine, for example, the C–O bond is shortened by 0.0021 Å, and the α-hydrogen bond is shortened by 0.0060 Å. The C–C' bond is shortened by electronegativity, and also by the Corey resonance (very small for the equatorial) between the bromine and oxygen, and by the hyperconjugation of the α-H with the oxygen. It is also lengthened by the steric effect between those two atoms. There is a net lengthening of 0.0060 Å.

When we look at the axial bromine compound, the Corey resonance causes a lengthening of the C–Br and C=O bonds, and a shortening of the C–C' bond. These effects can best be seen by comparing the axial values with those in the equatorial. The

α-HALO KETONE EFFECT

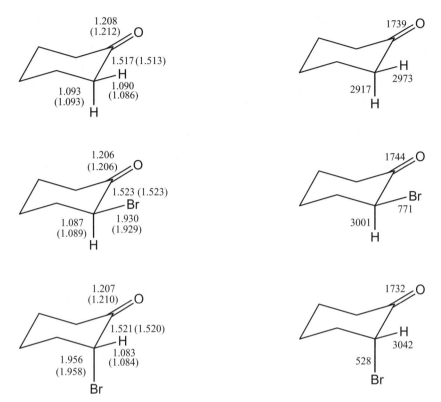

Figure 7.2. Bond lengths (Å) and vibrational frequencies (cm⁻¹) for cyclohexanone and α-bromo derivatives.

largest bond length difference between the two bromoketone diastereomers is the 0.026 Å longer C–Br bond for the axial from the Corey resonance.

The MM4 vibrational frequencies shown in the right column of Figure 7.2 are also informative. Other things being equal, the same type of bond with the shorter length has the higher vibrational frequency. Introducing the α bromine into cyclohexanone conformations causes the carbonyl bond lengths to be reduced by the electronegativity or inductive effect, and the axial one is additionally in part compensated by lengthening from the Corey resonance. Hence both the bond lengths and the carbonyl frequencies are reasonably in accord with what is expected. But what about the other vibrational frequencies? The hydrogens α to the bromine also have their bond lengths shortened by the inductive effect of the bromine, and hence bromocyclohexanone has the α C–H stretching frequencies very much raised. Cyclohexanone itself has these frequencies at 2917 and 2973 cm⁻¹ for axial and equatorial, respectively. The axial C–H is reduced by hyperconjugation, so its frequency is lower than that of the equatorial. The introduction of the bromine causes both of these C–H stretching frequencies to be markedly increased, and the equatorial one remains well above the axial (3042 versus 3001 cm^{-1}, as expected).

And what about the bromine itself? If we look first at cyclohexyl bromide, the equatorial C–Br stretching is at 702, while the axial is at 534 cm^{-1}. These numbers are less straightforward than they may appear. The bromine is so heavy that it really stretches very little relative to the molecule; rather the molecule stretches relative to the bromine. Hence, the coupling of the stretching vibration with other vibrations is large when compared to cases with a smaller halogen. But these numbers seem reasonable. When the carbonyl group is introduced into bromocyclohexane, the inductive effect strengthens all of the bonds in the neighborhood, including the C–Br bond. The axial and equatorial bonds in the bromoketone are thus expected to have their frequencies increased from this inductive effect. The equatorial one is increased markedly, from 702 in bromocyclohexane to 771 cm^{-1} in the bromoketone. But the axial one, again suffering from Corey resonance, also has its C–Br bond weakened. The corresponding stretching frequency barely changes, from 534 in the bromocyclohexane to 528 cm^{-1} in the bromocyclohexanone.

Thus, we see that the molecular mechanics model for α-bromocyclohexanone is reasonably accurate in all respects. It faithfully reproduces the geometries of the conformations, together with all of the expected trends in the related vibrational frequencies. This "explains" the structures within the framework of the resonance pictures of traditional organic chemistry. The other halo ketones have not been studied in such detail. However, superficial studies show the same general behavior as the bromo derivative.[34] Corey resonance is important for the halogens near the bottom of the periodic table, and much less so when one comes up to fluorine. All of the geometric, spectroscopic, and energetic changes that might have been qualitatively expected appear to have been found.

Molecular mechanics, in its most simple form, easily reproduces the energetics of the system (with the axial halogen being the more stable) by adjustment of the torsional potential. The bond length changes, however, do not come out of a simple molecular mechanics calculation (harmonic and diagonal) because hyperconjugation is not included at that level of calculation. These interactions can be represented with torsion–stretch interactions in the force constant matrix, however, so that the ab initio situation can be reproduced with molecular mechanics by including these terms. We have frequently made the point that there are a number of such "effects" in organic chemistry, the anomeric effect and the Bohlmann effect, being typical hyperconjugative interactions. In order for molecular mechanics to accurately reproduce molecular structures, these effects have to be explicitly taken into account. This "Corey resonance" is another such hyperconjugative interaction and is easily enough included in molecular mechanics a proper way once it is understood. It should be recalled, however, that when Corey proposed what we now call Corey resonance, hyperconjugation was still much under debate, and the general feeling by most organic chemists was that there appeared to be no such thing (see Hyperconjugation in Chapter 6). If there were no such thing with a proton, it did not seem that a bromonium ion would contribute very much to the situation. It has been only recently that it has been recognized how ubiquitous hyperconjugation is. It occurs everywhere, not only with

TABLE 7.7. MM4 Component Energies for Bromocyclohexanones (kcal/mol)[34]

	VDW	TORS	TOR/STR	TOR/BND	μ/μ	IND_μ	E_{TOT}
Ax-2-bromocyclohexanone	2.96	4.75	−0.38	−0.31	1.43	−1.30	8.13
Eq-2-bromocyclohexanone	3.50	4.38	−0.09	−0.16	2.70	−1.88	9.41
$\Delta\ (E_{eq} - E_{ax})$	+0.54	−0.37	+0.29	+0.15	+1.27	−0.58	+1.28

hydrogen, but also with alkyl groups, halogens, nitrogen, sulfur, phosphorus, oxygen, and so forth.

Energetics

Let us now consider axial going to equatorial α-bromocyclohexanone, as in Structure **8**. We know that the energy increases as the axial conformation goes over to the equatorial. We can divide this energy into several categories using the molecular mechanics headings. These are shown in Table 7.7. The total (MM4) energy difference is 1.28 kcal/mol (gas phase). Where, exactly, does this energy come from?

As Corey originally concluded, much of it comes from the electrostatics of the situation. According to MM4 the dipole–dipole repulsion difference is 1.27 kcal/mol for the isolated (unsolvated) molecules, but the induced dipoles lead to an energy difference in the opposite direction of −0.58 kcal/mol. Hence, of the total calculated 1.28 kcal/mol energy difference, only 0.70 comes directly from the electrostatics in this model.

There is some simple torsional energy (revealed by the bromoacetone work). The axial form is simply less stable torsionally than the equatorial form, by 0.37 kcal/mol. Then there is the Corey resonance energy. This is reflected in the torsion–stretch and torsion–bend interaction energies, which amount to 0.44 kcal/mol, favoring the axial halogen. It was early on recognized that an axial halogen is destabilized for van der Waals reasons, but in fact, the equatorial halogen is even more destabilized than the axial, for other van der Waals reasons. For the axial bromine, the oxygen–bromine distance is 3.61 Å, and the energy of that interaction is −0.08 kcal/mol. But for the equatorial, the distance is only 3.08 Å, and the repulsion energy is a very large 0.93 kcal/mol. Mainly because of this large repulsion, the total van der Waals energy of the molecule is 0.54 kcal/mol *less* in the axial conformation than in the equatorial. Adding all of these energies in Table 7.7, we find a total energy favoring the axial isomer by 1.28 kcal/mol.

The largest term in the energetics is indeed the dipole–dipole repulsion, but when the induced dipoles are allowed for, this electrostatic energy only amounts to 0.70 kcal/mol in the gas phase. And this number is much reduced in solution, in general agreement with experiments.[32] The Corey resonance on the other hand, while small in absolute terms (0.44 kcal/mol), is a substantial factor here. But a slightly larger factor is the van der Waals repulsion (0.54 kcal/mol), especially between the bromine and oxygen in the equatorial conformation. And a significant −0.37 kcal/mol comes simply from torsional energy.

Some general conclusions are possible here. All of the parameters required for the molecular mechanics treatment of 2-bromocyclohexanone and related compounds were derived from a study of small molecules, and transferability was assumed. Predictions from the model agree with experiment quite well, especially when one compares the shift of equilibrium constant with solvation in different solvents, and when one compares similar information for the different halogens fluorine through iodine.

Mathematicians are always happy when equations can be solved in "closed form." That is like when the diameter of a circle is compared with its radius, there is a factor of 2 difference. And it is exactly 2. Contrast this with the case when the diameter of the circle is compared with the circumference. The factor (π) comes out to be an "irrational" number. That is, the value of π cannot be written exactly with a finite number of digits.

In physical organic chemistry, one likes to say that this equilibrium is primarily a result of this or that effect. But here is a case where, according to the model, there are all different kinds of things happening, and we can calculate the sum of them rather well, to agree with both experiment and with quantum mechanical calculations. There are six different component energies that come into this equilibrium, some positive and some negative, which total 3.20 kcal/mol. The dipole–dipole repulsion is the largest single term, but it contributes just 40% of the total (gas phase) energy here, and much less in solution. So to state that the equilibrium results from an electrostatic interaction would be somewhat of an oversimplification in this case.

Summary

There is a point of general interest here. Corey's problem can be approached in three different ways. His approach was experimental and accompanied by such calculations as were feasible at the time. The experimental approach "solves the problem" in that it tells us what happens (axial conformation more stable than equatorial, changing the halogens down the series makes it more so, and the changes with solvation are clear). But it doesn't really tell us exactly why these facts are found. Quantum mechanics and molecular mechanics can be employed to help answer that latter question.

Quantum mechanics will give us the structural and energy differences between the conformations. The structures tell us something about what is happening (bond orders and electron densities), but some of this information can be difficult to interpret. Molecular mechanics, creating a general force field for the type of compound in question, tells us in the straightforward language of physical-organic chemistry why all of this happens, and it tells us quantitatively the contributions of each of the different effects on the overall energetics.

REFERENCES

1. (a) F. Bohlmann, *Angew. Chem.*, **69**, 641 (1957). (b) F. Bohlmann, *Chem. Ber.*, **91**, 2157 (1958).

2. (a) D. C. McKean, J. L. Duncan, and L. Batt, *Spectrochim. Acta*, **29A**, 1037 (1973).
   (b) D. C. McKean, *Spectrochim. Acta*, **31A**, 861 (1975). (c) D. C. McKean, J. E. Boggs, and L. Schäfer, *J. Mol. Struct.*, **116**, 313 (1984).
3. H. D. Thomas, K. Chen, and N. L. Allinger, *J. Am. Chem. Soc.*, **116**, 5887 (1994).
4. T. Laube and T. K. Ha, *J. Am. Chem. Soc.*, **110**, 5511 (1988).
5. J.-H. Lii, K.-H. Chen, and N. L. Allinger, *J. Phys. Chem.*, **108**, 3006 (2004).
6. K.-H. Chen, J.-H. Lii, Y. Fan, and N. L. Allinger, *J. Comput. Chem.*, **28**, 2391 (2007).
7. (a) P. Pulay, G. Fogarasi, F. Pang, and J. E. Boggs, *J. Am. Chem. Soc.*, **101**, 2550 (1979).
   (b) P. Pulay, G. Fogarasi, G. Pongor, J. E. Boggs, and A. Vargha, *J. Am. Chem. Soc.*, **105**, 7037 (1983). (c) J. Baker and P. Pulay, *J. Comput. Chem.*, **19**, 1187 (1998). The specific theory and basis set, B3LYP/6-31G**, was used for the frequency calculations because the scaling factors published by Pulay were obtained with this level of theory. For further information see J.-H. Lii, *J. Phys. Chem.*, **106**, 8667 (2002).
8. (a) N. L. Allinger and Y. Fan, *J. Comput. Chem.*, 18, 1827 (1997). (b) P. Fox, J. P. Bowen, and N. L. Allinger, *J. Am. Chem. Soc.*, **114**, 8536 (1992).
9. (a) D. H. R. Barton, *Experientia*, **6**, 316 (1950). (b) E. L. Eliel, N. L. Allinger, S. J. Angyal, and G. A. Morrison, *Conformational Analysis*, Wiley, New York, 1965.
10. (a) E. Juaristi and G. Cuevas, *Tetrahedron*, **48**, 5019 (1992). (b) Exo anomeric effect, p. 5061 in Ref. 10(a). (c) C. Ceccarelli, J. R. Ruble, and G. A. Jeffery, *Acta Crystallogr.*, **B36**, 861 (1980). (d) G. A. Jeffrey and R. Taylor, *J. Comput. Chem.*, **1**, 99 (1980). (e) R. J. Woods, *The Application of Molecular Modeling Techniques to the Determination of Oligosaccharide Solution Conformations, Reviews in Computational Chemistry*, Vol. 9, K. B. Lipkowitz and D. B. Boyd, Eds., Wiley, New York, 1996, p. 129.
11. J. R. Kneisler and N. L. Allinger, *J. Comput. Chem.*, **17**, 757 (1996).
12. G. A. Jeffrey, J. A. Pople, J. S. Binkley, and S. Vishveshwara, *J. Am. Chem. Soc.*, **100**, 373 (1978). Actually, these authors missed the stable gg' structure because they searched the potential surface at only 60° intervals. At the +60 −60 point, the molecule goes through an energy maximum, and the gg' conformation that is an energy minimum is noticeably skewed away from this exact geometry (Table 7.4). This structure is evident in Figure 7.1 with a torsion angle of about 260°.
13. J.-H. Lii, K.-H. Chen, K. A. Durkin, and N. L. Allinger, *J. Comput. Chem.*, **24**, 1473 (2003).
14. (a) E. E. Astrup, *Acta Chem. Scand.*, **27**, 3271 (1973). (b) E. E. Astrup, *Acta Chem. Scand.*, **25**, 1494 (1971).
15. N. L. Allinger, K.-H. Chen, J.-H. Lii, and K. Durkin, *J. Comput. Chem.*, **24**, 1447 (2003).
16. (a) B. Ma, J.-H. Lii, H. F. Schaefer III, and N. L. Allinger, *J. Phys. Chem.*, **100**, 8763 (1996).
    (b) B. Ma, J.-H. Lii, K. Chen, and N. L. Allinger, *J. Am. Chem. Soc.*, **119**, 2570 (1997).
    (c) B. Ma and N. L. Allinger, *J. Mol. Struct.*, **413–414**, 395 (1997).
17. L. Nørskov-Lauritsen and N. L. Allinger, *J. Comput. Chem.*, **5**, 326 (1984).
18. N. L. Allinger, M. Rahman, and J.-H. Lii, *J. Am. Chem. Soc.*, **112**, 8293 (1990).
19. (a) R. U. Lemieux, A. A. Pavia, J. C. Martin, and K. A. Watanabe, *Can. J. Chem.*, **47**, 4427 (1969). (b) J.-P. Praly and R. U. Lemieux, *Can. J. Chem.*, **65**, 213 (1987).
20. (a) A. J. Kirby, *The Anomeric Effect and Related Stereoelectronic Effects at Oxygen*, Springer, Heidelberg, 1983. (b) W. A. Szarek and D. Horton, *Anomeric Effects. Origin and Consequences*, ACS Symp. Ser. 87, Washington, D.C., 1979. (c) P. Des Longchamps, *Stereoelectronic Effects in Organic Chemistry*, Pergamon, New York, 1983.
21. N. L. Allinger, K. Chen, and J.-H. Lii, *J. Comput. Chem.*, **17**, 642 (1996).

22. E. J. Corey, *J. Am. Chem. Soc.*, **75**, 2301 (1953).
23. W. D. Kumler and A. C. Huitric, *J. Am. Chem. Soc.*, **78**, 3369 (1956).
24. J. Allinger and N. L. Allinger, *Tetrahedron*, **2**, 64 (1958).
25. (a) C. J. Cramer and D. G. Truhlar, *Continuum Solvation Models: Classical and Quantum Mechanical Implementations, Reviews in Computational Chemistry*, Vol. 6, K. B. Lipkowitz and D. B. Boyd, Eds., Wiley, New York, 1995, p. 1. (b) L. Dosen-Micovic, D. Jeremic, and N. L. Allinger, *J. Am. Chem. Soc.*, **105**, 1716 (1983). (c) A. V. Marenich, C. J. Cramer, and D. G. Truhlar, *J. Chem. Theory Comput.*, **5**, 2447 (2009).
26. N. L. Allinger and J. Allinger, *J. Am. Chem. Soc.*, **80**, 5476 (1958).
27. S. Winstein and N. J. Holness, *J. Am. Chem. Soc.*, **77**, 5562 (1955).
28. N. L. Allinger, J. Allinger, and N. A. LeBel, *J. Am. Chem. Soc.*, **82**, 2926 (1960).
29. E. J. Corey and H. J. Burke, *J. Am. Chem. Soc.*, **77**, 5418 (1955).
30. N. L. Allinger and H. M. Blatter, *J. Org. Chem.*, **27**, 1523 (1962).
31. N. L. Allinger, J. C. Tai, and M. A. Miller, *J. Am. Chem. Soc.*, **88**, 4495 (1966).
32. F. Yoshinaga, C. F. Tormena, M. P. Freitas, R. Rittner, and R. J. Abraham, *J. Chem. Soc., Perkin Trans. 2*, 1494 (2002).
33. Discussed in books on nuclear magnetic resonance spectroscopy, for example, J. B. Lambert and E. P. Mazzola, *Nuclear Magnetic Resonance Spectroscopy. An Introduction to Principles, Applications, and Experimental Methods*, Pearson Education, Upper Saddle River, NJ, 2004.
34. J.-H. Lii, unpublished.

8

ADDITIONAL STEREOCHEMICAL EFFECTS IN CARBOHYDRATES

Carbohydrates are biologically an important group of compounds. As it turns out, collectively they illustrate a number of different stereoelectronic effects, and hence this makes a convenient place to further study these effects.

The effects that we are going to discuss here apply in a general way to molecules and are represented by continuous mathematical functions. But here we will limit our discussion to the stable conformations of a relatively limited group of compounds, the hexopyranoses (six-carbon sugars consisting of a ring of five carbons and one oxygen, plus an exocyclic carbon).

GLUCOSE

Almost all chemists have studied at some point the problems associated with the structure and stereochemistry of glucose, but most chemists probably do not have this information at their fingertips, so we will take a page or two and review some pertinent facts here.

Glucose (including its polymers starch and cellulose) is certainly the most widespread organic compound found in living systems. It serves as the principle energy

---

Molecular Structure: Understanding Steric and Electronic Effects from Molecular Mechanics,
By Norman L. Allinger
Copyright © 2010 John Wiley & Sons, Inc.

source for biological activity in most plants and animals and, hence, is an exceedingly important organic compound.

Formally, glucose can be thought of as a hexanal, a 6-carbon aldehyde, that has five hydroxyl groups, one attached to each of the carbons other than the carbonyl. It is often written in the form of a *Fischer projection* (Structure **1**, center structure) to define its stereochemistry.

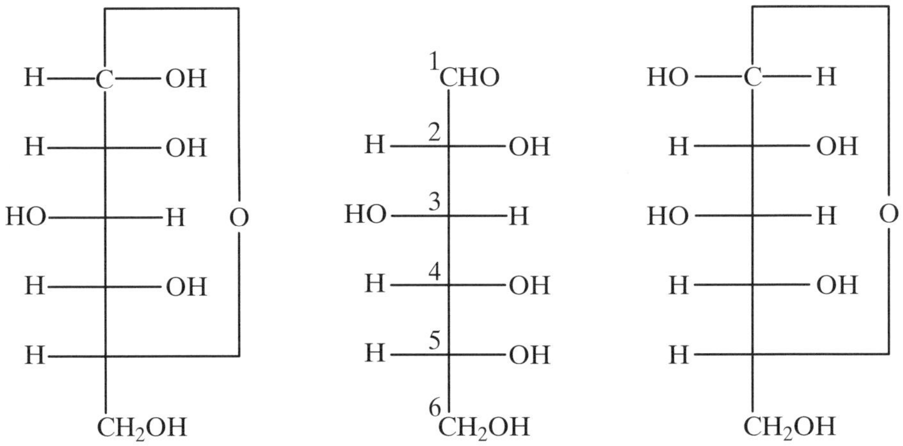

D-Glucose
(α-hemiacetal form)

D-Glucose
(aldehyde form)

D-Glucose
(β-hemiacetal form)

Fischer Projections

1

But actually, in water solution only a trace amount of the free aldehyde is an equilibrium with the cyclic hemiacetals that can be formed by the reaction of one hydroxyl group with the aldehyde function (right and left structures in Structure **1**). It is perhaps worth a moment to briefly discuss this equilibrium. The amount of free aldehyde is only about 0.02%. What is it that makes the cyclic forms so much more stable than the open chain aldehyde? The reaction is as shown in Reaction (1):

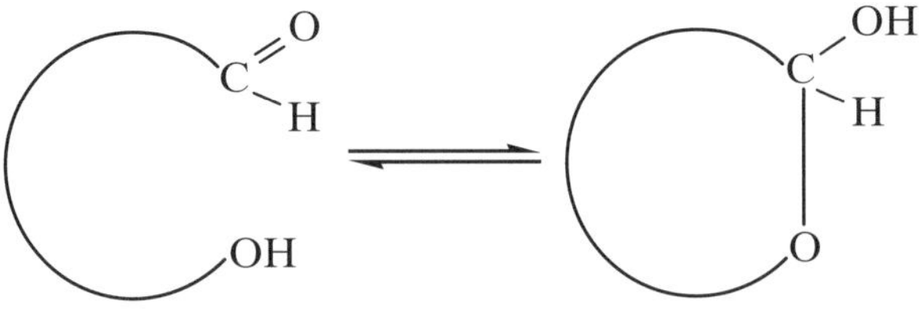

Cyclic hemiacetal formation (1)

In the ring closure, a C=O bond is converted into two C–O bonds. This has an energy change of approximately −5 kcal/mol (favorable). But, if we think about it a little further, in water solution, where the glucose normally exists, it would also be possible to hydrate the aldehyde with a water molecule [Reaction (2)]:

$$-\overset{O}{\underset{H}{C}} + H_2O \rightleftharpoons -\overset{OH}{\underset{H}{C}-OH} \qquad (2)$$

Open-chain hemiacetal formation

This would give us approximately the same change in bond energy. So why does the aldehyde not just hydrate, why does it cyclize? (And cyclize it does, very completely.) The reason is this. The water will tend to hydrate an aldehyde because there is a considerable driving force in terms of an energy (enthalpy) decrease. But, when the hydration is carried out with water, we also see that two molecules (one being the water) combine to give one molecule [Reaction (2)]. This has a very unfavorable entropy effect. The translational entropy of the water that is lost is about 20 entropy units (kcal/molT), so that leads to an increase in free energy of the hydrate of about 6 kcal/mol at room temperature. These entropy and enthalpy changes for the hydration approximately cancel out. Hence, hydration of an aldehyde (or of a sugar) by water is a thermodynamically marginal process; some do and some don't.

When glucose cyclizes (Reaction 1), one molecule goes to one molecule, so there is no loss of translational entropy. There is an entropy change due to the change in the shape of the molecule, but it is quite small. There is also an unfavorable change in entropy from the cyclization because the six-membered ring is pretty much confined to a single chair structure, although the hydroxyls are free to rotate. But in the open-chain form there are numerous gauche and anti forms for the carbon–carbon bonds and a substantial entropy of conformational mixing. This entropy of mixing is lost upon cyclization. This conformational entropy is, however, much smaller than the translational entropy of the water molecule. Accordingly, glucose exists almost completely in the cyclic form in aqueous solution, mainly because of these entropy effects.

The opening and closing of the ring to give the hemiacetal occurs slowly in neutral aqueous solutions and leads to an equilibrium mixture if the solution is allowed to stand. This reaction generates a new asymmetric center at what was the aldehyde carbon, so that the resulting hydroxyl can be in either of two forms, called α and β.* These two forms are shown in Structure **1**. These are two different isolable and stable isomeric compounds (diastereomers) with different melting points, optical rotations, and other physical properties.[1] The β form has a melting point of 150°C and an optical rotation at the sodium D line of +18.7°. The α form melts at 146°C and has an optical rotation

---

*The terms α and β are commonly used to differentiate diastereomers, particularly in carbohydrates and steroids. They mean, respectively, *down* and *up* with respect to the mean plane of the ring system (written in the conventional way). Thus, in Structure **2**, the orientation of the hydroxyl group at C-1 in glucose is indicated.

of +112.0°. In water the equilibrium mixture of the two has an optical rotation of +52.5° and corresponds to 64% β and 36% α. This equilibration process is called *mutarotation*, and it is easily followed by measuring the optical rotation (rotation of the plane of polarization of polarized light) of the solution. Whether the hydroxyl is written to the right or the left has stereochemical implications with Fisher projections. These projections were introduced by Emil Fischer about a hundred years ago to facilitate the discussion of stereochemistry. Now that we understand better the three-dimensional structures of molecules, it is more convenient to use ordinary stereochemical structures, as shown in Structure **2**, for this representation. The nature of the α- and β-hydroxyl groups at C-1 is evident in these pictures. The two diastereomers shown in Structure **2** have a special kind of relationship. They are called *anomers*. They differ in configuration at and only at C-1. They can be interconverted by simply standing in water. And this is possible because of the special properties of C-1, the aldehyde carbon. This carbon is called an *anomeric carbon*, and the anomeric effect has to do with special properties of these structures.

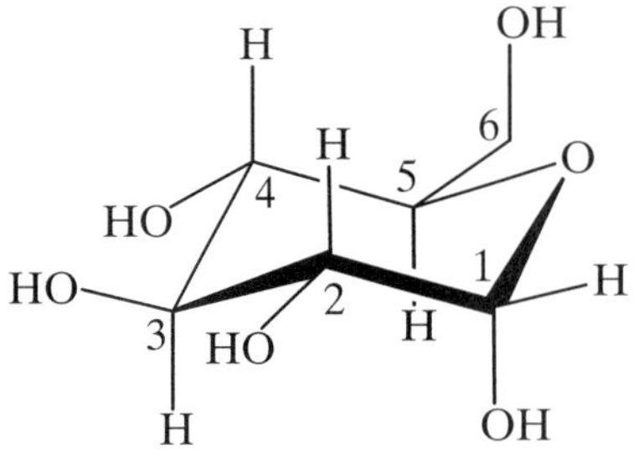

α-D-glucose 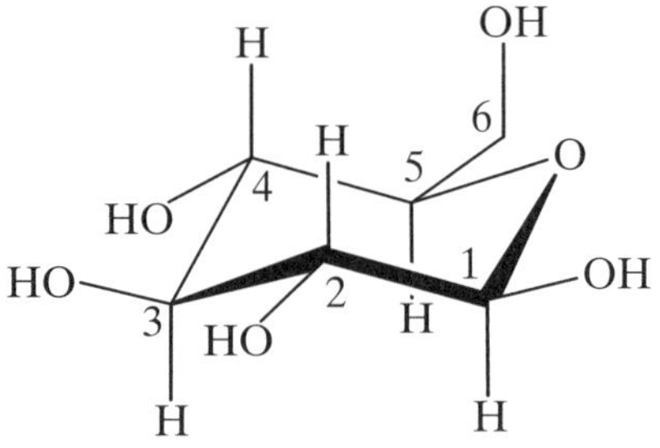   β-D-glucose

Conformational Formulation

2

The expression *anomeric effect* is used to describe the behavior pattern first reported by Edward in 1955,[2-4] which was observed with ordinary pyrannose forms of sugars (six-membered rings) and with their 1-methoxy analogs. It was well understood at that time that cyclohexanol and its methyl ether (and most other cyclohexane derivatives[5]) are found to preferentially have the substituent in the equatorial position. The corresponding axial form of cyclohexanol has a conformational energy of about 0.7 kcal/mol.[5] On the contrary, in the pyranose sugars the hydroxyl (or better its methoxyl equivalent, so as to reduce the possibility of complications from hydrogen bonding) at the 1-position is usually found to be more strongly in the *axial* position than that, and it has a lower than expected energy by about 0.5–1.0 kcal/mol.[1,6] Thus, the anomeric effect was the expression used to describe this energy difference. This energy is quite solvent dependent (whereas the cyclohexanol number is only slightly solvent dependent), with the equatorial form typically becoming increasingly stable in increasingly polar solvents. The large internal electric bond dipole moments of the 2-methoxytetrahydropyran type of structure would lead one to expect that qualitatively this would happen. The equatorial conformation of the 2-methoxytetrahydropyran has a substantially larger dipole moment than does the axial (by about 2D*), and the equilibrium is shifted by solvation in polar solvents toward the more polar form (here the equatorial) because of dipole solvation. The basic causes of the anomeric effect are two: negative hyperconjugation or Bohlmann effect and electrostatics. Since the latter is much reduced in water solution, the fact that the anomeric effect was discovered in carbohydrates attests to the strength of the hyperconjugation (Chapter 7).

GAUCHE EFFECT

Our MM4 molecular mechanical model for alcohols and ethers is a straightforward extension of the alkane force field described earlier (Chapter 4), and it already includes three important relevant structural effects: the electronegativity effect, hyperconjugation (Chapter 6), and the Bohlmann effect (Chapter 7). In carbohydrates, a particular example of the latter (the anomeric effect) is of special importance. In this case we have two or more hydroxyls (or equivalent) attached in a geminal (1,1) manner. Of further interest is what happens when there are two or more hydroxyls (or equivalent) arranged in a vicinal (1,2) manner because this arrangement is ubiquitous in carbohydrates, and it introduces still another stereoelectronic effect. The present section discusses this *gauche effect*, which refers to the unusual stability of a gauche conformation when there are two electronegative elements on adjacent carbon atoms.[7,8]

If we consider ethylene glycol, the central bond is a little more stable (in terms of energy rather than free energy) in the trans conformation than in the gauche. But before looking at this particular compound, it will be convenient to look at a simpler case, the corresponding difluoride, namely 1,2-difluoroethane. The effect that we wish to now discuss results from the electronegativity of the substituents, and it is larger and can be seen more clearly and unambiguously in this example. Specifically for the difluoro system, the gauche effect in 1,2-difluoroethane stabilizes the gauche conformation relative to the anti by about 1.5–2.0 kcal/mol. This is a sizable number, and we would like to know what causes it because we certainly need to include it in molecular mechanics.

According to MM4 the dipole–dipole interaction favors the anti form of 1,2-difluoroethane by 1.02 kcal/mol, and the induced dipoles disfavor it by 0.25 kcal/mol. The difference in the van der Waals energies favors the anti by 0.30, for a total energy difference here of 1.07 kcal/mol favoring the anti.[9] Experimentally, the gauche form is more stable by 0.6–0.9 kcal/mol,[7] leaving about 1.7–2.0 kcal/mol to be attributed to the gauche effect. The MM4 value for the (anti–gauche) energy is 0.66 kcal/mol. Thus, the MM4 energy of the gauche effect in this molecule is 1.73 kcal/mol. It is included in the F–C–C–F and H–C–C–F torsional potentials as will be described.

---

*The exact value varies, of course, according to the Boltzmann distribution between the conformations, which in turn is dependent upon solvent, dielectric constant, and temperature. This could all be taken into account if desired, but the overall changes would be small.

Interestingly, early Hartree–Fock calculations on 1,2-difluoroethane with small basis sets indicated (incorrectly, in disagreement with experiment) that the anti form was the more stable. As time went on, it was found that these calculational results improved with increasing basis set size. The calculated stability of the gauche form increased, and with very large basis sets, this form finally became the more stable one.[10] With small basis sets, MP2 correlation also significantly stabilized the gauche form. The importance of the correlation decreased somewhat as the basis set size increased, but the trends clearly show that quantum mechanics does predict a stable gauche form, but only when a reasonably large calculation is carried out. Wiberg has given an interpretation of this phenomenon,[10] based upon Bader's work[11] involving bond bending. In this interpretation, the electronegativity of the fluorine causes more p character to go into the CF bond, which in turn leaves more s character in the CH bonds, and causes the carbon–hydrogen part of the system to flatten. The C–C bond is bent in the process, and this destabilizes the trans conformation relative to the gauche. Alternatively, the effect has been interpreted in terms of hyperconjugation.[12]

Our model of the gauche effect[9,12] is straightforward and can be understood by examining Structure **3**.

$$
\begin{array}{cc}
\text{H} & \text{H} \\
| & | \\
\text{H}-\text{C}-\text{C}-\text{H} \\
| & | \\
\text{F} & \text{F}
\end{array}
\quad \longleftrightarrow \quad
\begin{array}{cc}
\text{H}^+ & \text{H} \\
& | \\
\text{H}-\text{C}=\text{C}-\text{H} \\
| & \\
\text{F} & \text{F}^-
\end{array}
$$

3

There is a hyperconjugative resonance between one hydrogen on the left carbon and the fluorine on the right carbon, as shown by the structure on the right. There is a similar resonance between the other hydrogen on the left carbon and the fluorine on the right carbon, not shown. And there is a similar set of resonances going in the opposite direction, involving the hydrogens on the right-hand carbon and the fluorine on the left. For this kind of resonance to be most effective, the hydrogen involved, the two carbons and the fluorine involved, must all lie in the same plane. The resonance will have onefold and twofold components because it is most effective when the hydrogen and the fluorine are in the same plane (so that their bonding orbitals overlap) and trans to one another. It is also effective, but less so, when they are cis to one another, and it goes to zero when they are at 90° to one another. This resonance does not specifically require a hydrogen, as an alkyl group will function in a qualitatively similar manner, as discussed in Chapter 6. With the fluorines the effect is pretty large, and the gauche conformation is more stable than the anti by 0.66 kcal/mol (MM4). The effect is a result of the electronegativity of the substituents; so it is similar but weaker when two oxygens replace the two fluorines. Thus, with MM4 we find[13] that if the two substituents are methoxyls the effect is much smaller. The electrostatics again favors the anti, similar to the difluoro compound, but here by 0.97 kcal/mol (MM4). The steric effect is negligible and the gauche effect reduces the energy to 0.20 kcal/mol favoring the anti. In carbohydrates, this gauche effect between hydroxyls is small but ubiquitous.

Polyoxyethylene (POE)

Polyethylene can be viewed as made up of a series of butane segments. In butane, the stable conformation is anti, and hence polyethylene is expected to have a stable all-anti conformation, and it does. What would be expected for polyoxyethylene $(OCH_2CH_2)_n$ is less clear, a priori. In methyl ethyl ether, the rotational profile about the ethyl C–O bond favors the anti conformation by a significant amount, about 1.4 kcal/mol according to MP2/B calculations, or by about 1.2 kcal/mol according to MM4.[13] (For definition of MP2/B, see under Jargon in the Appendix.) However, the gauche and anti conformations about the central C–C bond in 1,2-dimethoxyethane have essentially equivalent energies. (The anti is lower in enthalpy by 0.16 kcal/mol according to MM4 and 0.31 by MP2/B.) It was, therefore, of interest to see what MM4 would predict concerning the conformations of POE. This compound differs from the smaller model compounds in that longer range van der Waals and electrostatic forces are present, and they might easily change the overall conformational composition of the molecule somewhat from what might otherwise have been expected. Earlier experimental studies have indicated that the conformation about the C–C bond is mainly gauche in solution.[14]

For the MM4 calculations[15] our model was $CH_3–(OCH_2CH_2)_{10}–OCH_3$ (hereafter POE model). The forms studied are shown in Figure 8.1, and included the all-anti (AA) and the all-gauche (GG) conformations, and also the conformation that is anti at the C–O bonds and gauche at the C–C bonds (AG). While the AA molecule is linear, the latter two conformations both form helices in which the GG is more tightly wound.

From the energies of small model compounds, which do not include all of the longer-range interactions, one would predict that the all-anti conformation of the POE model would have the lowest energy, but only slightly so. The gauche C–C, anti C–O conformation should be a little less stable, while the conformations involving gauche C–O arrangements should be much higher in energy. (Entropy effects are ignored here.) Indeed, the all-anti conformation of the model is calculated by MM4 to be more stable than the all-gauche conformation by 41.65 kcal/mol (assuming the gas phase, which has an effective dielectric constant of 1.5 in MM4). On the other hand, the gauche C–C, anti C–O conformation has a calculated energy only 3.45 kcal/mol above that of the all-anti conformation. These energies are quite large numbers because we are looking at quite large molecules. When one looks at the component parts of the energies with MM4, it is seen that the electrostatic energy is very large, and, of course, these electrostatic quantities are dielectric constant dependent. We do not know the effective dielectric constant for polymer solutions (which would depend on the solvent and concentration), but we do know that the solvent will raise the dielectric constant relative to the gas phase. With a reasonable estimate of a dielectric constant of 5.0 (Table 8.1) the order of stabilities is changed, and the anti-gauche conformation becomes more stable than the anti-anti form by 4.16 kcal/mol. The gauche–gauche conformation remains high in energy, however, because the energy in this conformation is largely steric in nature.

So the prediction is that in solution at fairly low to high dielectric constant, the anti-gauche conformation is likely to be the stable one. But, when we consider the conformational equilibrium in this molecule, we expect to find that since gauche

196 ADDITIONAL STEREOCHEMICAL EFFECTS IN CARBOHYDRATES

POE Model, AA

(Anti c–o–c–c, Anti o–c–c–o)

POE Model, GG

(Gauche c–o–c–c, Gauche o–c–c–o)

POE Model, AG

(Anti c–o–c–c, Gauche o–c–c–o)

Figure 8.1. Polyoxyethylene model conformations.

TABLE 8.1. MM4 Conformational Energies (kcal/mol) of Polyoxyethylene Model (POE)[15]

Conformations	Steric Energy (Gas Phase)	Steric Energy (Solution, $D = 5.0$)
POE-AA	9.65	8.48
POE-AG	13.10	4.32
POE-GG	51.30	42.35

conformations have an $R\ln 2$ favorability compared to each anti due to the entropy, there are many possible conformations with 1 anti, 2 anti, and so forth. There will in fact be a complicated mixture (Boltzmann distribution). Molecular dynamics calculations[16] were carried out[15] on this molecule (gas phase). As might be expected for a molecule in which there were near equivalences between gauche and trans energies, the molecule tended to wad up to form a ball, rather than to be stretched out, both from entropy effects, and also because of the greater van der Waals attractions in the compact type of structure. This is as would be expected for the isolated molecule (gas phase). In solution, the van der Waals interactions between the molecule and the solvent would tend to suppress this tendency. It is also known that, at least with alkanes, gauche conformations tend to yield molecular structures that have smaller molar volumes and greater densities than the anti, and hence larger *intermolecular* van der Waals attractions.[17] Thus, the equilibrium concentration of a simple alkane shifts from a preponderance of anti conformations in the direction of increasing gauche conformations when the gas is condensed into a liquid. Presumably that would happen here also if the solution were in an indifferent solvent, which would tend to further stabilize the gauche conformations relative to the anti.

Our conclusion is that the AA conformation of POE itself, as with the POE model, is slightly more stable than the other conformations in the gas phase, in terms of energy. The AG conformation, however, becomes more stable in solutions, particularly at higher dielectric constant. Since these conformations are relatively close in energy, the entropy effect will be overwhelming so that POE itself will be a serious conformational mixture.

DELTA-TWO EFFECT

In a study published in 1950 by Reeves,[18] it was noted that when he examined a lengthy series of hexose structures he expected to have all sorts of combinations of axial and equatorial hydroxyl groups with the pyranose rings in chair forms. Mostly from previous work, he was able to eliminate boat forms as possible structures for these molecules and to establish whether a ring had a $^4C_1$ or a $^1C_4$ conformation in each case.* The α or β preference for the anomeric hydroxyl could be accurately determined by measuring the optical rotations of both the pure epimeric compounds and of their equilibrium mixture, obtained by allowing them to mutarotate (equilibrate). But one conspicuous peculiarity was found. When a lengthy series of compounds was examined (later), it was found that the relative stabilities of the α and β anomers appear to be more or less random if there was an equatorial hydroxyl at C-2, but the α anomer strongly predominated when there was an axial hydroxyl at C-2. For example, α-D-glucose could be equilibrated to a mixture of α and β epimers containing 36% of the α [Reaction (3)]:

---

*See Appendix under Carbohydrate Conformational Nomenclature for conformational definitions. The $^4C_1$ conformation is the usual one as is found for both diastereomers of D-glucose (shown) where the terminal –CH$_2$OH group is equatorial.

[Structures of α-D-Glucose ⇌ β-D-Glucose, $\Delta G°_{298} = -0.35$ kcal/mol]

(3)

But for α-D-mannose, with an axial hydroxyl at C-2, the equilibrium yielded 67% of the α epimer [Reaction (4)]:

[Structures of α-D-Mannose ⇌ β-D-Mannose, $\Delta G°_{298} = +0.70$ kcal/mol]

(4)

It seemed that there was something about the particular kind of conformational structure of β-mannose that destabilized it by about 1.0 kcal/mol. In other words, there was another effect, which Reeves called the *delta-two effect*, which was about the same size but in the opposite direction (in this case) from the normal anomeric effect. The conformations about the C-1–C-2 bond for mannose (or other compounds with a β-hydroxyl at C-2) can be written in Newman projections as in Structure **4**, looking from C-2 toward C-1:

[Newman projections: β Conformer and α Conformer]

β-D-Mannose

4

Note that in the β conformer in Structure **4**, the oxygen on C_2 is *between* two other oxygens on C_1. This particular conformation is destabilized by the delta-two effect.

We have previously discussed the anomeric effect and the gauche effect, where we have geminal (1,1) or vicinal (1,2) hydroxyl groups. Note that in the delta-two effect we have both of these arrangements at the same time (1,1,2). The delta-two effect is, if you will, a measure of the lack of additivity of anomeric and gauche effects.

The details of the delta-two effect were worked out for MM4 in a straightforward way. The model compound used was 1,1,2-trimethoxyethane. (The use of the trihydroxy derivative may have led to complications from hydrogen bonding.) The torsional profile about the 1,2 bond of this molecule was calculated at the BC level, and it was also calculated by MM4O, as has been described to this point (before inclusion of the delta-two effect). (For the details of MM4O, see under Jargon in the Appendix.) The torsional potential from the MM4O calculation did not match that from the BC calculation. Placing the oxygen on C-2 between the two oxygens on C-1 yielded an MM4 energy that was 1.36 kcal/mol too low at that torsion angle. Thus, the MM4 torsional potential for the 1,1,2-trihydroxyethane system was brought into agreement with the BC calculation by adding an appropriate, mainly V_2, torsion–torsion interaction.[19]

What is the cause of the delta-two effect? This question has been asked several times since Reeves initial study on the subject, but no very definitive answer has been forthcoming. It seems like "some kind of an electrostatic effect" is the consensus. But MM4 takes into account very well, as far as we are aware, the electrostatics of a situation like this, including both the dipole–dipole and the induced-dipole interactions. We find that the effect as shown by MM4 [which is the effect that comes from (gas-phase) quantum mechanics at the MP2/B level] is almost identical in magnitude to that found in sugars in water solution. This would indicate that it cannot have a large electrostatic component, or these numbers would be quite different. MM4 says that in a mechanical sense it is an interaction between two torsional degrees of freedom. Unfortunately, that does not convey much information in a chemical sense. Just what causes the delta-two effect is going to need, at a minimum, further thought.*

In order to get MM4 to reproduce the structures and energies of the aldohexapyranoses in an adequate way then, it was found that in addition to the usual force field required for alkanes and alcohols, one had to explicitly include three additional different effects: the anomeric effect, the gauche effect, and the delta-two effect. While all of these effects are known experimentally to occur in sugars, they were measured and studied only in aqueous solutions. The sugars are generally insoluble in solvents other than water, and they tend to decompose on heating. This leaves solvation as a problem still to be addressed.

---

*Interactions between close electronegative atoms in molecules have not been very well studied. There are, of course, both strong van der Waals and electrostatic interactions between them. These interactions may require potentials that are more accurate at short distances than those used in MM4. The r^{-8} or higher term in the dispersion interaction and/or the quadrupole interactions in the electrostatics may be significant here. Whatever this interaction may turn out to be, it seems to be accounted for fairly well by a torsion–torsion interaction in MM4.

GLUCOSE DIASTEREOMERS

Having developed from small molecules a suitable model for each of the effects previously described, we can now apply this model to glucose and related hexose molecules. Note that no parameterization of MM4 utilized any data from the carbohydrates themselves, only from the small model compounds as described earlier. The fit obtained in Table 8.2 of MM4 to the MP2/B data is thus a stringent test of the transferability of parameters in molecular mechanics. In Table 8.2 are listed the eight diastereomers of glucose, together with the energy differences of the β–α anomers ($E_\beta - E_\alpha$) from B3LYP/B calculations and similarly from the MP2/B calculations. The β isomer is usually higher in energy (gas phase). Also shown are the MM4 calculated values.

Let us first compare the two quantum mechanical calculations, MP2B and B3LYP. It is certain that the MP2 calculations, in general, give more accurate geometries than the B3LYP, and significantly so. It is less clear regarding the energies, but the B3LYP calculations do not include dispersion energy, and so if one has a set of compounds in which the dispersion energies vary, then the B3LYP results tend to give poor energies because of this incompleteness. The error from the incompleteness is always in the same direction, namely the MP2 calculations include the dispersion energy and stabilize the systems where this is important but have a less effect where it is unimportant. (There is evidence that at least in the case of hydrocarbons, MP2 tends to overestimate the effect of correlation, as discussed in Chapter 11 under Heats of Formation from Quantum Mechanics.) Other errors are expected to occur also, of course, and in either direction. The compounds in Table 8.2 differ only in the α or β orientation of the anomeric hydroxyl group. The α group is closer to the rest of the molecule than is the β, and hence the α energies on average are expected to be lowered relative to the β ones by the dispersion energies from MP2, relative to those from B3LYP. This means

TABLE 8.2. Anomeric Energy Differences ($E_\beta - E_\alpha$) for Glucose Isomers[a]

	MM4[19]	MP2B[20]	(MM4–MP2/B)	B3LYP/B[19]	(MM4–B3LYP/B)	
Glucose	1.16	1.29	−0.13	0.82	0.34	
Allose	0.31	0.16	0.15	−0.09	0.40	
Altrose	0.85	0.58	0.27	0.32	0.53	
Mannose	−0.39	0.92	−1.31	0.36	−0.75	
Gulose	−0.62	−0.41	−0.21	−0.63	0.01	
Idose	2.69	3.71	−1.02	3.07	−0.38	
Galctose	2.56	2.26	0.30	1.65	0.91	
Talose	1.56	0.74	0.82	0.44	1.12	
	1.02	1.16	−0.14	0.74	0.27	(AVG)
			0.68		0.65	(RMS)

[a]The O–C–C–O–H side-chain conformation is $^4C_1$GG (gauche,gauche) in all cases. The most stable hydroxyl conformation was determined[21] for each skeletal conformer by conformational search (Appendix) and was used for these calculations.

that the average difference ($E_\beta - E_\alpha$) should be greater for the MP2 calculations than for the B3LYP. And it is greater by 0.42 kcal/mol, a small but non-negligible amount.

Then the question is: How do the MM4 calculations compare with the MP2 and B3LYP calculations, overall? Taking the MP2/B calculations first, we can look at the column labeled (MM4–MP2/B) in Table 8.2. The average and rms discrepancies are given at the bottom, −0.14, 0.68 kcal/mol. If we then look at the column labeled MM4–B3LYP/B, we note the agreement is similar, with an average discrepancy of 0.27 kcal/mol and an rms of 0.65 kcal/mol.

These calculations are stringent tests of accuracy, for both MM4 and for the quantum mechanical calculations. The MP2B and B3LYP/B results are as they are. That is simply determined by the level of accuracy of those types of calculation. The MM4 values have been carefully determined, including the electronegativity, anomeric, gauche, and delta-two effects from small molecules. The assumption of transferability here is put to a severe test. Errors in hydrogen bonding energies could be a problem also (Chapter 9) because there are many hydrogen bonds, with all different kinds of geometries, in these molecules. We cannot draw further conclusions from the available information. But these numbers give us some idea as to what sorts of error might be expected in molecular mechanics and quantum mechanics calculations for molecules of this size and complexity. It is at present not known to what extent this transferability of parameters limits the agreement between quantum mechanics and molecular mechanics here, or the extent to which the limiting factor is just the accuracy of the quantum mechanical calculations themselves. We would expect that it is some of each.

The anomeric grouping O–C–O also has additional consequences affecting the structural unit to which it is attached. These become especially important in polysaccharides,[22] and an example will be discussed in the following section.

CELLOBIOSE ANALOG

Another example of the combined anomeric and delta-two effects can be illustrated using a molecule referred to as "the cellobiose analog," CBA (Structure **5**).[23] This is just cellobiose with all the hydroxyl groups replaced by hydrogens for computational simplicity.

Cellobiose Analog (CBA)

5

This study led to the discovery of still another of these "effects," which turns out to be important in carbohydrates. Here we would like to consider the torsional barriers about the external anomeric C–O bond for the ring on the left ($C_{1'}$–O_4), and the C–O bond at carbon 4, of the ring on the right (C_4–O_4). We will refer to these bonds as φ and ψ, respectively, as shown. Note that the angle φ ($C_{1'}$–O_4), is what is normally referred to as an *exo-anomeric* bond. That is to say, it is *internal* to an anomeric system, and *exo* to the carbohydrate ring (on the left). On the other hand, ψ is not internal to the anomeric system but is attached to it externally (analogous to the methyl group in dimethoxymethane). The earlier unrecognized *effect* that we wish to discuss in the following concerns this torsion angle ψ. Rotations about these two bonds (φψ) establish a potential surface that is important in the overall conformational behavior of disaccharides in particular and of polysaccharides in general.

In Figure 8.2 is shown a Ramachandran plot of the CBA φψ potential surface, as calculated by quantum mechanics (B3LYP). There is a large, roughly oval-shaped energy minimum, with a 1 kcal/mol contour near the center of the diagram (about φψ −75, −130) and four large mountain peaks in the four corners. There are two other minima at about φψ + 60, −105 and −75, −280, but these are 3–4 kcal/mol above the global minimum and hence of little importance. To compare these quantum mechanical calculations (which include correlation by the density functional theory method) with experimental information, the following was done. Numerous crystal structures (about

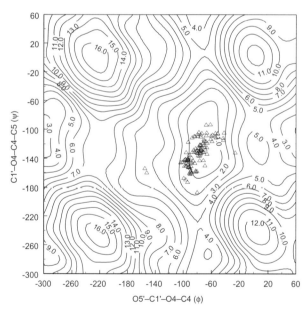

Figure 8.2. Ramachandran plot of the CBA potential surface by QM (B3LYP/6-311++G(2d,2p)//B3LYP/6-31G*) calculations. The numbers on the contours are their energies relative to the minimum at about Ψ −120°, Φ −70°. (From Lii, Chen, Johnson, French, and Allinger.[23] Copyright 2005 by Elsevier. Reprinted by permission of the authors.)

40) are known for molecules containing the CBA structure, which are substituted with a multitude of hydroxyl and other groups in diverse arrangements about the rest of the molecule. For these crystal structures, the torsion angles about the C–O bonds indicated (ϕ and ψ) are known. Hence, we can plot on the potential map from quantum mechanics the experimental points that are observed. Of course, the crystal packing may distort these ϕψ angles. Additionally, the various substituents in the molecules may lead to more distortion directly, and indirectly in that they will change the crystal packing somewhat. Hence, we don't expect to get an exact agreement between experiment and theory, but we should get a general idea of the torsional situation in question over a broad range of experimental measurements, and see how that compares with theory.

The small triangles shown in Figure 8.2 are the experimental points, and it can be seen that most of them lie within the contour represented by 2 kcal/mol of energy above the minimum. There is a somewhat systematic discrepancy in that the experimental points are slightly southwest of the center shown by the contour map. The accuracy of the quantum mechanical calculations is somewhat uncertain. However, it is known that the B3LYP correlation calculations exclude dispersion energy, and this will have some predictable consequences as we will discuss later.

After MM4 had been developed to the point so far discussed, we thought we could make a similar plot to that shown in Figure 8.2 of the surface predicted by MM4 and compare it with the same experimental points as are also shown in Figure 8.2. That was done, and the result is plotted in Figure 8.3. From this point on, we will define

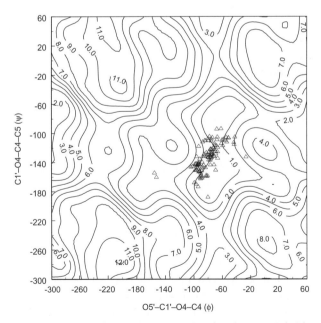

Figure 8.3. Ramachandran plot of the CBA potential surface by MM4O (without the external anomeric torsional effect). (From Lii, Chen, Johnson, French, and Allinger.[23] Copyright 2005 by Elsevier. Reprinted by permission of the authors.)

"MM4O" as the "original" MM4 program, as developed and described up to this point. The new current (further developed) version of MM4 will be referred to simply as "MM4." A comparison of Figure 8.3 (MM4O) with the experimental points, and with Figure 8.2 (QM), shows some serious discrepancies. The global minimum is, of course, the most important part of the surface. Accepting that the QM curve in Figure 8.2 is approximately correct, the global minimum in Figure 8.3 is distorted by splitting into north and south lobes. The experimental points do not similarly divide but are concentrated in the central region. There are the two higher energy minima in Figure 8.2, and those in Figure 8.3 are in approximately the right places, with reasonable depths. The four "mountain peaks" in the corners are much lower here than in the QM map (averaging 4–5 kcal lower). This is potentially a worrisome discrepancy because it is so large, even if it occurs in a portion of the map that is sufficiently high in energy that the molecules will not be located there in any case. It is worrisome because in other related molecules the parts of the surface that are of low energy here may be inaccessible (e.g., because of constraints imposed by rings). These normally high-energy regions might then become occupied, so we have to be concerned about describing them adequately.

The problem of the "mountain peaks" is easily understood, at least in part. We know that congested molecules will have energies that are too high when they are calculated by the B3LYP method. The reason for this is that with a normal van der Waals curve, when one has a close approach between the two atoms, the total van der Waals energy becomes quite high. But the dispersion energy (the r^{-6} part) also becomes quite negative. The B3LYP method omits this latter term in the energy calculation as an approximation. But the result is that a B3LYP calculation on a congested system will give energies that are too high. And if they are very high to start with, then they will become much too high. So what we see when we compare Figures 8.2 and 8.3 is that the mountain peaks at the four corners, take the southwest one, for example, are quite high by B3LYP (16 kcal/mol), and much lower by MM4 (12 kcal/mol). We will come back to this point later.

A more serious problem here is in the region of greater interest, namely near the potential minimum. There appears to be a peak in the MM4O $\phi\psi$ surface about 2 kcal/mol in height, centered at about $\phi\psi$ −100 −125, that is lacking in the QM plot, Figure 8.2. This discrepancy between the general shapes of the energy surfaces in Figures 8.2 and 8.3 is in fact mostly due to errors in MM4O calculations.

EXTERNAL ANOMERIC TORSIONAL EFFECT

Discussion in the literature regarding the exo-anomeric effect[24] has to do with the anomeric effect when it occurs with a pyranose ring, where one oxygen is in the ring and the other one is external (exo) to the ring. It was thought earlier that this particular arrangement led to unusual characteristics for the anomeric effect that seemed somewhat different from the ordinary anomeric effect found with simple molecules such as dimethoxymethane. Earlier,[19] we showed that the so-called exo-anomeric effect is really just an ordinary anomeric effect, but it may occur in more complicated molecules in such a way that various steric effects, hydrogen bonding, and the like also occur at the same time. Earlier authors did not always explicitly account for these other interactions and, hence, concluded that exo-anomeric effects might somehow be different from ordinary anomeric effects.[24] In the present case we do have an effect that occurs, not exo to a ring but external to the anomeric system. The ordinary anomeric effect involves the C–O bonds in the O–C–O unit. The effect that we now wish to discuss involves a C–O bond external to the anomeric unit, specifically the ψ bond in CBA. We considered using the prefix exo- to denote this effect since the effect is exo to the anomeric unit, but decided that that would be misleading. Hence, we have used the expression *external anomeric* effect, which should be clear. There is really no special exo-anomeric effect in the classical sense. Any anomeric effect that is exo to the ring system is similar to any other anomeric effect. Hence, while the separate classification of such an effect is unnecessary, it is sometimes a useful descriptor. The anomeric effect involves two ether oxygens and always operates simultaneously in two opposing directions between them. If the case under examination has a ring on one side only, as is normal when ordinary simple sugars are considered, it may be useful to call the effect in one direction *exo*-anomeric and in the other direction *endo*-anomeric but these adjectives are location descriptors only, not indicators of chemical differences.

The anomeric effect (also see earlier discussion in Chapter 7) has been quantitatively understood for a long time.[1–4] The delta-two effect was recognized long ago[18] but put on a quantitative basis only recently.[19] But when both of these things were taken into account, and the potential for the rotation about the φψ bonds in CBA was examined, we thought that the potential surface calculated by MM4O (Fig. 8.3) should match rather well with that obtained by comparison with known crystal structures. But, clearly, it does not. If we notice how the experimental points impact the MM4O calculated surface, there is a conspicuous problem. This was at first very puzzling. Everything known at the time was included in this calculation. The apparent explanation for this discrepancy is that there is another previously unrecognized "effect" that comes into play here. Not having accounted for it properly, the MM4O potential surface is distorted and is approximately 2 kcal/mol too high right in the midst of the experimental points in Figure 8.3.

While the anomeric effect has been studied for a very long time, and a great deal is known regarding the structures and energies that are influenced by this effect, almost the entire available literature on the subject is related to the bonds that occur *between* the two oxygen atoms. It has been previously noted (Chapter 7) that even in dimethoxymethane, the C–O bonds to the methyl groups are a little bit shorter than one might have expected or than MM4 calculates, but the differences are pretty small and they have generally been ignored. (See, however, Aped et al.,[25] where the geometric effects here have been studied in some detail.) When these methyl groups external to the anomeric system were looked at a little harder, it was found that considering them to be ordinary methyl groups (e.g., as would be found in methyl propyl ether) was not a very good approximation. It was found that a methyl in dimethoxymethane differs significantly from that in methyl propyl ether in the sense that the presence of the second (more distant) oxygen affects the properties of the methoxy group that is further away from it. The C–O bond length shortening is now attributed in MM4 to the β electronegativity shortening from the second oxygen. But now we come to the problem.

In Structure **6**[23] are shown the pertinent conformers of dimethoxymethane. A capital letter is used to indicate the bond about which the conformation is of particular interest (the 3,4 bond).

Structure 6. Conformers of dimethoxymethane.

In Table 8.3 are shown methyl rotational barriers for dimethoxymethane in the aA and aG conformations.[23] The barriers about the 4–5 bond were calculated both by the B3LYP/B and MP2/B methods. The calculated methyl rotational barrier height for aA was found to be 2.22 kcal/mol by B3LYP and 2.39 kcal/mol by MP2, while the MM4O value was 2.60 kcal/mol. Thus, the MM4O value is a little high, but not alarmingly so.

On the other hand, if we look at the aG calculation, the B3LYP value is much smaller than it was for the aA conformation, 1.52 versus 2.22 kcal/mol, respectively. Similarly, the MP2 calculation gives a much smaller barrier (1.66 kcal/mol), while the MM4O calculation gives 2.57 kcal/mol, substantially unchanged from the aA value. Thus, we conclude that while the MM4O value for the gauche conformation is "normal," in view of what was previously known, it is actually about 1 kcal/mol too high. This is a pretty significant number in terms of chemical accuracy. And this reduction in barrier height came about because of the presence of the second (more distant) oxygen in the molecule. Hence, there appeared to be an external anomeric effect here on the rotational barrier that had not previously been recognized.

Where does this reduction of the barrier for the methyl rotation in dimethoxymethane come from? Structure **7** shows the aG conformation in which this barrier reduction is important.

TABLE 8.3. Methyl Rotational Barriers (C_4–C_5) in Dimethoxymethane (kcal/mol)

Conformation	Calculation			
	B3LYP/B	MP2/B	MM4O[a]	MM4[b]
aA	2.22	2.39	2.60	2.24
aG	1.52	1.66	2.57	1.52

[a]Without the external anomeric or Bohlmann torsional effects.
[b]With the external anomeric and Bohlmann torsional effects.

EXTERNAL ANOMERIC TORSIONAL EFFECT

Structure 7. External anomeric torsional effect: aG dimethoxymethane.

The resonance form shown on the right is helpful in understanding the barrier reduction. What is shown is the normal anomeric effect type resonance, which puts double-bond character into the C_3–O_4 bond. And, of course, if the double-bond character in the 3,4 bond is increased, then the hydrogens typified by hydrogen-13 in Structure **6** will have their torsional energy lowered at the eclipsed form. This is analogous to the case in propene where a methyl group hydrogen eclipses the double bond in the ground state. Here, this lowers the energy of the eclipsed conformation relative to that in other conformations where this resonance is less important. And this resonance is less important here in the aA conformation. Hence, we can see that there is a simple, readily understandable interpretation for the lowering of the barrier to the methyl group rotation in dimethoxymethane, relative to what it would be in methyl propyl ether.

There is an earlier example of what amounts to this same effect, which is known in hydrocarbon systems. In the ground state of 2-methyl-1,3-butadiene, the methyl group is oriented with a hydrogen eclipsing the 1,2 bond not the 2,3 bond, and this is a result of the same effect. The methyl group tends to eclipse the bond of higher bond order (see discussion of Structure **1** in Chapter 5).

Calculations were similarly carried out for the 4,5 bond in ethoxymethoxymethane (Structure **8**) to see what the change would be in the barrier height in this group. As one might surmise, the barrier lowering was similar to, but somewhat larger in the ethyl than in the methyl case.

Structure 8. External anomeric torsional effect: aaG methoxyethoxymethane.

In order to reproduce this external anomeric torsional effect in MM4, a torsion–torsion interaction (between the 2,3,4,5 and 3,4,5,6 bonds in Structure **8**) was used.[19] It was worked out so that the results on dimethoxymethane and on methoxyethoxymethane as obtained by MMP2 calculations were reproduced with MM4. (It is

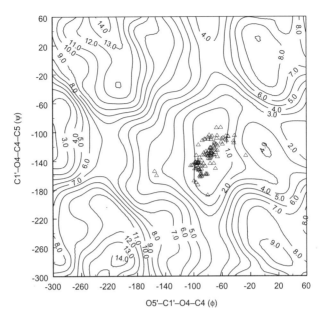

Figure 8.4. Ramachandran plot of the CBA potential surface by MM4 (with the external anomeric torsional effect included). (From Lii, Chen, Johnson, French, and Allinger.[23] Copyright 2005 by Elsevier. Reprinted by permission of the authors.)

necessary in the parameterization only to obtain numbers for hydrogen and for carbon. As usual in molecular mechanics, the rotational barriers for isopropyl, *t*-butyl, etc. are just the result of taking each of the C–H and C–C barriers into account one by one.)

A Ramachandran plot of the CBA φψ potential surface by MM4 (with this external anomeric torsional effect now included) was prepared, and the crystal data were plotted on that potential surface as previously described.[23] The result is shown in Figure 8.4, and this may be compared with the result in Figure 8.3, which is the same surface plotted by MM4O where the external anomeric torsional effect was not included. The result of this barrier lowering is quite dramatic. The double-well minimum that was present in Figure 8.3 is gone in Figure 8.4. The north and south halves of the potential well have collapsed into a single minimum, into which almost all of the experimental points now fall. The overall MM4 surface in Figure 8.4 certainly agrees much better with experiment than that in Figure 8.3. If we take the difference between these two MM4 potential surfaces [Fig. 8.3 (MM4O) – Figure 8.4 (MM4)], we get what is shown in Figure 8.5.[23] Clearly, there is a hill about 2 kcal/mol high present in the potential surface of Figure 8.3 that led to the observed problem near the energy minimum that has disappeared in Figure 8.4.

We can also make a comparison of Figure 8.4, the MM4 plot of these data, with Figure 8.2, which plots the corresponding potential surface directly from QM [B3LYP/6-311++G(d,p)//B3LYP/6-31G(d)] calculations. Superficially, the two plots are seen to be quite similar. The MM4 plot shows an eliptical shaped global minimum, with a

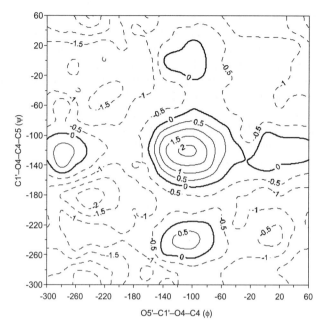

Figure 8.5. Ramachandran difference plot of the CBA potential surface (MM4O–MM4). (From Lii, Chen, Johnson, French, and Allinger.[23] Copyright 2005 by Elsevier. Reprinted by permission of the authors.)

north–south orientation, centered at approximately −80, −120°. This minimum is just slightly southwest of the one from the corresponding QM plot (−75, −130). Somewhat more of the experimental points fall within the 1-kcal/mol contour for the MM4 plot than for the QM plot. There are still the four large peaks in the four corners of each diagram, from 12 to 16 kcal/mol high from QM, and 8–14 kcal/mol from MM4. The latter values are believed by the present author to be more accurate as discussed earlier. Note also that when the external anomeric torsional effect is included in the MM4 calculation (Fig. 8.4), the hill shown in the center of Figure 8.5 is removed, and the zero of the energy point in both of these figures is accordingly lower. This also has the effect of raising somewhat the heights of the four peaks in Figure 8.4 relative to those in Figure 8.3 by 1–2 kcal/mol. The potential surface shown in Figure 8.4 appears to be rather good, giving a somewhat better fit to the experimental points than that obtained by QM. (The latter calculations could, of course, be improved by using an MP2 or other better correlation calculation in place of B3LYP, and by optimizing the geometry at the same time, instead of a single point, but that was not feasible at the time this work was done.)

Some further comments regarding the accuracy of Figure 8.4 appear to be in order at this point. This figure is a fairly good example of the accuracy and transferability of potential functions that can be obtained with molecular mechanics. Note that in the parameterization of the MM4 force field for alcohols, ethers, and related compounds, a large amount of data, both experimental and from quantum mechanics, had to be considered and converted into a general force field. Besides the usual terms that come into an alcohol force field, there are four different known chemical effects that are important if one wishes to calculate the structural and energetic characteristics of CBA correctly. These four effects are the anomeric effect, the gauche effect, the delta-two effect, and the external anomeric torsional effect. Those effects were each quantitatively worked out for small molecules (dihydroxymethane, 1,2-dihydroxyethane, 1,1,2-trihydroxyethane, and related ethers). It might be pointed out that no data from CBA itself, or anything closely related to it, were used in this parameterization. The parameterization came only from small molecules, and the fit with the potential surface shown in Figure 8.4 clearly attests to the accuracy to which the transferability of parameters exists.

Oligosaccharides (those containing more than two or three sugar units, but smaller than polysaccharides such as starch) are of major importance in biology, and an understanding here requires their study in aqueous solution. A general introduction to, and discussion of, the subject of the conformational analysis of these compounds has been given by Woods.[22] Whether or not additional effects will be found in oligosaccharides beyond those known in disaccharides has yet to be established.

REFERENCES

1. Described in books on carbohydrate chemistry, for example, J. F. Stoddart, *Stereochemistry of Carbohydrates*, Wiley-Interscience, New York, 1971.
2. J. T. Edward, *Chem. Ind.*, 1102 (1955).
3. E. Juaristi and G. Cuevas, *Tetrahedron*, **48**, 5019 (1992) for a review. Also see The Anomeric Effect in Chapter 7.
4. (a) A. J. Kirby, *The Anomeric Effect and Related Stereoelectronic Effects at Oxygen*, Springer, Heidelberg, 1983. (b) W. A. Szarek and D. Horton, *Anomeric Effects. Origin and Consequences*, ACS Symp. Ser. 87, Washington, D.C., 1979. (c) P. Des Longchamps, *Stereoelectronic Effects in Organic Chemistry*, Pergamon, New York, 1983. (d) L. Nørskov-Lauritsen and N. L. Allinger, *J. Comput. Chem.*, **5**, 326 (1984). (e) N. L. Allinger, M. Rahman, and J.-H. Lii, *J. Am. Chem. Soc.*, **112**, 8293 (1990). (f) J.-H., Lii, K.-H., Chen, K. A. Durkin, and N. L. Allinger, *J. Comput. Chem.*, **24**, 1473 (2003). (g) R. U. Leimuex, *Molecular Rearrangements*, P. de Mayo, Ed., Interscience, New York, 1964.
5. (a) D. H. R. Barton, *Experientia*, **6**, 316 (1950). (b) J. Hirsch in *Topics in Stereochemistry*, Vol. I, N. L. Allinger and E. L. Eliel, Eds., 1967, p. 199.
6. E. L. Eliel, N. L. Allinger, S. J. Angyal, and G. A. Morrison, *Conformational Analysis*, Wiley, New York, 1965.
7. E. L. Eliel and S. H. Wilen, *Stereochemistry of Organic Compounds*, Wiley, New York, 1993.
8. S. Wolfe, *Acc. Chem. Res.*, **5**, 102 (1972).
9. K.-H. Chen, J.-H. Lii, G. A. Walker, Y. Xie, H. F. Schaefer, III, and N. L. Allinger, *J. Phys. Chem.*, **110**, 7202 (2006).
10. K. B. Wiberg, M. A. Murcko, K. E. Laidig, and P. J. MacDougall, *J Phys. Chem.*, **94**, 6956 (1990).

11. G. Runtz, R. F. W. Bader, and R. R. Messer, *Can. J. Chem.*, **55**, 3040 (1977).
12. N. L. Allinger, D. Hindman, and H. Honig, *J. Am. Chem. Soc.*, **99**, 3282 (1977).
13. J.-H. Lii, K.-H. Chen, T. B. Grindley, and N. L. Allinger, *J. Comput. Chem.*, **24**, 1490 (2003).
14. (a) A. Abe and J. E. Mark, *J. Am. Chem. Soc.*, **98**, 6468 (1976). (b) M. Ohsaku and A. Imamura, *Macromolecules*, **11**, 970 (1978). (c) E. L. Eliel, *Acc. Chem. Res.*, **3**, 1 (1970).
15. J.-H. Lii, unpublished.
16. (a) J. A. McCammon and S. C. Harvey, *Dynamics of Proteins and Nucleic Acids*, Cambridge University Press, Cambridge, 1987. (b) J. M. Haile, *Molecular Dynamics Simulation: Elementary Methods*, Wiley, New York, 1992. (c) D. Frenkel and B. Smit, *Understanding Molecular Simulation*, Academic Press, San Diego, CA, 1996.
17. (a) N. L. Allinger, M. Nakazaki, and V. Zalkow, *J. Am. Chem. Soc.*, **81**, 4074 (1959). (b) E. L. Eliel, N. L. Allinger, S. J. Angyal, and G. A. Morrison, *Conformational Analysis*, Wiley, New York, 1965, p. 172.
18. R. E. Reeves, *J. Am. Chem. Soc.*, **72**, 1499 (1950).
19. J.-H. Lii, K.-H. Chen, and N. L. Allinger, *J. Comput. Chem.*, **24**, 1504 (2003).
20. J.-H. Lii, unpublished. Some preliminary results were published in Ref. 19.
21. B. Ma, H. F. Schaefer, III, and N. L. Allinger, *J. Am. Chem. Soc.*, **120**, 3411 (1998).
22. R. J. Woods, *The Application of Molecular Modeling Techniques to the Determination of Oligosaccharide Solution Conformations, Reviews in Computational Chemistry*, Vol. 9, K. B. Lipkowitz and D. B. Boyd, Eds., Wiley, New York, 1996, p. 129.
23. J.-H. Lii, K.-H. Chen, G. P. Johnson, A. D. French, and N. L. Allinger, *Carbohydr. Res.*, **340**, 853 (2005).
24. (a) R. U. Lemieux, A. A. Pavia, J. C. Martin, and K. A. Watanabe, *Can. J. Chem.*, **47**, 4427 (1969). (b) J.-P. Praly and R. U. Lemieux, *Can. J. Chem.*, **65**, 213 (1987). (c) G. A. Jeffrey and R. Taylor, *J. Comput. Chem.*, **1**, 99 (1980). (d) K. N. Houk, J. E. Eksterowicz, Y.-D. Wu, C. D. Fuglesang, and D. B. Mitchell, *J. Am. Chem. Soc.*, **115**, 4170 (1993).
25. P. Aped, Y. Apeloig, A. Ellencweig, B. Fuchs, I. Goldberg, M. Karni, and E. Tartakovsky, *J. Am. Chem. Soc.*, **109**, 1486 (1987).

9

LEWIS BONDS

HYDROGEN BONDS

When chemists refer to "acids," they are normally referring to compounds that contain a proton that dissociates easily if the acid is strong (such as sulfuric acid) or maybe not so easily with weaker acids (such as acetic acid). A hydrogen bond is an "almost" acid–base reaction where the acid shares, rather than donates, its proton. But there are also what are called "general acids," or *Lewis acids*. These are compounds, of which trimethylboron is a familiar example, that react with "general bases," or *Lewis bases*. While a strong proton acid ionizes to furnish a proton when reacting with a base, a substance that forms a hydrogen bond does not lose its proton, but rather shares it with a Lewis base. In this case the base is providing electrons that partially delocalize into the antibonding orbital of the X–H bond. In a general acid example, the electrons from the base delocalize into whatever low-lying orbital is available from the acid. A familiar example would be trimethylboron acting as a Lewis acid, and trimethylamine acting as a Lewis base, to form compound (**1**) in Structure **1**. Here the lone pair on nitrogen is delocalized into an empty 2p valence orbital on boron. In a hydrogen bond (**2**), shown here between water and ammonia, the lone pair on nitrogen is delocalized into an H–O σ^* orbital of the water.

Molecular Structure: Understanding Steric and Electronic Effects from Molecular Mechanics, By Norman L. Allinger
Copyright © 2010 John Wiley & Sons, Inc.

$$\underset{(1)}{\overset{\overset{CH_3}{|}\overset{CH_3}{|}}{CH_3-\overset{\delta+}{N}\cdots\overset{\delta-}{B}-CH_3}}\underset{(2)}{\overset{\overset{H}{|}\overset{H}{|}}{H-\overset{\delta+}{N}\cdots\overset{\delta-}{H}-O}}$$

1

We will here refer to the kind of bond formed in (**1**) as a *Lewis bond*. They are also commonly called *coordinate-covalent bonds* or *semipolar double bonds*. The hydrogen bond (**2**) is also a specific type of Lewis bond, and is the type of Lewis bond that is most commonly encountered in organic molecules.

The most familiar example of hydrogen bonding is in liquid water, which has an unusually high boiling point for a molecule of its molecular weight, due to the hydrogen bonds holding the liquid together. We will discuss here that particular kind of Lewis bond first.

Latimer and Rodebush appear to have published the first definitive reference on hydrogen bonding in 1920,[1] and the subsequent history of the subject was lengthy and complicated. Many chemists appear to have contributed to the concept. The expression "hydrogen bond" was apparently first used by Pauling in 1931 and later was discussed by him in his well-known book.[2] After a somewhat slow start, the hydrogen bond as we know it today became generally accepted, and in 1960 that led to a book on the subject by Pimentel and McClellan.[3]

The importance of hydrogen bonding in biological structures was realized[4] long before crystallographic structure determinations of proteins and nucleic acids were reported. The Watson–Crick[5] base-pair hydrogen bonding has subsequently been shown to be a feature of all known double-helical structures of naturally occurring nucleic acids, and is part of the basis for genetic coding in all known living organisms. Although hydrogen bonds are weak compared to covalent bonds, they are the most important forces determining the three-dimensional folding of proteins. Both the α-helix and the β-pleated sheet, two of the most common polypeptide structural schemes in the secondary structures of proteins,[6] are stablized by hydrogen bonding.

It was initially hard to understand that the relatively weak (about 3–5 kcal/mol) hydrogen bonds are of prime importance for living organisms, so much so that life as we know it would be impossible without them.[3] This is because many biological processes involve intermolecular recognition that has to be rapid. This recognition, consequently, requires a weak interaction that allows very fast association and dissociation, so that in a short time many possible combinations can be checked before the correct partners associate. Stronger interactions, with bonding energies much greater than those attained by hydrogen bonding would seriously hinder the flow of biological information and events. For important summaries concerning hydrogen bonds, general references are available.[2–4]

A *hydrogen bond* describes the attractive force that arises between the proton donor covalent pair **A–H**, with its significant bond dipole, and the noncovalently bound nearest-neighbor electronegative hydrogen acceptor atom **B** (Structure 2).

$$\overset{-\delta}{\text{A}}\text{───}\overset{+\delta}{\text{H}}\text{ⅡⅠⅠⅠⅠⅠⅠⅠⅠⅠⅠ}\overset{-\delta}{\text{B}}$$

Structure 2. Classic electrostatic model for hydrogen bond.

A hydrogen bond is the difference (lowering) in the energy of the system when A–H and B are brought together from infinity to the energy minimum. It is in part the Coulombic interaction of the **A–H** dipole with the excess electron density (or atomic dipole) at the proton-acceptor atom that forms the hydrogen bond interaction. The strength of the hydrogen bond formed is believed to be best correlated with acidity of the hydrogen atom and basicity of the atom **B**,[7–11] although electrostatic interactions are also important. Unless the acidity of the hydrogen and basicity of the acceptor atom are sufficient, any hydrogen bonds formed are usually too weak to be significant. Of course, if the hydrogen atom is too acidic or if the acceptor atom is too basic, the hydrogen will be transferred as a proton to the acceptor atom in a simple acid–base reaction.

Although the concept of a hydrogen bond has been generally accepted since Latimer and Rodebush[1] published the first definitive reference on hydrogen bonding, the importance of the role of the lone-pair electrons at basic site **B** was not understood until 1950. Kasha[12] found that lone-pair electrons are affected by hydrogen bonding by observing shifts in $n \rightarrow \pi^*$ transitions in ultraviolet spectra. Neutron and X-ray scattering experiments showed that the hydrogen bonding hydrogens usually tend to be situated on or near the lone-pair orbital axis (**B–LP**).[13,14] In 1979 Newton et al.[15] applied ab initio molecular orbital calculations to a study of the geometries of hydrogen bonds, which further delineated the roles of lone pairs.

A successful theory (or model) for the hydrogen bond has to be able to provide useful quantitative information and explain the many important properties of hydrogen-bonded complexes that are available from extensive experimental evidence. These include when the hydrogen bond, **A–H ... B**, is formed:

1. The molecules concerned come much closer together than the sum of the van der Waals radii of the nearest atoms would otherwise allow.
2. The lengths of **A–H** bonds are somewhat increased as hydrogen bond strength increased.
3. As a result of item 2, the infrared stretching frequencies of **A–H** bonds are shifted to lower frequencies.
4. The integrated infrared intensities of the **A–H** stretching bands are enhanced.
5. The association energy of attraction between the molecules concerned increases substantially.

6. The polarities of the molecules concerned are increased. The dipole moment of the hydrogen-bonded complex is larger than the vectorial sum of the monomer moments.
7. The linearity of the **A–H ... B** segment tends to increase as the strength of the hydrogen bond increases.
8. The protons of hydrogen bond donors usually tend to be situated on or near the hybrid lone-pair orbital axes of proton acceptors.

Before 1957, theoretical studies of a hydrogen bond used either classical electrostatic models or approximate quantum mechanical treatments of fragments of the hydrogen bonded system. The classical electrostatic model was the first proposed for a hydrogen bond because of the fact that all known hydrogen bonds are between electronegative elements (**A** and **B** in Structure **2**). Typical calculations utilizing this model were made by Lennard-Jones and Pople.[16,17] With this model, the theoretical chemists successfully calculated the hydrogen bond energy of the water dimer, 5.95 kcal/mol, in good agreement with the experimental values. However, the model failed to give correct predictions of hydrogen bond lengths and other properties of hydrogen bonds, such as the increase of molecular polarity and intensity of the **A–H** stretching frequency. The simple electrostatic model was an incomplete description of the hydrogen bond because it did not take into account three other essential effects: polarization, delocalization, and electron exchange between bonded molecules. Therefore, a better model was needed.

QUANTUM MECHANICAL DESCRIPTION OF A HYDROGEN BOND

A better model depended on a better understanding of the hydrogen bond. (The definition of the hydrogen bond energy here and in quantum mechanics generally is as given previously. Dimer energy minus twice the energy of the monomer, with the sign changed to positive.) Hydrogen bonding has been studied in some detail by quantum mechanical methods,[17–21] and a well-developed theoretical framework for it exists. We will use the water dimer as an example in the following discussion (Structure **3**):

Structure 3. Water dimer (stereographic projection).

To gain the necessary understanding, several research groups[7,22,23] partitioned the total hydrogen bond energy into individual contributions similar to those proposed by Coulson[24] in 1957. Among them, Umeyama and Morokuma[25] presented a more detailed scheme of energy partitioning within the framework of molecular orbital theory. Of course, when one solves the Schrödinger equation for a hydrogen bond, one obtains an energy and a wave function. But to improve our understanding of what really happens here, as usual, it is desirable to have a model that will relate this information to a classical electrostatic/mechanical model. There is no unique way of doing this, but Umeyama's and Morokuma's scheme proceeds in a clear and straightforward way. In their scheme the energy of the hydrogen bond, ΔE_{HB}, is given by Eq. (9.1):

$$\Delta E_{HB} = ES + EX + CT + PL + DISP + MIX \qquad (9.1)$$

where ES is the electrostatic interaction energy. This is the interaction energy between the unperturbed nuclei and electron clouds of two monomers, $\mathbf{M_a}$ and $\mathbf{M_b}$. This contribution includes the interactions of all permanent charges and multipoles, such as charge–charge, charge–dipole, dipole–dipole, dipole–quadrupole, and so on. This interaction may be either attractive or repulsive depending on the orientation of the charges and multipoles. This is the dominant interaction when the two monomers are far apart.

EX is the electron exchange repulsion energy. This is the interaction energy caused by the exchange of electrons between monomers $\mathbf{M_a}$ and $\mathbf{M_b}$. The interaction is a short-range repulsion due to the overlap of electron clouds of $\mathbf{M_a}$ and $\mathbf{M_b}$.

CT is the charge transfer or electron delocalization interaction energy. It results from the interaction caused by electron transfer from the highest occupied molecular orbital (HOMO) of $\mathbf{M_a}$ to the lowest unoccupied molecular orbital (LUMO) of $\mathbf{M_b}$, and from the HOMO of $\mathbf{M_b}$ to the LUMO of $\mathbf{M_a}$, and higher-order coupled interactions. This interaction is always attractive and highly directional. In hydrogen-bonded structures, a net electron transfer always occurs from the molecule with the nonbonded (lone-pair) electrons, or weakly bonded (π) electrons to the molecule with a highly polarized A–H bond.

PL is the polarization interaction energy. This energy results from the interaction between $\mathbf{M_a}$ and the distorted (polarized) electron cloud of $\mathbf{M_b}$ by $\mathbf{M_a}$, or vice versa. The contribution includes charge-induced dipole, dipole-induced dipole, quadruople-induced dipole, and so on. This is always an attractive interaction.

DISP is the dispersion energy. This interaction is caused by simultaneous and correlated excitations of electrons in both molecules and leads to correlation of the electron motions and to general net stabilization of the complex. The correlation contribution is found to be important for interactions between nonpolar molecules, but relatively unimportant for those of small polar molecules, as in hydrogen bonding.

MIX is the sum of the higher-order interaction energies among the ES, EX, PL, and CT components.

The results of a hydrogen bond energy partitioning are very basis set dependent. This is clearly shown by a comparison of the results by Morokuma and Winick,[26] Kollman and Allen,[22] and Morokuma.[7] Morokuma points out that the decomposition of the hydrogen bond energy into components as above is not rigorous. It is an approximate method for breaking down the interaction into component parts for purposes of qualitative comparisons. But he also issues "a word of warning" about overinterpreting the energy results.

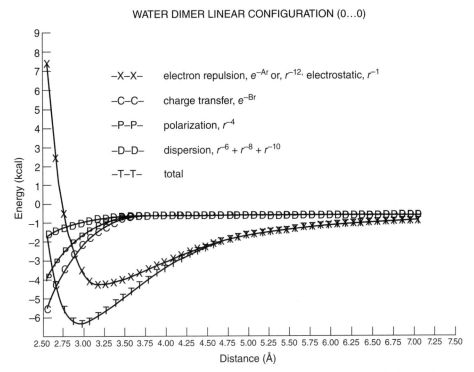

Figure 9.1. Morokuma decomposition of an ab initio molecular orbital calculation for the water dimer.[3b] (From Jeffrey and Saenger.[4] Copyright 1991 by Springer. Reprinted by permission of Springer.)

Normally, we want to know the hydrogen bond energy at the specific geometry that corresponds to the energy minimum. But we can calculate this energy as a function of distance (or other geometric variables). In Figure 9.1[3b] is shown a decomposition of an ab initio molecular orbital calculation for the water dimer. The major part of the energy here comes from the repulsion between electron clouds at very short distances, and is best represented by an exponential. This repulsion is largely Pauli repulsion. There is also a Coulomb term, which is proportional to r^{-1}, and dominating at long distances. There are also attractive polarization and dispersion terms (r^{-4} and r^{-6}) and importantly, a charge transfer term, e^{-Ar}. These terms are negligible at long distances but become significant attractions once the oxygen–oxygen distance gets below about 3.5 Å.

Note that electrostatics in quantum mechanics tends to be exaggerated with calculations that are in current wide use, and even more so for earlier calculations. For example, amines[27] and alkyl fluorides[28] (which have both been studied in detail) are calculated by MP2/B to have dipole moments that are rather uniformly too large (by about 15%), as compared with experiment and MM4. (For a definition of MP2/B, see under Jargon in the Appendix.) This is a basis set/correlation truncation problem and can be rather accurately corrected for in simple cases.

HYDROGEN BONDING MODELS IN MOLECULAR MECHANICS

The question to be asked here is: How do we wish to represent hydrogen bonding in molecular mechanics? The MM4 model for hydrogen bonding is shown in Figure 9.2.[29-31] The formulation of hydrogen bonding in molecular mechanics necessarily differs somewhat from that used in quantum mechanics since the method considers atoms rather than nuclei plus electrons. It is clear at the outset, however, that electrostatics will be quite important here. The A–H dipole has the hydrogen end positive, and B is an electronegative atom and the negative end of a dipole. Therefore, electrostatics is going to give us more or less a reasonable hydrogen bonding geometry without further refinements. But it is only reasonable—it is really not very accurate. Obvious things about the hydrogen bond that we have to be concerned with include the distance R. How close does the hydrogen actually come to B? We are also concerned about the depth of the potential well. And we know that hydrogen bonds tend to be linear; so we are going to have to worry about the angle θ_{H-A-B}. It was not apparent to all in the early days, but the hydrogen clearly bonds to a lone pair attached to atom B rather than to atom B in general.[21,32] We must, therefore, define somehow the location of that lone pair. It is not centered at the nucleus but is out away from the nucleus.

When we try to formulate an equation that describes the above items, as illustrated in Figure 9.2, we also recognize from quantum mechanics that the potential well of the hydrogen bond has a shape that is very similar to that of a van der Waals curve. Hence, the energy of the hydrogen bond in molecular mechanics can be represented, at least to a pretty good approximation, by a van der Waals curve with suitable parameters, which will be different from those normally used for the atoms A, B, and H, and including angular terms as suggested by Figure 9.2. The model in Figure 9.2 is described by Eq. (9.2), which is of the general form of a van der Waals equation (Chapter 4), with added angular terms:

$$E_{HB} = \sum \varepsilon_{HB}\{276,000\exp[-12.0R_{HB}/R_{HB}^\circ] - F(\theta_{H-A-B}, \theta_{H-B-LP}, \ell_{AH})2.25(R_{HB}/R_{HB}^\circ)^6\}/D$$

(9.2)

where R_{HB}° is the minimum energy bonding distance for the H–B bond, and D is the dielectric constant. $\Gamma(\theta_{H-A-B}, \theta_{H-B-LP}, \ell_{AH})$ means a function of the indicated variables, and LP identifies the lone pair of electrons. The lone pair is not treated as such. Rather,

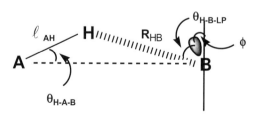

Figure 9.2. Geometric parameters for the MM4 hydrogen bond model.

it is identified only by the location of a line drawn from B along the formal axis of the lone pair. The function F has the form given in Eq. (9.3):

$$F(\theta_{H-A-B}, \theta_{H-B-LP}, \ell_{AH}) = E[\cos(\theta_{H-A-B})+1]^n[\cos(\theta_{H-B-LP})+1]^n(\ell_{AH}/(\ell^\circ_{AH}), \quad (9.3)$$

where E is a normalization factor, and n is related to the number of lone pairs. The directionality of the line representing the lone pair in Figure 9.2 is determined by the angle ϕ, which is the angle between lone pair and the reference axis. In the case where atom B is bound to only one atom X, the reference axis is the B–X bond itself. If B is bound to two or more atoms, such as BXY or BXYZ, the reference axis is the line that connects atom B to the midpoint of the line X, ... , Y or the centroid of the plane XYZ. The ϕ value is determined according to the atom type of B. It is 70° for oxygen (whether it is a carbonyl, alcohol, ether, or water oxygen) and 65° for the halogen atom, and 0° for amine nitrogen atom. With this method of defining the lone pair directionality in the MM4 equation for the oxygen case, the hydrogen bond interaction anywhere on the conically shaped surface around the reference axis is the same (no specific location of lone pair is defined). The exact resulting hydrogen bond geometry will then be determined by the steric and electrostatic effects from the attached atoms.

The MM4 molecular mechanics model of the hydrogen bond can be seen in general to be similar to that from quantum mechanics. The potential curve that describes the water dimer energy as a function of the O–O distance is shown in Figure 9.3. If we compare the curve (with hydrogen bond) in this figure with that from the ab initio calculation as given by Jeffrey[3b] (Fig. 9.1), we can see that the MM4 hydrogen bond energy is about 5 kcal/mol at an O–O distance near 3.0 Å. The curve shown in Figure

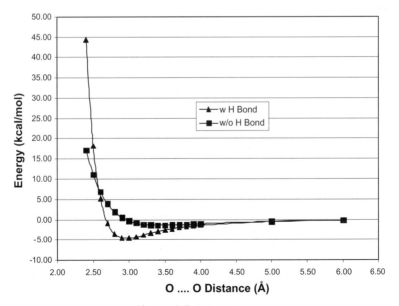

Figure 9.3. Water dimer.

9.1 is about 1 kcal/mol too strong, as explained earlier. The electrostatics also tends to be exaggerated in the quantum mechanical calculation in the distances going outward from 3.0 Å. But the figures are otherwise comparable. Also shown in Figure 9.3 is the same information regarding the energy but using the ordinary van der Waals parameters (without hydrogen bond). Thus, for the water dimer, the increase in the epsilon value accounts for about 40% of the increase in the well depth, but the greater part of the increased depth comes from the reduced value of the effective van der Waals distance. Because the energy minimum is moved inward from about 3.5 to less than 3.0 Å, the other (attractive) quantities that affect the well depth become larger. This is an important point, which we have noted elsewhere with hydrogen bonds, and it seems to be fairly general. That is, the increased well depth that is associated with the hydrogen bond as given directly by the van der Waals parameter (ε) accounts for only a part of the hydrogen bond energy. Much of the energy, and the greater portion of the well depth lowering, comes from the fact that the hydrogen bonding atoms move quite a lot closer together, and so the attractive interaction terms become significantly more important. This is also found to be true with Lewis bonds as discussed later in this chapter.

The electrostatics in the MM4 calculation (e.g., the dipole–dipole interaction in the water dimer) is not adjusted in any way to allow for the hydrogen bonding. The MM4 dipoles are present as usual, and they interact, and the energy of that interaction (the electrostatics in MM4) accounts for roughly half of the total hydrogen bond energy. However, if no further adjustment is made, the hydrogen does not come close enough to the oxygen to which the hydrogen bond is formed, and the bond is not strong enough. In other words, in the molecular mechanics model there appears to be required a considerable modification of the van der Waals interaction (parameters) between the hydrogen and the oxygen when they form a hydrogen bond, from the values that they would otherwise have.

There are various ways with which this problem might have been dealt. What we chose to do was to simply assign the van der Waals energy and distance parameters (ε and l_0) between the hydrogen and the oxygen special values for the hydrogen bonded pair, so as to allow the two atoms to come together to an appropriate distance and energy minimum. The ordinary van der Waals function has the same general shape as needed for a hydrogen bond, in going to zero at infinite distance, and coming downhill in energy with decreasing distance to an appropriate point, whereupon the curve turns sharply upward. It was expedient, therefore, to simply assign van der Waals parameters appropriately and treat the hydrogen bond (in a mathematical sense) as though it were just an unusually strong van der Waals interaction. And it turns out that this scheme works quite well. When the energy/distance curves from MM4 are compared from quantum mechanical calculations, the agreement is acceptable. This does not really mean that we are assigning the hydrogen bond as a van der Waals interaction. Rather we are assigning it as an interaction that fits the same energy versus distance function as does the van der Waals interaction, after the electrostatics, dispersion, and polarization are each accounted for. However, the parameter values are quite different from normal van der Waals values.

The molecular mechanics model that was developed, mainly from the quantum mechanics model, began in a simple fashion with MM2 and became more complex with time, as it was found that the simple examples were not completely general, and/or were of limited accuracy. Only the final result is given here, as the earlier work has been adequately reviewed elsewhere.[31] A few simple examples will outline the general accuracy of the MM4 model. This model has been applied to simple cases such as the hydrogen fluoride dimer, the water dimer, and the methanol dimer, each of which will be briefly discussed. It is found that these simple cases are generally modeled well. A more interesting case, frequently found in organic molecules, will then be discussed, namely the conformational problem presented by ethylene glycol.

Hydrogen Fluoride Dimer

Since quantum mechanics tells us that hydrogen bonds are mainly electrostatic in nature, one might have expected that the HF dimer would have the two molecules side by side in an antiparallel geometry, as this would be the most stable electrostatic arrangement for two dipoles. However, that is not what is found. If the hydrogen bond were strong enough to overcome the dipole–dipole electrostatics, then one might expect to find a linear head-to-tail dimer since this geometry can yield a reasonably good arrangement for the dipoles, and the hydrogen bond is linear. This was what MM3 calculated. However, both experiment[33] and MP2/B calculations[31] show that the hydrogen fluoride dimer has neither the linear nor the parallel geometry that might have been expected. Instead, the dimer *1* has a planar V shape (see figure in Table 9.1). The hydrogen-bonded hydrogen is off of the fluorine–fluorine line by 10° experimentally, by 6° from the BC calculation, and by 6.1° from MM4. The other hydrogen forms an angle with the fluorine–fluorine line of 117° experimentally, 111.5° BC, and 120.2°

TABLE 9.1. Geometry and Energy of Hydrogen Fluoride Dimer[a,b]

I. (HF)$_2$ (linear, C_s), **1**	Exp.[33]	MP2/6-311++G(2d,2p)	MM4
E(uncorrected)		−4.87	
BSSE		−0.97	
E(corrected)	−5.0 to −7.0	−3.90	−5.52
Dipole moment	2.988	3.569	3.000
F...F dist.	2.78	2.762	2.777
F...H dist.		1.846	1.865
<H–F ... F	10(2)	6.0	6.1
<F ... F–H	117(6)	111.5	120.2

[a]Experimental geometry is microwave r_0; MM4 r_g. Energies are in kcal/mol; distances are in Å; angles are in degrees.
[b]All four atoms are in the same plane.

Source: From Lii and Allinger,[31] Copyright 2008 by the American Chemical Society. Reprinted by permission of the American Chemical Society.

TABLE 9.2. Geometries and Energies of Water Dimer[a]

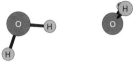

I. $(H_2O)_2$ (linear, C_s), 2	Exp.[34]	MP2/6-311++G(2d,2p)	MM4
E(uncorrected)		−5.36	
BSSE		−0.88	
E(corrected)	−5.2(1.5)	−4.48	−4.59
Dipole moment	2.6	2.783	2.494
O … O dist.	2.976	2.917	2.956
O … H dist.		1.956	1.987
<H–O … O	6(20)	4.6	2.8
<O … O … X[b]	123(10)	123.7	123.0

[a]Experimental geometry is microwave r_0; MM4 geometry r_g. QM values are approximate r_e values (without correction); energies are kcal/mol; distances in Å; angles are in degrees.
[b]O … X is the bisector of the angle H–O–H in the H-bond receptor molecule.

Source: From Lii and Allinger,[31] Copyright 2008 by the American Chemical Society. Reprinted by permission of the American Chemical Society.

MM4. The binding energies are in the 5–7 kcal/mol experimentally, 3.9–4.9 by quantum mechanics, and 5.5 by MM4.

Water Dimer

Table 9.2 shows that the MM4 hydrogen bonding potential gives a significant improvement in the dimer geometry, relative to earlier calculations. Unlike MM3, which predicted the $O_1 … O_2 … X$ ($O_2 … X$ is the bisector of $H_2–O_2–H_2$ angle, Table 9.2) angle of the linear water dimer 2 to be 160.3°, MM4 now calculates the angle to be 123.9°, which is very close to both ab initio and experimental[34] results [123.7° and 123°(10) respectively]. Simultaneously, the dipole moment of the dimer is also improved due to the improvement of the dimer geometry.

The total hydrogen bond energy of the water dimer is about 4.5 kcal/mol by BC calculations. Earlier calculations with smaller basis sets, and without BSSE corrections* usually gave a value of about 6 kcal/mol.[20] The experimental value is 5.2 ± 1.5 kcal/mol. [This might be a good place to point out that hydrogen bond energies were routinely calculated by quantum mechanics to be too strong by about 1–1.5 kcal/mol in most of the studies in the literature prior to about 1999.[30] With the 6-31G* basis sets (or smaller) commonly used prior to about that time, the BSSE correction is of the order of 1.5 kcal/mol. If this correction is not made, the hydrogen bond is calculated to be quite a bit too strong. If appropriate diffusion functions are added to the basis set (required for an

---

*Basis set superposition errors (BSSE) result when small basis sets are used to describe intermolecular interactions. See Basis Set Superposition Error in the Appendix.

accurate result), the hydrogen bond is calculated to be somewhat weaker, and the BSSE correction is also reduced. It is presumed that the larger calculation gives the better result. The direct experimental value for the hydrogen bond in the water dimer is not known with high accuracy, and it is not, therefore, known with certainty which of the quantum mechanical values is to be preferred. However, there is a large quantity of data available in carbohydrate chemistry that says that pyranose structures, in general, are more stable than the corresponding furanose structures. The BC calculations give this order correctly, while the small basis set calculations give it the other way around.[30]]

Methanol Dimer

No experimental geometry exists for the methanol dimmer; so the comparison is made between the MP2/B and MM4 calculations (see Table 9.3).[31] The MM4 results show both geometries and binding energies that seem reasonable.

Ethylene Glycol

Ethylene glycol contains a key moiety (O–C–C–O) required for modeling studies of carbohydrates. It is important to get the ethylene glycol potential surface right so that one can transfer the alcohol/ether parameters to carbohydrate calculations. And the ethylene glycol potential surface is dependent on, among other things, the hydrogen bond interaction between the two vicinal hydroxyl groups. The conformational problem here was studied as early as 1949 by Bastiansen,[35] using electron diffraction. He could detect only the gauche OCCO skeletal conformation, indicating that this one was far

TABLE 9.3. Geometries and Energies of Methanol Dimers[a]

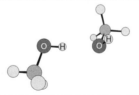

I. $(CH_3OH)_2$ (linear, C_1)	MP2/6-311++G(2d,2p)	MM4
E(uncorrected)	−6.25	
BSSE	−1.06	
E(corrected)	−5.19	−5.45
Dipole moment	2.800	2.794
O ... O dist.	2.851	2.902
O ... H dist.	1.895	1.936
<O ... O–H	6.6	4.0
<O ... O ... X[b]	123.3	126.6

[a]MM4 geometry is r_g. Energies are in kcal/mol; distances in Å; angles in degrees.
[b]O ... X is the bisector of the angle C–O–H.

Source: From Lii and Allinger,[31] Copyright 2008 by the American Chemical Society. Reprinted by permission of the American Chemical Society.

more stable than the trans. He presumed that there was considerable stabilization from hydrogen bonding, which is true. However, the gauche effect (Chapter 8) also plays a part in stabilizing that particular structure. When Kazerouni and co-workers[36] repeated Bastiansen's electron diffraction experiment in 1986, they used two different temperatures. They found a composition containing a mole fraction of only 0.08 T conformations at 376 K, which increased to 0.18 at 733 K. Earlier microwave work[37] had revealed an extremely rich and complicated spectrum. It was quite different from that expected for a rigid-rotor spectrum and indicated that there was a large-amplitude tunneling of the hydrogens between the two hydroxyl groups. This led to considerable further studies.[38]

Twelve conformations of ethylene glycol were studied (Fig. 9.4). They are 10 stable conformers and 2 rotational transition states as shown. The conformational

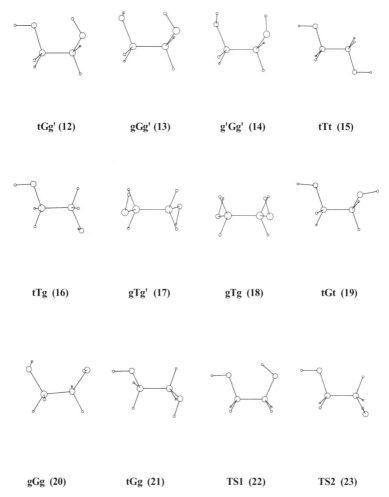

Figure 9.4. Ethylene glycol conformations. (From Lii and Allinger.[31] Copyright 2008 by the American Chemical Society. Reprinted by permission from the American Chemical Society.)

TABLE 9.4. Conformational Energies of Ethylene Glycol[31]

Conformations[a]	MP2/B	B3LYP/B	MM3	MM4
tGg′, 12	0.00	0.00	0.00	0.00
gGg′, 13	0.55	0.33	0.98	0.56
g′Gg′, 14	1.14	0.84	NA[b]	0.97
gTg′, 15	2.92	2.45	3.40	3.96
tTt, 16	2.66	2.46	2.04	2.27
tTg, 17	2.86	2.52	2.93	3.33
gTg, 18	3.13	2.67	3.98	4.24
gGg, 19	3.28	2.80	NA[b]	2.45
tGt, 20	3.21	2.95	2.62	2.99
tGg, 21	3.78	3.36	3.62	3.16
TS1, 22[c]	5.65	4.70	5.81[d]	5.82[d]
TS2, 23[c]	6.96	6.19	6.97[d]	6.50[e]

[a]The three dihedral angles are H_1–O_1–C–C, O_1–C–C–O_2 and C–C–O_2–H_2, respectively; g,G: gauche(+); g′,G′: gauche(−); t,T: trans (see Fig. 9.4).
[b]Stable conformation could not be found on the MM3 potential surface.
[c]TS1 (with O–C–C–H eclipsed) and TS2 (with O–C–C–O eclipsed) are rotational transition states (see Fig. 9.4).
[d]One imaginary frequency.
[e]Two imaginary frequencies.

energies are given in Table 9.4. As one can see from the table, the MM4 results are comparable to those from MP2/B calculations. It is particularly important that the low-energy conformations have their energies well calculated, as they correspond to the conformations that will actually be most often found in more complex systems. The three lowest energy conformations are, in increasing order, **12**, **13**, and **14**, by MP2/B (and also by B3LYP/B), and the MM4 values are within about 0.2 kcal/mol of the MP2/B values in each case. The high-energy conformations are generally fairly well calculated, but there are two discrepancies of about 1.3 kcal/mol.

Looking at the MM4 energies for the conformations of ethylene glycol, the difference between the most stable trans and the most stable gauche is 2.27 kcal/mol. The MM4 determination of the Boltzmann distribution over all of the conformations was carried out, and it indicated that the total percentage of trans conformations would only be 1.5% at 376 K. The calculation was also carried out at 460°C, the temperature used by Hedberg, but because of the low rotational barriers for the hydroxyls, the calculated values of entropies of the conformations (which are obtained using the harmonic approximation) would be quite inaccurate, and the corresponding free energy results would not be accurate. However, for the record, the value is 13.0%. A higher value would be expected in reality, for the reasons mentioned. Of course, there is an experimental problem here too. The temperature of the nozzle from which the gaseous compound expands into the vacuum can be measured, but the temperature of the expanded gas is somewhat uncertain.

TABLE 9.5. Geometry and Energy of Ethylene Glycol (tGg′)[31]

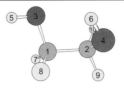

12

	R_e			R_g	
	MP2/6-311++G(2d,2p)	MM3	MM4	E.D[a]	MM4
$E(tTt)-E(tGg')$	2.66	2.04	2.27		
C1–C2	1.508	1.514	1.507	1.517	1.514
C1–O3 (H acceptor)	1.434	1.426	1.424	1.424$_{ave}$	1.430
C2–O4	1.422	1.424	1.421		1.427
O3–H5	0.958	0.930	0.952	0.961$_{ave}$	0.967
O4–H6 (H donor)	0.962	0.932	0.955		0.97
C2–C1–O3	105.9	108.2	107.1	109.3 (4)$_{ave}$	109.5$_{ave}$
C1–C2–O4	111.2	109.2	111.2		
C1–O3–H5 (H acceptor)	108.8	108.2	108.7	105.8 (27)$_{ave}$	106.9$_{ave}$
C2–O4–H6 (H donor)	106.0	106.2	106.4		
H6–O4–C2–C1	−167.3	179.5	179.2		
O3–C2–C1–O4	62.0	61.7	62.9	60.7 (18)	63.2
C2–C1–O3–H5	−51.8	−50.5	−48.2		

[a] The r_g structure by ED is at 376 K.[36]

Conformations **12**, **13**, and **14** can internally hydrogen bond, with a favorable XGg′ arrangement. These arrangements between the oxygens and the hydrogens that form the hydrogen bonds are optimum (for an ordinary 1,2-diol) and are lacking in all other conformations. These conformations have MM4 hydrogen bonds that are of sufficient strength as to lower their stabilities by at least 1 kcal/mol below those of the remaining conformations.

The MM4 geometry of the most stable conformation (tGg′, **12**) was compared with the experimental electron diffraction structure (r_g)[36] and with the MP2/B calculation[31] (Table 9.5). Some important values cannot be uniquely determined from the diffraction pattern, and are only reported as average values with similar quantities. The MP2/B values can be compared with MM4 item by item. The agreement seems to be satisfactory.

Upon examining other atoms or groups that are hydrogen bond acceptors, such as nitrogen, halogens, double bonds, and the like, we find that we can mimic the results of strong or weak hydrogen bonding, as it is known to occur, by simply substituting the special hydrogen bonding (van der Waals type) parameters into the normal MM4 calculation, for each different proton donor–acceptor combination. The special van der

Waals parameters here contain what in quantum mechanics is the charge transfer term, and dispersion is always included in the van der Waals attraction anyway. In MM4, polarization is specifically included as a separate term throughout, so nothing special is needed for hydrogen bonding.[39]

It is convenient that the distance dependence of the hydrogen bond can be represented adequately using a simple van der Waals function plus the ordinary electrostatics. This means that for practical purposes, wherever there is a hydrogen bond (an interaction that differs significantly from the normal van der Waals interaction expected between the two elements in question), we can simply insert a pair of special van der Waals parameters for the explicit atoms involved to override the normal van der Waals calculation. This seems to work adequately for ordinary hydrogen bonds and has the additional advantage that it can also be generalized to cover interactions of the Lewis acid/Lewis base type (discussed later). Thus, it seems that hydrogen bonding should be extendable to various atom combinations in a reasonably straightforward manner. And, of course, as in the rest of molecular mechanics, these terms have to be adequately transferable from molecule to molecule. This appears to be the case as far as we know at present.

OTHER LEWIS BONDS

Acids were recognized since the early days of chemistry as compounds that were proton donors. Bases were the corresponding proton acceptors. Lewis generalized this phenomenon by defining what he referred to as *general acids* and *general bases*, which now are often referred to as Lewis acids and bases. These are compounds wherein what is being transferred or bonded is not a proton but is a general acid (an electrophile), being bound to a general base (a nucleophile). An example would be the trimethylamine, trimethylboron complex (Structure **4**):

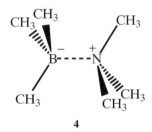

4

Here the boron is a general acid (accepting a lone pair as a proton would) and the nitrogen is a general base (donating a lone pair, as it would to a proton). As with an ordinary hydrogen bond, the B⋯N distance is appreciably shorter than the sum of the van der Waals distances, and the energy is lower than it would be for an ordinary van der Waals interaction (i.e., the bond is stronger), while the bond stretching function is described adequately near the minimum by a van der Waals function.

Amine–Carbonyl Interactions

Amines (Lewis bases) react with carbonyl groups (Lewis acids) by carbonyl addition reactions. Indeed, a substantial fraction of organic chemistry is concerned with carbonyl addition reactions of which this is a typical example.[40] What we wish to discuss here is both their *interaction* and also their *reaction*. What, exactly, happens when these two groups come together, and how do they interact or react?

Let us consider the general problem of the *reaction coordinate diagram* of this reaction. As a simple example, consider the reaction with ammonia and formaldehyde [Reaction (1)]:

(1)

It is known that in such a reaction, the ammonia attacks the carbonyl carbon of the formaldehyde by coming in with its electrons pointing toward the carbon, and approximately perpendicular to the plane of the carbonyl system, as shown in Structure **5**:

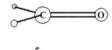

5

The geometry of the initial carbonyl system is trigonal, planar, and approximately sp² hybridized at carbon. The addition product shown for Reaction 1 would be approximately tetrahedral and sp³ hybridized.

If one wishes to study a reaction theoretically, the most simple example is usually where one starts, which would be with the ammonia/formaldehyde reaction in the present case. But when studying a reaction experimentally, it is often convenient to start with much more complicated examples. The latter approach was followed historically, and we will discuss it here in historical order.

To obtain a reaction coordinate diagram for an addition of an amine to a carbonyl group, the actual approach was used as follows. Consider the series of structures related to 1-methyl-1-aza-cyclooctan-5-one (**7**) (Structure **6**):

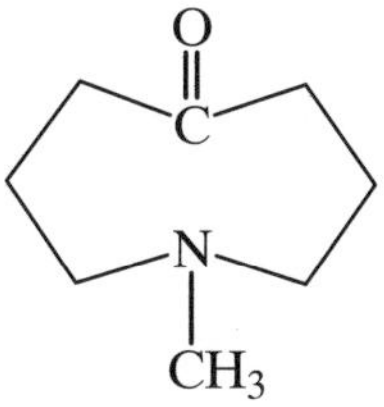

(structures 7, 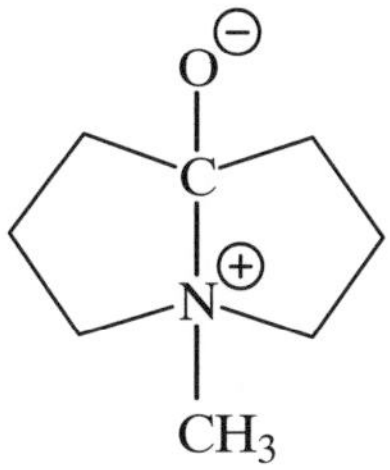 8, 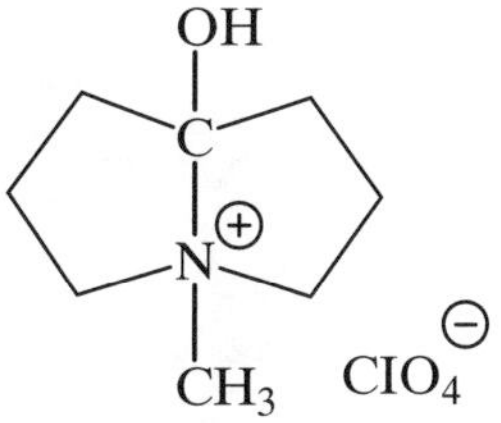 9)

6

In this molecule we have an amino group and a carbonyl group held in close proximity, and they can orient in the correct geometry for the nitrogen to add to the carbonyl group, which would give (**8**). This does not happen very well in the gas phase, and for the same reason that even strong acids such as HCl do not ionize in the gas phase. The energies involved with a charge separation like this are very large. On the other hand, when (**7**) is placed in aqueous solution, it does convert to (**8**) because the charges can be solvated by water. If (**7**) is treated with one equivalent of perchloric acid, it does not form the protonated (on nitrogen) ammonium salt of (**7**) that might have been expected, but rather forms compound (**9**). Leonard[41] discusses the extensive evidence concerning interactions and reactions of this sort, which can be detected by infrared (IR) and ultraviolet (UV) spectra, dipole moments, optical rotatory dispersion, O NMR, and ^{13}C NMR spectra. So this kind of interaction/reaction has been long known in cyclic systems such as (**7**), and it is quite dependent on solvation. Leonard points out that the functional group combinations thus created would not exist without the benefit of ring constraints. The reason that simple open-chain systems do not show this type of reaction is an entropy effect analogous to that discussed for the hydration of glucose (Chapter 8). Putting together two small molecules such as formaldehyde and ammonia to make one molecule causes a total entropy loss ($T\Delta S$) of about 5–10 kcal/mol at room temperature. The binding energy between a simple amine and a ketone is smaller than that. Hence, although the energy may favor this kind of reaction, it is quite small and it is overwhelmed by the loss in entropy of having two molecules combine to form one. This problem, of course, is avoided when a transannular reaction occurs.

The compound 11-methyl-11-azabicyclo[5.3.1]undecane-4-one is a molecule representative of the kind of system that we will here examine. It is shown in Structure **7**:

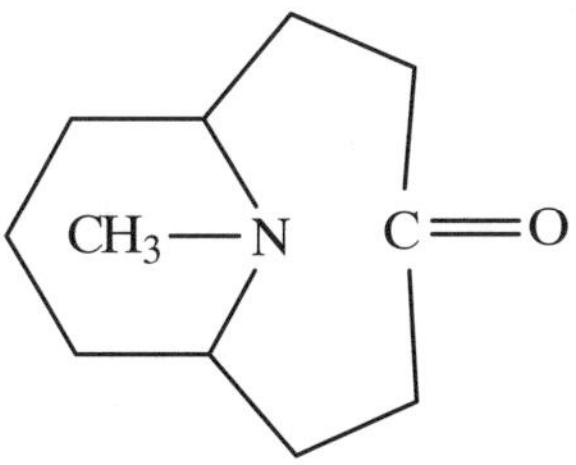

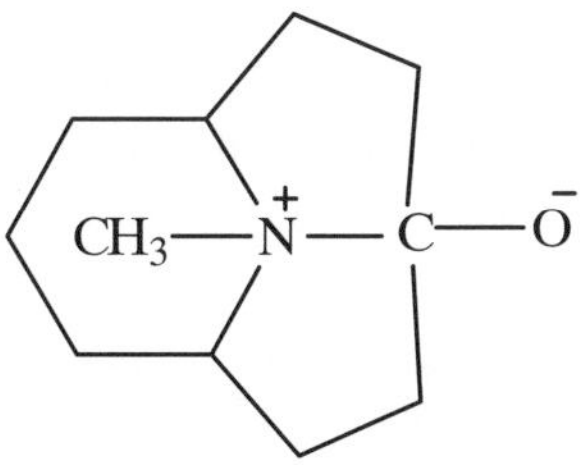

7

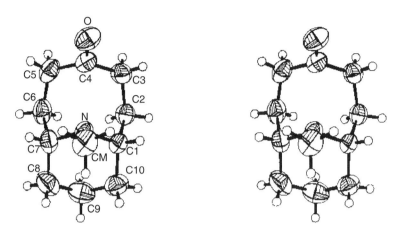

Figure 9.5. 11-Methyl-11-azabicyclo[5.3.1]undecan-4-one: Stereographic projection of molecule, showing vibrational ellipsoids at the 50% probability level.

It can be seen that there is a tertiary amino group in position where it can rather easily add to the carbonyl group, and in fact it is literally being forced into the carbonyl group. This can be seen quite clearly in the stereographic crystal structure of Kaftory and Dunitz[42] (Fig. 9.5).

What was actually done by Bürgi and Dunitz[42b] was to use compounds (referred to as A–F in the following*) related to (**7**), which were perturbed in various ways by the addition of rings and substituents to give a whole group of related structures. The crystal structure of each of these compounds was determined by X-ray crystallography, so the geometry of each compound in the vicinity of the carbonyl group was accurately known in each case. Because of the presence of the various substituents, the distances between the carbonyl carbon and the nitrogen varied considerably from one molecule to the next. These structures were then arranged in order according to this distance, with the nitrogen far from the carbonyl, and then closer to the carbonyl, and then very close to the carbonyl. This yielded in effect the reaction coordinate diagram for the addition of a tertiary amine to a carbonyl group.

As will be evident from the foregoing paragraph, this ingenious experiment was a rather difficult problem to carry out in practice, designing (or imagining) the appropriate compounds, the synthesis (or acquisition) of all of those compounds, and then the X-ray crystallography required to provide the information. And how did this come out? It came out very well indeed. And we will outline this work here in some detail.

It was determined just what the distance between the amino group and the carbonyl carbon was for each molecule A–F from the crystal structures. These distances are shown in Figure 9.6 (the scale in angstroms is shown at the right). For each molecule,

---

*The compounds actually used by Bürgi were as follows: A, methadone; B, cryptopine; C, protopine; D, clivorine; E, retusamine; F, n-brosylmitomycin A. Their exact structures are not required for the following discussion. The interested reader is referred to the original work and the review by Nelson Leonard for additional information and references.[41]

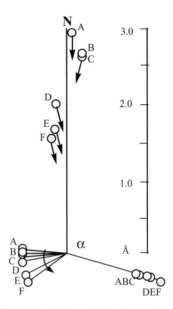

Figure 9.6. Nucleophilic addition to a carbonyl group.

the nitrogen position is represented by a circle, and the lone-pair direction is represented by the downward arrow. Thus, A–C are at long distances (2.5–3.0 Å), while D–F [which are better represented as ammonium salts analogous to (**9**)] are sequentially closer, down to 1.5 Å.

One can also examine the angle that the incoming amino group makes with the C=O bond. This angle turns out to remain fairly constant at $107 \pm 5°$. Note that depending on the distance between the amino and carbonyl groups, the initially planar trigonal system becomes increasingly tetrahedral, as shown for the positions of the rest of the molecule below the carbonyl plane by the symbols A–F in the lower left corner. At the same time, the carbonyl oxygen moves to the lower right, as the bond stretches, increasingly through the series A–F. Taken all together, the diagram represents a reaction coordinate diagram for an amino group adding to a carbonyl group.[42b]

One would also expect, and it is found, that as we go through the series of compounds A–F the carbonyl bond goes from strongly double to almost single. Since the bond becomes weaker through the series, the carbonyl stretching frequency is also greatly reduced. Birnbaum[43] found that for four different alkaloids of known crystal structures where the C⋯N distances varied from 3.0 to 1.8 Å, the corresponding carbonyl stretching frequencies varied from about 1700 to 1600 cm^{-1}.

There are often practical problems with spectroscopic frequencies, especially of polar molecules such as ketones, which can depend upon solvation and concentration. But the qualitative results are clear. The longer N⋯C distance corresponds to a normal ketone, while the shorter distance corresponds to a carbonyl that has undergone a large fraction of the addition and is a much weaker (tending toward single) bond.

The only thing about which the crystallographic study did not provide much information was the energy of the amine–carbonyl interaction. This is a problem that is well suited to investigation by ab initio calculations. We can easily calculate the reaction coordinate for carbonyl additions between simple molecules, and simultaneously obtain the energy of the system as a function of distance and orientation between the molecules.

The experimental work so far described gave us a reasonably good understanding of the interaction between amino groups and carbonyl compounds. Many years later Kulkarni[44] was able to carry out an ab initio study of this type of interaction, utilizing the powerful methods that were developed subsequent to the experimental work. In this case it was expedient to return to the simple molecules to gain an understanding of what the energies were like during the course of the reaction described. MP2/6-31G** calculations[45] were utilized for this work. We will return to the most simple case previously mentioned of the ammonia–formaldehyde interaction.

To carry out these calculations the ammonia nitrogen was fixed at a given distance from the carbonyl carbon, and the ammonia was allowed to orient to the arrangement of minimum energy. (The orientation came out to be as shown in Structure **5** and is in accord with that given earlier by Bürgi et al.[42b]) The distance N⋯C was then incrementally reduced, and the calculation repeated several times, until the curve shown in Figure 9.7 was obtained. The lower curve shown gives an energy minimum of about 4.2 kcal/mol at a distance of 2.85 Å. But we know that this type of calculation overestimates the energy of the interaction because of basis set superposition error (BSSE; see Appendix for additional information), so that needed to be corrected for.[46] When that

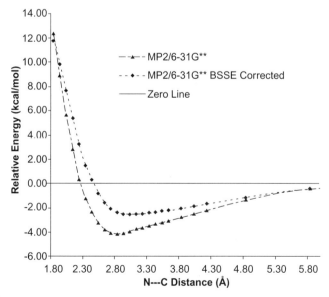

Figure 9.7. Effect of basis set superposition error (BSSE) correction on the ab initio energy profile for the ammonia–formaldehyde pair.[44]

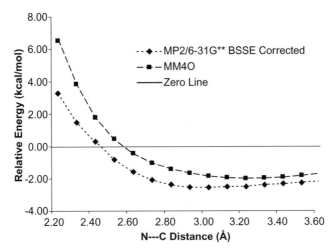

Figure 9.8. Comparison of the MM4O energies without the Lewis bonding parameters, with the BSSE corrected ab initio energies for the ammonia—formaldehyde pair.[44]

was done, the interaction energy is reduced to about 2.6 kcal/mol, and the minimum energy distance is increased somewhat, to about 3.00 Å (Fig. 9.7, upper curve). Note just how small this energy difference is (only about 5% as strong as would be expected for an actual covalent bond). But the actual Lewis bonding interaction must be still smaller. After all, there are electrostatic and van der Waals interactions that also lower the energy of the system, and the energy of those, plus the Lewis bonding energy all together yield a total energy lowering of only 2.6 kcal/mol. To resolve out the Lewis bond energy, we can plot the MM4O interaction energy for the two molecules on the same graph. (MM4O is the original MM4 program, which does not include Lewis bonding. For details, see under Jargon in the Appendix.) The raw MM4O energy includes the van der Waals and electrostatic components of the dimerization but not the Lewis bond energy, so the latter is given by the difference between the two curves (Fig. 9.8).[44] As expected, adding the Lewis bond energy to the electrostatic and van der Waals attractions shown by MM4 pulls the two molecules closer together (from 3.23 to 3.03 Å), and deepens the potential well (by 0.54 kcal/mol). Again, the effect is clear enough, but it is quite small.

Formaldehyde and ammonia are not exactly a typical case for a carbonyl–amine interaction, but this result certainly suggests that whatever Lewis bonding occurs in more typical cases will likely be pretty small. The more typical case is shown in Figure 9.9, which is for the interaction of acetone with trimethylamine.

In this case MM4O gives an interaction energy between the molecules of −2.57 kcal/mol at a distance of 3.38 Å (without Lewis bonding), while the BSSE-corrected QM calculation lowers the energy to −3.22 kcal/mol at the same distance. Again, the Lewis bonding energy appears to be quite small, 0.65 kcal/mol.

We know now, of course, that large basis sets with more complete correlation will give better results, but such calculations were not really feasible at the time of Kulkarni's

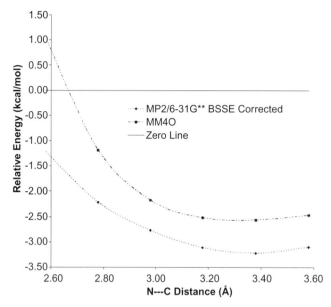

Figure 9.9. Comparison of MM4O energies calculated without use of the Lewis bonding parameters, with the BSSE corrected ab initio energies for the trimethylamine–acetone pair.[44]

work. And we also know that the standard BSSE treatment tends to overcorrect the actual error. Accordingly, he used the above information as a starting point and looked at some real-world examples.

In Figure 9.10 are shown some organic molecules in which amine–ketone intramolecular interactions of the type discussed are known to occur. In each of these molecules there is an amino group near the carbonyl carbon of a ketone, and a Lewis bond between them is anticipated. In each case there is also available a crystal structure, and hence the N⋯C=O bond distance, for each molecule. It is expected that the X-ray position of the amino group is nearer to the carbonyl carbon by about 0.005 Å than the nuclear positions (as the X rays measure the locations of electron density (Chapter 2). But the carbonyl carbon is located by X rays to be about 0.005 Å further from the amino group because of the arrangement of its bonding electrons (Chapter 2). Since these numbers are similar in magnitude and in direction, we will assume that they will approximately cancel, and the experimental N⋯C distances can be compared directly to the MM4 calculated internuclear distances.

In Table 9.6 (column 1) are shown the X-ray values of the N⋯C distances for compounds 1–6, and it can be seen that they range from about 2.47 to 2.76 Å. (These include the shortest distances that could be found where the ketone oxygen remained unprotonated. Of course, the crystal itself provides a medium of relatively high dielectric constant, and tends to "solvate" and stabilize the zwitterionic form of the adduct.) The MM4 values for the sum of the van der Waals radii of these two atoms is 3.80 Å. It is clear that the observed distances are closer than that, which is suggestive of a

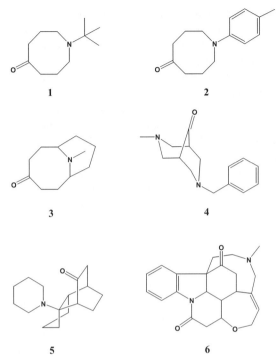

Figure 9.10. Experimental (X-ray) structures of compounds 1–6.

TABLE 9.6. Comparison of N–C Distances (Å) before (MM4O) and after (MM4) the Lewis Bonding Parameterization with the Experimental Observations for Compounds 1–6

Compound	Exptl.	N–C Distance (Å)			
		MM4O	MM4O-Exptl.	MM4	MM4-Exptl.
1	2.692	2.850	0.158	2.673	−0.018
2	2.764	2.974	0.210	2.802	0.038
3	2.472	2.734	0.262	2.556	0.084
4	2.514	2.567	0.053	2.503	−0.011
5	2.765	2.846	0.081	2.774	0.009
6	2.466	2.631	0.165	2.499	0.033
Averaged signed difference			+0.155		+0.022
RMS difference			0.187		0.045

significant attraction between the nitrogen and the carbonyl carbon. Note, however, that in an alkane crystal where there are only van der Waals attractions holding the crystal together, the hydrogens on adjacent molecules come together to within 2.4 Å, or 74% of their van der Waals distance. Here the carbonyl carbon and the nitrogen approach one another to within 65% of their van der Waals distance, which suggests only a modest attraction beyond normal van der Waals attraction in a crystal.

In the second column of Table 9.6 are shown the corresponding N⋯C distances as calculated by MM4O (the *original* MM4 program) *without* any Lewis bonding special interaction, but the normal van der Waals and electrostatic interactions are present. Note that here the MM4O distances range from 2.57 to 2.97 Å for the different compounds. These values are all larger than the experimental distances. The difference (MM4 − Exper.) for each of the compounds is given in column 3 of Table 9.6. The numbers are all positive (+0.05 − +0.26 Å), with the average MM4 distance being +0.16 Å greater than the average experimental distance.

Our interpretation of these results is as follows. The MM4 N⋯C distances are on average approximately 0.16 Å larger than the experimental values. This means that the discrepancy is mainly systematic, and it is larger than one might ordinarily expect when MM4 distances are compared with experiment. The experimental distances are smaller than the sum of the van der Waals radii of the atoms (by 1.0–1.5 Å), as is expected qualitatively from the van der Waals attractions between molecules in the crystal, which leads to a compression of these distances that might be expected to be about 0.8 Å (Chapter 4). There is also an electrostatic attraction that is included in the MM4O calculation. But there also seems to be a significant attraction that is observed experimentally, but is not shown by MM4O. This extra attraction is most reasonably attributed to Lewis bond formation, which was subsequently included in the MM4 program.

The next question to ask is just how big is this Lewis bond force or energy? The force needs to be whatever is required to yield the observed bond distances, and the energy cannot be determined directly from the crystal structure. For that we will return to the ab initio calculation plot shown in Figure 9.9. Here is shown the ab initio energy (with BSSE correction) versus the N⋯C distance when trimethylamine approaches acetone, and the total interaction (attraction) energy is roughly 3.0 kcal/mol at an energy minimum distance of about 3.4 Å. The corresponding MM4 curve is also shown in Figure 9.9. It roughly parallels the ab initio curve and is about 0.5 kcal/mol higher in energy.

The MM4 Lewis bond energies were determined as those necessary to move the MM4O energies down so as to superimpose the curves shown in Figure 9.9, and the results applied to the compounds in Figure 9.10 are included in Table 9.6. The MM4 distances are considerably reduced (last column) and the signed average discrepancy from experiment is now only +0.02 Å, vs. +0.16 Å in MM4O where Lewis bonding was omitted.

Our conclusion is that in the amine-carbonyl interaction there is a Lewis-bond energy, similar to a hydrogen-bond energy, small but easily detectable, about 0.5 kcal/mol vs. 5.0 kcal/mol for the hydrogen bond in water. It seems likely that for the most part such Lewis bonds will be of minimal significance, with possibly some exceptions. When the interaction involves two molecules, as in the acetone-trimethylamine case, the entropy loss upon formation of the Lewis-dimer (roughly 6 kcal/mol at room temperature) will be so great that it will overwhelm the bond formation energy. In that case the interaction will have a negligible effect. The intramolecular case avoids this entropy problem. However, the Lewis-bonded dimer contains a greater amount of charge separation in the general direction of the dipole moment of the complex. Hence the complex is expected to have a larger solvation energy than the component monomer molecules. This greater solvation energy will lead to greater charge transfer, which leads to still greater solvation, and so on. Accordingly, what is a small effect in the gas phase may become more important in solution in polar solvents.

Because of the flatness of the van der Waals function near the energy minimum, rather small energy differences can lead to sizable shifts in the location of such minima along the distance axis. In the average complex in Table 9.6, while the Lewis bond energy amounts to only about 0.5 kcal/mol, the N$\cdots$C distance is shortened by 0.14 Å.

REFERENCES

1. W. M. Latimer and W. H. Rodebush, *J. Am. Chem. Soc.*, **42**, 1419 (1920).
2. (a) L. Pauling, *J. Am. Chem. Soc.*, **53**, 1367 (1931). (b) L. Pauling, *The Nature of the Chemical Bond*, Cornell University Press, Ithaca, NY, 1939.
3. (a) G. C. Pimentel and A. L. McClellan, *The Hydrogen Bond*, Freeman, San Francisco, 1960. (b) G. A. Jeffrey, *An Introduction to Hydrogen Bonding*, Oxford University Press, Oxford, 1997.
4. G. A. Jeffrey and W. Saenger, *Hydrogen Bonding in Biological Structure*, Springer, Berlin, New York. 1991.
5. J. D. Watson and F. H. C. Crick, *Nature*, **171**, 737 (1953).
6. L. Pauling, R. B. Corey, and H. R. Branson, *Proc. Natl. Acad. Sci.*, **37**, 205 (1951).
7. K. Morokuma, *J. Chem. Phys.*, **55**, 1236 (1971).
8. W. Gordy and S. C. Stanford, *J. Chem. Phys.*, **8**, 170 (1940).
9. L. P. Hammett, *J. Chem. Ed.*, **17**, 131 (1940).
10. S. C. Stanford and W. Gordy, *J. Am. Chem. Soc.*, **63**, 1094 (1941).
11. C. Curran, *J. Am. Chem. Soc.*, **67**, 1835 (1945).
12. M. Kasha, *Disc. Faraday Soc.*, **9**, 14 (1950).
13. I. Olovsson and P. G. Jonsson, *Hydrogen Bond*, **2**, 393 (1976).
14. I. Olovsson, *J. Z. Phys. Chem.*, **220**, 963 (2006).
15. M. D. Newton, G. A. Jeffrey, and S. Takagi, *J. Am. Chem. Soc.*, **101**, 1997 (1979).
16. J. Lennard-Jones and J. A. Pople, *Proc. R. Soc.*, **A205**, 155 (1951).
17. J. A. Pople, *Proc. R. Soc.*, **A239**, 550 (1957).
18. K. Kitaura and K. Morokuma, *Int. J. Quant. Chem.*, **10**, 325 (1976).
19. K. Morokuma, *Accts. Chem. Res.*, **10**, 294 (1977).
20. U. C. Singh and P. A. Kollman, *J. Chem. Phys.*, **83**, 4033 (1985).
21. J.-H. Lii, Hydrogen Bonding: in *The Encyclopedia of Computational Chemistry*, Vol. 2, P. v. R. Schleyer, N. L. Allinger, T. Clark, J. Gasteiger, P. A. Kollman, H. F. Schaefer, III, P. R. Schreiner, Eds., Wiley, Chichester, UK, 1998, p. 1271.
22. P. A. Kollman and L. C. Allen, *Theor. Chim. Acta.*, **18**, 399 (1970).
23. M. Dreyfus and A. Pullman, *Theor. Chim. Acta.*, **19**, 20 (1970).
24. C. A. Coulson, *Research (London)*, **10**, 149 (1957).

25. H. Umeyama and K. Morokuma, *J. Am. Chem. Soc.*, **99**, 1316 (1977).
26. K. Morokuma and J. R. Winick, *J. Chem. Phys.*, **52**, 1301 (1970).
27. K.-H. Chen, J.-H. Lii, Y. Fan, and N. L. Allinger, *J. Comput. Chem.*, **28**, 2391 (2007).
28. K.-H. Chen, J.-H. Lii, G. A. Walker, Y. Xie, H. F. Schaefer, III, and N. L. Allinger, *J. Phys. Chem.*, **110**, 7202 (2006).
29. J.-H. Lii and N. L. Allinger, *J. Phys. Org. Chem.*, **7**, 591 (1994).
30. (a) B. Ma, H. F. Schaefer, III, and N. L. Allinger, *J. Am. Chem. Soc.*, **120**, 3411 (1998). (b) J.-H. Lii, B. Ma, and N. L. Allinger, *J. Comput. Chem.*, **20**, 1593 (1999).
31. J.-H. Lii and N. L. Allinger, *J. Phys. Chem.*, **112**, 11903 (2008).
32. B. R. Brooks, R. E. Bruccoleri, B. D. Olafson, D. J. States, S. Swaminathan, and M. Karplus, *J. Comput. Chem.*, **4**, 187 (1983).
33. (a) B. J. Howard, T. R. Dyke, and W. Klemperer, *J. Chem. Phys.*, **81**, 5417 (1984). (b) A. S. Pine, W. J. Lafferty, and B. J. Howard, *J. Chem. Phys.*, **81**, 2939 (1984).
34. J. A. Odutola and T. R. Dyke, *J. Chem. Phys.*, **72**, 5062 (1980).
35. O. Bastiansen, *Acta Chem. Scand.*, **3**, 415 (1949).
36. (a) M. Kazerouni, S. L. Barkowski, L. Hedberg, and K. Hedberg, *Eleventh Austin Symposium on Molecular Structure*, University of Texas at Austin, 1986, p. 33. (b) M. R. Kazerouni, L. Hedberg, and K. Hedberg, *J. Am. Chem. Soc.*, **119**, 8324 (1997).
37. K.-M. Marstokk and H. Møllendal, *J. Mol. Struct.*, **22**, 301 (1974).
38. H. Møllendal, Recent Gas-Phase Studies of Intramolecular Hydrogen Bonding, in *Structures and Conformations of Non-Rigid Molecules*, Ed. by J. Laane, M. Dakkouri, B. van der Veken, and H. Oberhammer, Eds., NATO ASI Series, Series C: Mathematical and Physical Sciences, Kluwer Academic, Dordrecht, The Netherlands. Vol. 410, 1993, p. 277.
39. B. Ma, J.-H. Lii, and N. L. Allinger, *J. Comput. Chem.*, **21**, 813 (2000).
40. Discussed in textbooks on elementary organic chemistry.
41. N. J. Leonard, *Acc. Chem. Res.*, **12**, 423 (1979).
42. (a) M. Kaftory and J. D. Dunitz, *Acta Cryst.*, **B31**, 2914 (1975). (b) H. B. Bürgi, J. D. Dunitz, and E. Shefter, *J. Am. Chem. Soc.*, **95**, 5065 (1973).
43. G. I. Birnbaum, *J. Am. Chem. Soc.*, **96**, 6165 (1974).
44. S. C. Kulkarni, Molecular Mechanics (MM4) Study of Lewis Type Amine-Ketone (N→C=O) Interactions, Thesis, Department of Chemistry, University of Georgia, 1998.
45. M. J. Frisch, G. W. Trucks, H. B. Schlegel, P. M. W. Gill, B. G. Johnson, M. W. Wong, J. B. Foresman, M. A. Robb, M. Head-Gordon, E. S. Replogle, R. Gomperts, J. L. Andres, K. Raghavachari, J. S. Binkley, C. Gonzalez, R. L. Martin, D. J. Fox, D. J. Defrees, J. Baker, J. J. P. Stewart, and J. A. Pople, Gaussian92/DFT, Revision F.2, Gaussian, Inc., Pittsburgh, 1993, and subsequent versions.
46. (a) S. F. Boys and F. Bernardi, *Mol. Phys.*, **19**, 553 (1970). (b) N. R. Kestner and J. E. Combariza, *Basis Set Superposition Errors: Theory and Practice, Reviews in Computational Chemistry*, Vol. 13, K. B. Lipkowitz and D. B. Boyd, Eds., Wiley, New York, 1999, p. 99.

10

CRYSTAL STRUCTURE CALCULATIONS

After a force field has been developed that is suitable for the calculation of molecular structures, it is fairly straightforward to extend it to cover assemblages of molecules, be they in the gas, liquid, or crystalline phase. We will start with the simplest cases first. If we just put two acetone molecules together in the gas phase, with the general planes of the molecules parallel, so that the carbonyl groups are more or less one on top of the other and pointing in opposite directions, a molecular mechanics (or quantum mechanics) calculation will show that the two molecules are held together at substantially less than the van der Waals distances. The individual nonbonded interactions between the two molecules in MM4 calculations are the same (dipole–dipole and van der Waals) as those that would occur intramolecularly. The fact that they are between two different molecules really makes no difference. So if one wants to study gas-phase assemblages of molecules with MM4, in principle one does not have to do anything differently from the way it is done with a study of a single molecule, other than put the molecules together, and then optimize the total structure by minimizing the energy in the usual way. In practice, it may not be quite that simple.

For even small assemblages of molecules, such as dimers and trimers, the *multiple minimum* problem may arise. That is, there may be different ways that one can put the molecules together and obtain minimum energy structures. One will probably want to

---

Molecular Structure: Understanding Steric and Electronic Effects from Molecular Mechanics, By Norman L. Allinger
Copyright © 2010 John Wiley & Sons, Inc.

know all of these structures, and their relative energies. For larger assemblages, the situation becomes increasingly complicated.

In the liquid phase, the studies are necessarily more complicated because many molecules must be assembled in order to represent a liquid-phase structure. To be of practical use for any given problem, one must actually study a very large number of possible molecular arrangements using Monte Carlo methods.[1] A number of bulk properties of liquids can be well calculated in this way, such as the densities and heats of vaporization, for example.

CRYSTALLINE PHASE

The structures of most molecules in crystals are quite similar to their structures when the molecules are in isolation. (The principle exception is that crystals usually contain a single conformation, while fluid phases are likely to contain a Boltzmann distribution of conformations.) So for the most part, MM4 studies on crystals have led to rather few surprises. There are, however, things that can be learned by molecular mechanics from a study of molecules packed into crystals that cannot necessarily be easily learned in other ways. There is some introductory discussion of the usefulness and limitations of X-ray crystallography in determining molecular structures in Chapter 2. If the reader is unfamiliar with the technique, it may be helpful to refer to that discussion and the references cited therein.

When one determines the structure of a molecule by X-ray crystallography, one also determines the structure of the crystal, that is, the details of the way in which the molecules are packed into the crystal. (A nice introduction, much more comprehensive, to material covered in this paragraph can be found in the book edited by Gavezzotti.[2]) There are 230 *space groups*[3] that define just how identical objects can be packed into repetitive three-dimensional arrays. If we have an object of a general arbitrary shape, multiple copies of that object can be packed into a regular lattice only if the lattice corresponds to one or another of these space groups. Thus, in crystallographic studies of molecular structure, determining the space group of a given molecular crystal is an important part of determining the crystal structure. The size and shape of the *unit cell* is also important. If a crystal is looked at the way we look at a brick wall, there are similarities and differences. The bricks are generally of high symmetry, whereas the molecules are not, although they are identical to one another. Bricks are normally packed in a certain specific way, but molecules can pack in 230 different ways. The size and shape of the unit cell of the crystal corresponds to the size and shape of the brick (if we ignore the mortar). However, when one who is not a crystallographer reads the crystallographic literature, it is quickly found that the unit cell is not always pictured to be exactly the way most chemists would have chosen it. Thus, instead of one molecule being considered the unit cell, sometimes the cell is considered as two half molecules, separated by some space, where the right half of the molecule is in the left half of the cell, and the disconnected left half of the molecule is in the right half of the cell. The cell can be chosen with considerable arbitrariness since it needs only to be the entire repeating unit (in three dimensions). It can start at any arbitrary point, and the ways crystallographers choose to picture the cells are frequently as described. There may be just one molecule in the unit cell for a given crystal, but there may be two, or more, however many it takes to make a repeating unit.

Anticipation of Unit Cell

If the unit cell parameters are known, it is often easy to find a possible arrangement of the molecules within the unit cell. Kitaigorodski[4,5] demonstrated this long ago using a mechanical apparatus. This approach should be readily extendable to force-field methods. It requires that the structure of the molecules in the crystal be known in advance, and hence it is a much easier problem than that normally solved in crystallography, where one must determine both the structure of the molecule and that of the crystal at the same time. The structure of the crystal alone should be found relatively easily. Numerous methods for dealing with this problem have become available recently, and are being explored.[2,6]

A Priori Calculations of Crystal Structures

There has been quite a lot of recent interest in the possibility of predicting the crystal structures of molecules using computational methods.[7,8] Some substances are rather nonpolar, and their crystal structures can be treated fairly well with sufficiently good van der Waals potentials in a force field. Other structures are highly polar, and those can be treated fairly well only if one has a sufficiently good representation of the electrostatics in a force field.[2,7] Mooij et al.,[8] for example, have studied with care the crystal structure of acetic acid (strongly polar and hydrogen bonded), and it seems that the crystal can be well described using multipole expansions within a molecular force field. Work by many others has led to trials wherein crystal structures are predicted by computational methods,[2,8] both molecular mechanics and quantum mechanics. The results have gotten better with time, and it looks like this is a problem that is solvable in many cases at present and may be solvable for most cases at some point.

Some molecules (or substances) have more than one stable crystal form. Such a molecule is said to be *polymorphic*. These different crystal forms can usually be identified by different physical properties, such as melting points. They may have the molecules present in different conformations.

Usually, the conformation of a molecule in the crystal is the lowest energy conformation that the molecule can take up in the gas phase. Thus, for example, the *n*-alkanes tend to crystallize in the all-anti conformation. Occasionally, a crystal is found in which a slightly less stable conformation in the gas phase becomes the only conformation in the crystal phase. For this to happen, the less stable gas-phase molecule must form a crystalline lattice that is more stable than the one formed by the anticipated conformation, and by more than the energy difference between the conformations. Or, it may be an unstable polymorph, which will go over to the stable polymorph (and the stable conformation) under thermodynamic conditions.

It is also found that whatever forces lead to stability in molecular dimers will typically lead to stability in crystals. Thus, the dipoles tend to align themselves in an anti

MOLECULAR MECHANICS APPLICATIONS TO CRYSTALS

With MM3 and MM4 there is a useful subsidiary program (CRSTL[9,10]) that is pertinent here. One begins with the MM3 or MM4 structure of the molecule of which one wishes to study the crystal. The molecule is initially treated as a rigid body. It is expedient to actually carry out the molecular mechanics calculations of the crystal by starting with the optimized gas-phase structure of the molecule since the structure of the molecule in the crystal is usually very close to the structure in the gas phase. The CRSTL program will pack the molecules according to any designated space group for any given number of molecules per unit cell. The size and shape of the unit cell (three distances and three angles) are then optimized, keeping the internal coordinates of the molecules fixed.[11] The optimized unit cell coordinates are then fixed, and the internal coordinates of one of the molecules are reoptimized (using MM3 or MM4) in the force field generated by a surrounding block of unit cells. The reoptimized structure is then replicated in the block of unit cells. This cycle is then repeated until the structure corresponding simultaneously to the energy minima for both the unit cell and for the internal coordinates of the molecule is found. These can be compared with experiment, and the agreement is usually found to be good.

It is sometimes of interest to compare the structure found in the crystal lattice with that that exists in the gas phase. The lattice forces in a crystal are fairly strong, so molecules usually distort somewhat in the crystal environment. The force constants for stretching or compressing bonds are roughly 10 times larger than the force constants for bending bond angles. Rotational barriers in saturated systems usually correspond to even smaller force constants. So what one finds in general is that bond lengths do not change much upon packing the molecules into crystals, typically by 0.002 Å or less. Bond angles, on the average, change more, but for the most part, still not very much, typically maybe 0.5°, which is roughly the experimental error in the crystal work and also the calculational error in molecular mechanics. However, in special cases, for example, a large molecule that has a methyl group sticking out of it, sometimes such a methyl may be bent by up to 5° or so away from its gas-phase position by lattice forces. The sizes of torsional barriers vary from negligibly small (as in dimethylacetylene) to quite large (as in ethylene), and hence the torsion angles of the molecules may change by negligible to very substantial amounts, depending on the case.

Comparison of X-Ray Crystal Structure with Calculated Structures

Recall (Chapter 2) that there are a number of differences between an X-ray structure and the same structure as determined by any other method, experimental or calculated. Depending on the problem at hand, these differences between the same structure determined by two different methods may be quite important, or negligibly so. X-ray structures differ from the others in some important ways. First, the X-ray structure is determined by the positions of the electron densities corresponding to the atoms. All other structural methods that we have discussed depend upon the nuclear positions, and this difference may be important. X-ray structures also suffer from what is called *rigid-body motion*. This has little effect upon angles but can cause bond lengths to artifically appear shortened by 0.005–0.015 Å if determined at room temperature (less if determined at low temperatures). Vibrational amplitudes and, hence, things like the unit cell size and refractive index are temperature dependent.

In molecular mechanics it is normally assumed that the system is at 25°C. It is possible to work at other effective temperatures, but this is not normally done. The comparison between quantum mechanical calculations and X-ray crystal structures tends to be even more difficult. The ordinary quantum mechanical structure reported in the literature refers to a point on the potential energy surface (the equilibrium point, r_e), corresponding to the nuclear positions of minimum energy. The accuracy to which that point is located depends in part on the size of the basis set and the amount of electron correlation included in the calculation. Sometimes potential energy surfaces are nearly flat over a substantial area, containing undulations that are of the order of 0.3 kcal/mol in amplitude or less. What this means in reality is that the lowest vibrational level probably lies above the maxima of the waves; hence, the minima do not exist as actual physical points. Rather, the structure is simply vibrating back and forth across the entire area (or delocalized over the area in quantum mechanics). It may be that one is interested in such a surface, but for the most part, things like this should be considered as simple artifacts to be worked around, and they are rarely anything with which to be concerned. In the earlier literature, however, this situation led to meaningless discussions about how many conformations exist for a given molecule. In the following discussion, these distractions may be occasionally noted where they are of interest, but they need to be kept in mind at all times, so that one is not misled in any given comparison.

Benzene Crystal

This crystal is particularly instructive. If we have two benzene molecules and allow them to come together in the gas phase to form a dimer, one might have anticipated that they would stack like two plates, in such a way as to maximize the van der Waals interactions between them, as in (**A**) in Structure **1**. But this is not what is found. Rather, if one benzene ring defines a plane, the second benzene is found to be perpendicular to that plane, as in (**C**) or (**D**). (This is known because the dimer has a dipole moment.[12])

Structure 1. Benzene Dimers.

In terms of molecular mechanics, the reason for this is that each C–H bond in benzene has a bond dipole, with the hydrogen positive, the carbon negative, and a magnitude of about 0.6 D. If one stacks the benzene rings together like two plates as in (**A**), one on top of the other, these dipoles are aligned in an unfavorable way, with all of the positive hydrogens in one molecule near to the positive hydrogens of the other molecule, and all of the negative carbons similarly near to each other. The electrostatic situation can be improved somewhat by rotating one of the benzene rings 60° so that the hydrogens on one are in between the hydrogens on the other one. But still better, one can slide the ring in such a way as to move the positive hydrogens away from each other, and nearer to the negative carbons, as in (**B**). Alternatively, one ring can become perpendicular to the other, so that positive hydrogens are close to negative carbons, as in (**C**) or (**D**). These things have been well studied, both by quantum mechanics[12,13] and by molecular mechanics.[14]

When we look at the benzene crystal, we find that it does not have a structure corresponding to a stack of plates (**A**), as would be indicated by the maximization of the van der Waals attractions. (In fact, Williams showed as early as 1966 that van der Waals interactions alone do not lead to the observed crystal structure.[15]) We can imagine the stack of plates leaning in the sense that if we slide each plate a little ways in the same direction relative to the one below it [from (**A**) to (**B**) and continuing the process with a tall stack of rings], then we will get a leaning stack in which the negatively charged atoms on one plate are nearer to the positively charged atoms on the neighboring plates and as is consistent with the dimers. But, if we push the stack a little too far, then one benzene ring can go over the edge and kind of around the one below it [as in (**D**)]. If we look at the actual crystal structure (Fig. 10.1), that (**D**) type of structure is what we find.[16]

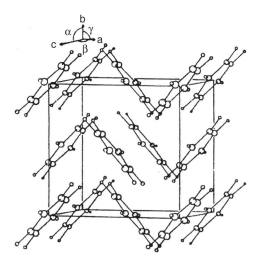

Figure 10.1. Benzene crystal structure. (From Allinger and Lii.[14] Copyright 1987 by John Wiley & Sons. Reprinted by permission of John Wiley & Sons.)

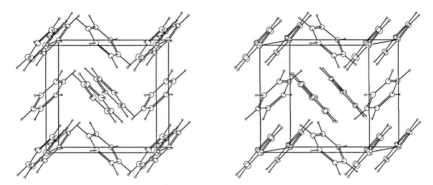

Figure 10.2. Benzene crystal structure (stereographic projection).

The rings are each perpendicular to a neighboring ring, with the positive hydrogens near the negative carbons, and this structure is replicated in three dimensions. And not surprisingly to us now, the structure of liquid benzene[17] was said to be "surprisingly similar to that of the solid."[17a] A stereographic projection of the benzene crystal is also shown in Figure 10.2. While it is more difficult to see (at least for this author), it shows some finer nuances of the structure that cannot be seen in Figure 10.1.

Interestingly, the crystals of many other aromatic hydrocarbon molecules such as naphthalene and biphenyl also have a benzene-like structure and for similar reasons. Namely, the basic structure is determined largely by the electrostatics, rather than by the van der Waals forces, and there is a similar repetitive arrangement of bond dipoles throughout the series. But also quite common are crystal structures like that of hexamethylbenzene, which are not like that of benzene but rather of the "leaning stack of plates" type (**B**).[18] Here the positive substituents (the methyl groups) are much further from the negative ring carbons because of the larger physical size of the methyl in this molecule than of the hydrogen in benzene. The balance between the electrostatics and van der Waals forces here is an intermediate case. And so the (tilted) stack of plates type structure is actually found (Fig. 10.3) in hexamethylbenzene, while in benzene itself the stack has tipped over and rearranged.

All of the above is what one finds from the MM3[10] and MM4[19] calculations, and it is also what is found experimentally. From the MM4 calculations one can also correctly calculate both the crystal structures and the heats of sublimation for both the benzene and hexamethylbenzene crystals. We conclude that the MM4 model here is a good one. And the details of all of the electrostatic and van der Waals interactions are also available from the molecular mechanics calculations.

The benzene crystal also gives us an example of an "experiment" that can be carried out by calculation but that cannot be carried out in the laboratory! (Such experiments are sometimes said "to have been carried out *in silico*.") One would predict that if there were no bond dipoles, the benzene crystal structure of type (**A**) would be more stable than that of (**B**), and molecular mechanics calculations on the dimers agree. If we start with a vertical "stack" of plates for the benzene structure, we can study the

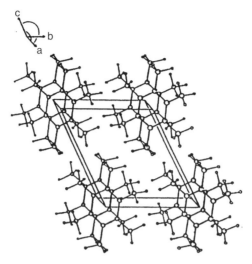

Figure 10.3. Hexamethylbenzene crystal structure. (From Lii and Allinger.[10] Copyright 1989 by the American Chemical Society. Reprinted by permission of the American Chemical Society.)

effect that varying the magnitude of the C–H bond dipole has on the structure. When the C–H bond dipole was changed over a range of test values in MM4, and the energy minimization calculation repeated for each, (**A**) was indeed found to be more stable than (**B**) when the C–H bond dipole had the starting value of zero. As the dipole was increased (0.1 C, 0.2 D, etc.) and the calculation was repeated, the plates start to slide away increasingly to new minimum energy geometries, yielding structures like (**B**). Eventually, when the bond moments become large enough, the top plate slips over the side of the bottom plate to give a (**D**) type of structure (which in three dimensions becomes the structure shown in Fig. 10.1). Thus, we find that the MM4 (and experimental) bond dipole of the C–H bond in benzene (0.60 D) happens to be somewhat more than enough to give benzene the crystal structure that is observed.

Biphenyl

This is a well-studied case that might be mentioned. In the gas phase the biphenyl molecule has the two rings rotated with a torsion angle of about 42° between them.[10,20] The biphenyl molecule is apparently planar in the crystal structure (Fig. 10.4), although there has been a lengthy and perhaps still not completely settled disagreement about this.[21]

The barriers to rotation of biphenyl are 1.8 and 2.5 kcal/mol at 90° and 0°, respectively.[19] The planar structure (0°) is clearly resonance stabilized, but when the molecule is planar, the ortho hydrogens on one ring come rather close to the ortho hydrogens on the other ring, and there is a substantial van der Waals repulsion between them. So one set of forces (resonance) is trying to planarize the molecule (torsion angle 0°), while

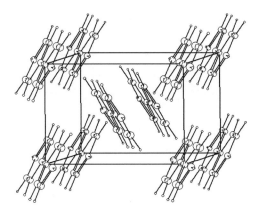

Figure 10.4. Biphenyl crystal structure.

the other set (van der Waals repulsions between *ortho*-hydrogens) is tending to push the torsion angle toward 90°, and the result comes out somewhere in between. But if the molecules are to be packed into a crystal structure, one can imagine that the planar biphenyl molecules can and do easily pack in a way that is quite analogous to the crystal packing of benzene itself (Fig. 10.1). It's not so clear what to expect with the nonplanar molecule, but it cannot optimally pack both rings at the same time into a benzene-like lattice. In order to overcome the torsional energy and planarize, the heat of sublimation of the planar biphenyl crystal would have to be increased (the energy lowered) by more than the height of rotational barrier to planarity, or the molecule would remain nonplanar in the crystal. The heats of sublimation of the crystalline biphenyl were also calculated by MM4, for both the planar and (constrained) nonplanar conformations.[19] The value for the planar conformation was greater than that for the nonplanar by 1.90 kcal/mol, so clearly the extra energy of crystal packing was sufficient to overcome the barrier to planarity in the molecule. The MM4 calculated heat of sublimation for biphenyl is 18.42 kcal/mol versus 16.52 for the nonplanar form (18.5 ± 0.5 experimental).[22]

Ditrityl Ether

Let us consider the specific example of di-trityl ether[23] (Tr-O-Tr), a stereographic picture of which is shown in Figure 10.5. The C–O–C angle in this molecule is opened rather widely in the crystal, 135.8°, because of the substantial steric repulsion between the two trityl groups.[24] (The usual value for the C–O–C angle in a dialkyl ether is only about 111.5°.[23]) But, of course, this is an X-ray value, so the lone pairs on oxygen make the oxygen appear to be about 0.005 Å further away from the attached carbons than the nuclear positions. If we correct the X-ray value for this, the C–O–C angle opens slightly to 136.2 Å. The MM4 (gas phase) value is 132.9°. We worried about the reason for this sizable discrepancy (3.3°) and, accordingly, carried out the MM4 calculation on the crystal of the molecule. It was found that indeed the angle was squashed by the crystal lattice, and it opened up to a value of 135.3° in the MM4 calculations on the crystal,

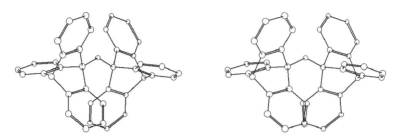

Figure 10.5. Ditrityl ether (stereographic projection). The hydrogens have been deleted for clarity.

in reasonable agreement (0.9°) with experiment for such a large distortion. The difference between the MM4 values calculated for the crystal and for the gas phase (2.4°) shows a potential hazard of comparing a gas-phase structure with a crystal structure. One might have thought that since the C–O–C bond angle has quite a large bending force constant (it is already opened from the usual value of a dialkyl ether by over 25°!) that it would be very resistant to further opening. But nonetheless, it opens as stated, indicating that these lattice forces are indeed very strong in this case.

In principle, one can find the unit cell of lowest energy for any molecule by optimizing the structure (minimizing the energy) for each space group and, thereby, predict the crystal structure and lattice energy of the molecule.[7] We tried to interest crystallographers in pursuing this problem after we developed the CRSTL program (1985)[9,10] but failed to convince anyone to undertake it. The general difficulty that was foreseen was that for most compounds there will be many different space group structures found that correspond to energy minima, and typically many of the energies of the various structures will lie within a range of a few tenths of a kcal/mol of one another. The molecular mechanics calculations were not expected to be of sufficient accuracy with respect to these energies so as to really say which of these would actually be the most stable, and quantum mechanical calculations for such problems were still much too primitive in 1985. But if one knows the preferred crystal structure, then MM4 will calculate that structure (in numerous cases studied) with reasonably good accuracy. Once one has the structure, and the unit cell dimensions (determined by minimizing the energy of an assemblage), one can in principle also calculate other crystal properties, such as the heat of sublimation of the crystal. For crystalline hydrocarbons (the only class of compounds so far studied in any detail) that are not too large, these heats generally range from perhaps 5–30 kcal/mol over a wide range of typical molecules. Their heats can be calculated with an accuracy of perhaps 0.5–1 kcal/mol. The different space groups will give different crystal structures for a molecule, and not all of them will correspond to energy minima. Sometimes one space group will have its structure go over to that of another during the energy minimization, unless symmetry constraints are used in the calculation. In other cases there are small or large barriers that prevent or at least hinder this. It is well known that many substances, probably most, are *polymorphic*. That is, they typically form different kinds of crystals, with different space groups, and with different melting points. If a substance is polymorphic, then there must be a barrier separating the corresponding different space group lattices. Such barriers are often pretty small, and sometimes one polymorph goes over to another simply upon standing or more usually upon warming.

More of the [18]Annulene Story

Recall that as of about 1970 there were two structures available for [18]annulene (Chapter 5). One had approximately a D_{6h} structure, roughly planar with equivalent benzenoid bond lengths, and the other had a slightly non-planar alternating bond structure (approximately D_3), essentially that of a polyene. Not unexpectedly, the [18]annulene molecules pack into the crystal in the motif of the "stack of pancakes," similar to hexamethylbenzene (Fig. 10.3). There was crystallographic evidence for the D_6 structure that seemed correct, proper, and definitive, for the structure of the molecule in the crystal.[25a] Since the symmetry of the crystal is different from the symmetry of the molecule, the packing forces on different atoms that would be equivalent in the isolated molecule differ somewhat in the crystal. Thus, the structure is not expected to be exactly D_6 or D_{6h} because of the packing deformations, but the difference from D_6 symmetry should be (and is) small. But there was also evidence from quantum mechanical calculations of the electronic transitions in the ultraviolet region (gas phase) that agreed quite well with the D_3 structure, and not at all with the D_6 structure as measured experimentally in solution. The problem remained in that state for many years.

In 1995 Gorter et al.[25b] redid the earlier X-ray structure using modern instrumentation and methods. Their temperature was a little higher than the original work, however (111 versus 80 K). In the more recent work it was found that there was some disorder in the crystal, where about 14% of the molecules were rotated 30° about their principle axis, relative to the rest of the molecules. This was properly dealt with and corrected for. They also reexamined the original experimental data from 1965 and found that in that work there was a 6% disorder that was not recognized originally, which they also could and did correct for. (The disorder is apparently temperature dependent, and only 6% at the lower temperature had little effect on the results.) But for our purposes, the overall conclusions were really the same from both pieces of work, namely the bond lengths were approximately equivalent and not alternating. That is, the structure was approximately D_6.

The explanation usually offered until 1995 for this discrepancy in structures between the crystal and the solution/gas phase was that structures were simply different and phase dependent. This meant that the lattice forces in the crystal deformed the structure away from D_3 to D_6 (or from alternating bond lengths to equivalent bond lengths). This is conceivable, and such deformation might be suspected for a molecule in which there were large charges or the possibility for charge transfer existed. But it would not be expected with a simple polyene. See, however, the work by Ermer[26] wherein a suggestion is offered as to how this might occur. In 1971, another important experiment was reported.[27] It showed that the proton NMR was a single line at room temperature but separated into two lines, one at high field (6 protons) and one at low field (12 protons), at low temperatures (in solution). These lines are widely separated and showed the presence of a strong ring current, which is usually equated with aromaticity.

Quantum mechanical calculations on the structure of [18]annulene were carried out repeatedly over the years and are worth a little discussion here. In the 1960s and earlier, one was limited to assumed geometries and π-system calculations using very small basis sets. As time went by, all electron calculations using larger basis sets could be carried out, and geometry optimization became possible. The accuracy of the calculations generally increased a great deal. But [18]annulene was, and still is, a relatively large molecule for quantum mechanics. Already in the 1960s, it was found that Hartree–Fock calculations with relatively small basis sets gave the D_3 structure as the most stable. But when electron correlation was included in the calculation, it stabilized both the D_3 and the D_6 structures, of course, but it stabilized the D_6 structure much more. And when sufficient correlation was included, the D_6 structure became more stable.[28] In 2004, fairly detailed all-electron quantum mechanical calculations were reported[29], but did not really settle the question. Depending on the details of the correlation calculation, the range of energies obtained favored the C_2 structure by up to 20.45 kcal/mol, or disfavored it by as much as 15.06 kcal/mol. So the calculated energy differences here are very large, and the basis set/correlation calculations are still at such a level that the results are not convincingly definitive. But what they do currently show is that the D_{6h} structure is a transition state (one imaginary vibrational frequency at 1218 cm^{-1}), between two ground-state structures with alternating bond lengths, but only 2.1 kcal/mol higher in energy. This means that the molecule has alternating bond lengths in the ground state, and that the double bonds all shift at the same time.

Both conformations (equal and alternating bond lengths) correspond to stationary points on the energy surface, and it is now possible to calculate fairly accurate chemical shifts for the two conformations.[29] It was found that the alternating bond length structure gave chemical shifts in reasonable agreement with the experimental values (in solution). The equivalent bond lengths, on the other hand, gave very poor agreement with the observed chemical shifts. These chemical shifts are huge, and the protons within the ring are, of course, shifted in the opposite direction from those outside the ring by the ring current. These chemical shift data obtained experimentally and by calculation are shown in Table 10.1.

(For the moment, consider the C_2 structure in Table 10.1 to be the same as a D_3 structure. It has alternating bond lengths.) Note that the low field frequencies calculated for the D_{6h} structure and the X-ray structure are similar (11.5 and 10.2) and that calculated for the C_2 (9.0) is lower and in better agreement with experiment (9.3). The high field absorptions for the QM D_{6h} and X-ray structures are −10.9 and −12.5, while the QM C_2 value (−2.4) differs markedly from both of those but is in good agreement with experiment (−3.0).

There is no question but that the agreement of the QM calculations with the experiment is only consistent with the molecule having alternating bond lengths in the gas phase and in solution. Again, this is the same conclusion that was reached from the ultraviolet spectra in 1969, namely the alternating bond structure is the stable one in the gas phase and in solution.[30] Chemical shift data of the molecule in the crystal are not available at this writing. So does this really mean that the structure of the molecule

TABLE 10.1. NMR Chemical Shifts for Proton Absorptions in [18]Annulene[a]

Structure		Field	
		High (6)	Low (12)
Experimental		−3.0	9.3
Calculated			
QM[b]	D_{6h}	−10.9	11.5
X-ray		−12.5	10.2
QM	C_2	−2.4	9.0

[a]Relative to tetramethylsilane (TMS) with a computed proton shielding of 31.98. The relative band areas are given in parentheses.
[b]QM here indicates a 6-311+G** basis with a wide range of correlation calculations.

is different in the crystal from that which exists in solution? In this author's view, that is not likely. In the 1969 calculation[30] it was estimated that the D_6 structure had an energy approximately 10 kcal/mol above that of the D_3 structure. The best current number is 2.1 kcal/mol.[29] Either way, on the X-ray timescale (a matter of perhaps an hour, or a day for the earlier work), the structure would average to D_6, consistent with the crystallographers' conclusions. But if one looks on a very short timescale (subsecond, as in the NMR and ultraviolet spectra), one sees the properties that correspond to a D_3 structure. One needs to note that there are two structures for the D_3 molecule that are otherwise indistinguishable, but in which the alternating long and short bonds change places. For this to happen, since the difference in length between a long and short C–C bond is only 0.10 Å, each carbon needs to move by only about 0.07 Å. The barrier to this motion appears to be pretty small. Accordingly, the structure over the X-ray timescale would not have one carbon (of the 18 total) at the center of the ellipsoid for thermal motion as described by the (Bregman/Gorter) crystal structure,[25] but rather it would have two half-carbons (36 total), one on each side of the central position of the ellipsoid, separated by about 0.07 Å. The thermal ellipsoid has a length along the major axis of about 0.5 Å, and these motions of the carbons, officially a disorder in crystallography, are simply buried in those ellipsoids. This complete disordered structure can probably be squeezed out of the X-ray data with further work and is expected (by the present author at least) to show that the structure does in fact have alternating long and short bonds. Or alternatively, the two structures (the degenerate D_3 and the D_6) will fit the data equally well. So in this view, the structure is in fact a function of the time required to do the experiment by which the structure is determined.

It needs to be added before we leave this topic that Schleyer's quantum mechanical calculations[29] indicate that the structure is not actually D_3 but C_2. This structure differs slightly from the D_3 structure (and still has alternating bond lengths), and not only has a lower energy, but it fits the NMR data better. It is quite similar to the D_3 structure, but while the long bonds are always long, and the short bonds always short, the long bonds differ slightly from one another and similarly for the short ones. The result is C_2 rather than D_3 symmetry. The difference between these two structures is quite small.

If the two structures were accurately drawn and placed side by side, the difference is too small to see by just looking at them. They are really similar but not identical.

So is [18]annulene aromatic? In this case one can definitely say yes or no, depending on which of the available definitions of aromaticity one chooses. The heat of formation of the compound is known to be 124.0 kcal/mol experimentally,[31] and the MM4 value is 123.4 kcal/mol.[32] Since the experimental error is reported as 6.0 kcal/mol, this means that the MM4 value is in agreement with experiment. There are different ways that one can calculate numerical values for aromaticity. We use basically the Dewar scheme,[33] as appropriately modified to include in molecular mechanics calculations.[34] (Other methods generally give systematically larger but otherwise similar numerical values.) Additionally, we require that the aromaticity be calculated only for a planar system.*

At any rate, [18]annulene is substantially planar, the small amount of out-of-plane bending will have little effect for present purposes. In the Dewar approach to aromaticity, one takes as the reference point ordinary linear planar trans polyenes and defines them to have a resonance energy of zero. The energy of the π system for any conjugated hydrocarbon molecule is then calculated relative to planar polyenes. This method gives benzene a resonance energy of about 20 kcal/mol whereas older methods (designed for hand calculations, and using a less reasonable zero of reference) usually placed the resonance energy of benzene about 30 kcal/mol. On the Dewar scale, when the MM4 resonance energy of [18]annulene is calculated, it comes out to be –0.8 kcal/mol. A minus number here means antiaromatic or less stable than the reference polyene. In other words, the magnitude of the aromaticity is insignificant to within the error of the calculation, and that which is calculated to exist is actually slightly antiaromatic. It leads to a (negligible) destabilization rather than to a stabilization of the molecule.

So the thermochemical definition of aromaticity is not met by [18]annulene, and by that definition, the compound is not aromatic. But note also that it is not significantly antiaromatic either. It is just a polyene, and the chemical behavior of the compound is consistent with this statement. For example, it reacts with air and polymerizes (turns brown) on standing.

Turning to the ring current definition of aromaticity,[35] the chemical shift calculations for different structures for [18]annulene can now be carried out in a straightforward manner by quantum mechanics. These calculations[29] predict quite a large ring current, and the chemical shifts are substantially larger for the inner protons if there is bond alternation than if the bonds were to be "equivalent." (By equivalent here is meant "similar to those in benzene and to each other," and the quantum mechanical stationary

---

*If one tries to calculate aromaticity of a nonplanar system, the energy effect due to aromaticity is mixed together with various kinds of strain energies that are a function of the details of the particular system under examination, and these energies can be separated only based on some sort of model. While this can lead to useful information, we here (and usually) limit our discussion of aromaticity to strictly planar systems because the results otherwise may become too model dependent. (Utilizing MM4, this normally presents no problem because the system can be "planarized" by removing the direction cosine terms from the calculation and just squashing the molecule into a plane. One can then calculate whatever properties one wants for that planar system and discuss those, and one needs not to worry about what happens when the system is deformed out of planarity, although that is separately calculated.

point was used for this purpose.) What was found was that the ring current is substantial in any case, but even larger when the ring has alternating rather than equivalent bond lengths. By the NMR criterion then, we might conclude that [18]annulene is aromatic, and even more so if the bond lengths alternate in length than if they are the same. This latter fact is in direct conflict with the thermochemical conclusions, and also with the idea that aromatic means equal (or similar) bond lengths.

Thus, while we really understand quite well the "aromaticity" of [18]annulene, we do not at this point have a generally agreed upon way to define it. In the present author's view, the molecule is not aromatic since it has a negligible resonance energy, and it does have bond alternation, *and* a large ring current. If a molecule has a large ring current, this tells us that the molecule has a large ring current. It does not tell us about bond length alternation, thermochemical stability, or aromaticity.

There are some other points of interest that need to be made here before we leave this subject. X-ray crystallography is an extremely powerful tool, as is now well known. However, as with all other tools, there are limitations. One limitation is that in any actual experiment the X rays used have a certain wavelength (0.70 Å for the molybdenum X rays used by Gorter.[25b] The other widely used X-ray source, copper, has an X-ray wavelength of 1.5 Å, and for the occasionally used silver, the value is 0.5 Å). When one uses electromagnetic radiation to "look at" something, one wants if possible to use a radiation wavelength that is *smaller* than the object being viewed. Preferably quite a lot smaller, as this gives better resolution. As far as X-ray crystallography is concerned, what this means as "a rule of thumb" is stated by Glusker and Trueblood[36] in the following way: "Since the wavelength of the X-ray beam is generally about 1 Å, one will not be able to resolve objects that are closer together than about 0.5 Å."

Most bond lengths are longer than 1 Å, and so bound atoms come close together, but not too close for X-rays to be useful for investigating them. But in this case the carbons are approximately 0.07 Å apart (i.e., the two half-carbons that will be seen before and after moving). This means that these atoms will not be resolved by X rays in the usual sense. In other words, following the standard crystallographic procedures as used by Bregman and Gorter,[25] one would expect to obtain a structure as they did (with equivalent bond lengths), no matter whether the structure actually is that or whether it is the alternating bond structure, and the bonds are either exchanging or packed rigidly, but in a random way. Hence, the conclusion appears to be that the X-ray experiment currently does not differentiate the two structures in question (D_3 or D_6). Since everything else points to a D_3 structure, Occam's razor says that we should choose the simplest alternative where the bond lengths are alternate.

It might be possible to resolve the crystallographic structure problem in either of two ways. One way would be to simply treat the molecule as 1:1 disordered as is frequently done in X-ray work. (Gorter[25b] considered doing this but did not for reasons that do not seem compelling to this author.) Or one might start with a model with each atom represented as two half-atoms, with the distance between them held fixed during the refinement. Either of these constraints should, in principle, refine to a structure of lower R value (or, in practice, at least just as good) as was obtained from the previously used refinement procedure.

As mentioned by Ermer,[26] even the crystal structure of benzene could be in doubt for the same reason as in the [18]annulene case. We know for sure that the crystal structure of benzene has, on the time average, equal bond lengths. But what about the instantaneous structure? There also exist electron diffraction and quantum mechanical structures, but again, all of the same problems discussed for [18]annulene could also present difficulties with the structure of benzene. On the other hand, quantum mechanics at a sufficiently high level tells us that [18]annulene has alternating bond lengths, while benzene has equal bond lengths at all levels of calculation (to date).* And what about chlorophyll and its relatives (Chapter 5)?

REFERENCES

1. (a) W. L. Jorgensen, *J. Phys. Chem.*, **90**, 1276 (1986). (b) R. Q. Topper, D. L. Freeman, D. Bergin, and K. R. LaMarche, *Computational Techniques and Strategies for Monte Carlo Thermodynamic Calculations, with Applications to Nanoclusters, Reviews in Computational Chemistry*, Vol. 19, K. B. Lipkowitz, R. Larter, and T. R. Cundari, Eds., Wiley, New York, 2003, p. 1.
2. A. Gavezzotti, Ed., *Theoretical Aspects of Computer Modeling of the Molecular Solid State*, Wiley, New York, 1997.
3. *International Tables for Crystallography*, T. Hahn, Ed. *Vol. A, Space Group Symmetry*, Reidel, Boston, 1996. This topic is also discussed in introductory books on crystallography.
4. A. I. Kitaigorodskii, *Izw. AN SSSR. Otd. khim. n.* **6**, 587 (1946).
5. A. I. Kitaigorodskii, *Organic Chemical Crystallography*, Consultants Bureau, New York, 1961.
6. W. Linert and F. Renz, *J. Chem. Inf. Comput. Sci.*, **33**, 776 (1993).
7. (a) R. J. Gdanitz, in *Theoretical Aspects and Computer Modeling of the Molecular Solid State*, A. Gavezzotti, Ed., Wiley, New York, 1997, p. 185. (b) P. Verwer and F. J. J. Leusen, *Computer Simulation to Predict Possible Crystal Polymorphs, Reviews in Computational Chemistry*, Vol. 12, K. B. Lipkowitz and D. B. Boyd, Eds., Wiley, New York, 1998, p. 327.
8. (a) W. T. M. Mooij, Ab Initio Prediction of Crystal Structures, Ph.D. Thesis, University of Utrecht, The Netherlands, 2000. (b) W. T. M. Mooij, B. P. van Eijck, S. L. Price, P. Verwer, and J. Kroon, *J. Comput. Chem.*, **19**, 459 (1998). (c) W. T. M. Mooij, B. P. van Eijck, and J. Kroon, *J. Phys. Chem.*, **A103**, 9883 (1999). (d) W. D. S. Motherwell, H. L. Ammon, J. D. Dunitz, A. Dzyabchenko, P. Erk, A. Gavezzotti, D. W. M. Hofmann, F. J. J. Leusen, J. P. M. Lommerse, W. T. M. Mooij, S. L. Price, H. Scheraga, B. Schweizer, M. U. Schmidt, B. P. van Eijck, P. Verwer, and D. E. Williams, *Acta Cryst., Section B, Structural Science*, **B58**(4), 647 (2002). (e) W. T. M. Mooij, B. P. van Eijck, and J. Kroon, *J. Am. Chem. Soc.*, **122**, 3500 (2000).
9. J.-H. Lii, Ph.D. Dissertation, University of Georgia, Athens, GA, 1987.

---

*It is generally agreed that quantum mechanics will give the "right answers," if the calculations are carried out to the Schrödinger limit. Just how close to that limit "current-day" calculations actually come has varied over time but is still, perhaps, less close than is commonly assumed. Note that various types of widely used large basis set calculations incorrectly showed benzene to be seriously nonplanar as recently as 2006.[37]

10. J.-H. Lii and N. L. Allinger, *J. Am. Chem. Soc.*, **111**, 8576 (1989).
11. (a) W. H. Press, B. P. Flannery, S. A. Teukolsky, and W. T. Vetterling, Downhill Simplex Method in Multidimensions and Linear Programming and the Simplex Method, §10.4 and 10.8 in *Numerical Recipes in FORTRAN: The Art of Scientific Computing*, 2nd ed., Cambridge University Press, Cambridge, England, 1992, pp. 402–406 and 423–436. (b) A. Forsgren, P. E. Gill, and M. H. Wright, Interior Methods for Nonlinear Optimization, *SIAM Rev.*, **44**, 525–597 (2002).
12. K. C. Janda, J. C. Hemminger, J. S. Winn, S. E. Novick, S. J. Harris, and W. Klemperer, *J. Chem. Phys.*, **63**, 1419 (1975).
13. J. B. Hopkins, D. E. Power, and R. E. Smalley, *J. Phys. Chem.*, **85**, 3739 (1981).
14. N. L. Allinger and J.-H. Lii, *J. Comput. Chem.*, **8**, 1146 (1987).
15. D. E. Williams, *J. Chem. Phys.*, **45**, 3770 (1966).
16. (a) E. G. Cox, D. W. J. Cruickshank, and J. A. S. Smith, *Proc. Roy. Soc. (London)*, **A247**, 1 (1958). (b) G. Karlström, P. Linse, A. Wallqvist, and B. Jönsson, *J. Am. Chem. Soc.*, **105**, 3777 (1983).
17. (a) A. H. Narten, *J. Chem. Phys.*, **48**, 1630 (1968). See especially p. 1634. (b) L. S. Bartell, L. R. Sharkey, and X. Shi, *J. Am. Chem. Soc.*, **110**, 7006 (1988).
18. L. O. Brockway and J. M. Robertson, *J. Chem. Soc.*, 1324 (1939).
19. N. Nevins, J.-H. Lii, and N. L. Allinger, *J. Comput. Chem.*, **17**, 695 (1996).
20. (a) O. Bastiansen and M. Traetteberg, *Tetrahedron*, **17**, 147 (1962). (b) R. Barrett and D. Steele, *J. Mol. Struct.*, **11**, 105 (1972). (c) A. Almenningen, O. Bastiansen, L. Fernholt, B. N. Cyvin, S. J. Cyvin, and S. Samdal, *J. Mol. Struct.*, **128**, 59 (1985).
21. (a) G. P. Charbonneau and Y. Delugeard, *Acta Crystallogr.*, **B32**, 1420 (1976). (b) J. L. Baudour, L. Toupet, Y. Delugeard, and S. Ghemid, *Acta Cryst.*, **C42**, 1211 (1986). (c) A. T. H. Lenstra, C. Van Alsenoy, K. Verhulst, and H. J. Geise, *Acta Cryst.*, **B50**, 96 (1994).
22. R. S. Bradley and T. G. Cleasby, *J. Chem. Soc.*, 1690 (1953).
23. N. L. Allinger, K.-H. Chen, J.-H. Lii, and K. Durkin, *J. Comput. Chem.*, **24**, 1447 (2003).
24. H. Iwamura, T. Ito, H. Ito, K. Toriumí, Y. Kawada, E. Osawa, T. Fujiyoshi, and C. Jaime, *J. Am. Chem. Soc.*, **106**, 4712 (1984).
25. (a) J. Bregman, F. L. Hirschfeld, D. Rabinovich, and G. M. Schmidt, *Acta Cryst.*, **19**, 227 (1965). (b) S. Gorter, E. Rutten-Keulemans, M. Krever, C. Romers, and D. W. J. Cruickshank, *Acta Crystallogr., Sect. B*, **51**, 1036 (1995).
26. O. Ermer, *Helv. Chim. Acta*, **88**, 2262 (2005).
27. (a) J. F. M. Oth, *Pure Appl. Chem.*, **25**, 573 (1971). (b) C. D. Stevenson and T. L. Kurth, *J. Am. Chem. Soc.*, **122**, 722 (2000). (c) C. D. Stevenson, personal communication to P. v. R. Schleyer.
28. H. Bauman, *J. Am. Chem. Soc.*, **100**, 7196 (1978).
29. C. S. Wannere, K. W. Sattelmeyer, H. F. Schaefer, III, and P. v. R. Schleyer, *Angew. Chem., Int. Ed.*, **43**, 4200 (2004).
30. F. A. Van-Catledge and N. L. Allinger, *J. Am. Chem. Soc.*, **91**, 2582 (1969).
31. J. F. M. Oth, J. C. Bünzli, and Y. de J. de Zelicourt, *Helv. Chim. Acta*, **57**, 2276 (1974).
32. Robert Lii (unpublished).
33. (a) M. J. S. Dewar, *The Molecular Orbital Theory of Organic Chemistry*, McGraw-Hill, New York, 1969, p. 176. (b) L. J. Schaad and B. A. Hess, Jr., Dewar Resonance Energy, in

Chemical Reviews, Vol. 101, P. v. R. Schleyer, Ed., American Chemical Society, Washington, D.C., 2001, p. 1465.

34. N. L. Allinger, F. Li, L. Yan, and J. C. Tai, *J. Comput. Chem.*, **11**, 868 (1990).
35. R. H. Mitchell, Measuring Aromaticity by NMR, in *Chemical Reviews*, Vol. 101, P. v. R. Schleyer, Ed., American Chemical Society, Washington, D.C., 2001, p. 1301.
36. J. P. Glusker and K. N. Trueblood, *Crystal Structure Analysis: A Primer*, 2nd ed., IUCr, Oxford Press, Oxford, 1985.
37. D. Moran, A. C. Simmonett, F. E. Leach, III, W. D. Allen, P. v. R. Schleyer, and H. F. Schaefer, III, *J. Am. Chem. Soc.*, **128**, 9342 (2006).

11

HEATS OF FORMATION

The heat of formation of a compound is, arguably, its most important property. Hence, if molecular mechanics is to furnish a good model of molecules, it must be able to calculate heats of formation. The *heat of formation* of a compound is its heat content relative to the elements therein in their standard states. It is very difficult to accurately calculate these quantities by purely ab initio quantum mechanics.[1] The major problem is that one must first be able to accurately calculate the heats of ionization of all of the electrons for each of the atoms concerned and then calculate the bonding energies between atoms for the molecule of interest, as well as for the standard states of the elements involved. The heat of formation then is just the difference between these two numbers. The basic difficulty stems from the fact that the energies of typical molecules are enormous numbers. Thus, chemical accuracy requires them to a great many decimal places. To illustrate, the HF/6-31G** energy of *n*-octane is −196,684.96 kcal/mol. Chemical accuracy would require numbers like this to about seven significant figures. This would require a very large basis set and a very extensive treatment of electron correlation. This is just not possible by straightforward calculation at present. [The "focal-point analysis" procedure (Chapter 3, under Ab Initio Methods), where one does the calculation with large basis sets and extrapolates to the Schrödinger limit is

---

Molecular Structure: Understanding Steric and Electronic Effects from Molecular Mechanics,
By Norman L. Allinger
Copyright © 2010 John Wiley & Sons, Inc.

becoming increasingly feasible and useful here, however (later).] Molecular mechanics, however, starts with atoms not with electrons and nuclei, so that the numbers to be calculated are orders of magnitude smaller. Equally important, whereas ab initio quantum mechanics has to calculate binding energies very accurately (again large numbers), molecular mechanics takes various kinds of bonds between atoms as standards too (elements in their standard states), so that only the relative energies of those bonds (much smaller numbers) have to be calculated.

The current situation is that while quantum mechanics cannot calculate accurate heats of formation for molecules containing more than a few atoms by strictly ab initio methods, much of the difficulty can be circumvented empirically (later).

Molecular mechanics, on the other hand, has traditionally calculated heats of formation to chemical accuracy for many (but certainly not all) compounds, but only if a sizable number of accurate *experimental* numbers are available for appropriate model compounds.

BENSON'S METHOD

How do we proceed? The calculation of heats of formation has a long history and was quite successful in limited areas long before molecular mechanics was developed. Thus, we will initially follow a historical path in this discussion. Brief, clear earlier reviews include those written by Wiberg[2a] and by Fan.[2b]

Organic chemists knew since the earliest days of the subject about the similarities of compounds containing identical functional groups, and it was easy to make the approximation that all C–H bonds are the same, and all C–C single bonds are the same with respect to their energy contents. Also, all C=O bonds might be assumed the same, and so on; but let us concentrate first on the most simple case, the saturated hydrocarbons. So then can we just assign a bond energy to each different kind of bond, add up how many bonds of each kind are in a molecule, and find the heat of formation of the molecule? Well, this sort of works, but the numbers obtained are not of chemical accuracy. Why? For several reasons. Early studies of simple molecules showed that branched-chain hydrocarbons tend to be more stable than straight chains. For example, isopentane and neopentane are about 2 and 5 kcal/mol more stable than *n*-pentane, respectively. Thus, one needs not only to add *bond energies* but also *structural energy increments*, primary, secondary, and so forth.

The heat of formation (H_f°) then can be calculated from Eq. (11.1):

$$H_f^\circ = \sum_i n\mathrm{BE}_i + \sum_i n\mathrm{ST}_i \tag{11.1}$$

where BE is a constant assigned for the energy of a particular type of bond, and ST is a structural energy increment assigned to such functions as primary, secondary, and so forth. Each of these terms has to be multiplied by the number of times (n) that it appears in the molecule.

Then, if one adds up these numbers for many (but not all) hydrocarbon molecules, one can get acceptable values for heats of formation. And indeed, if one looks at a series of ketones, for example, using a similar scheme one can calculate correctly many (but not all) of their heats of formation. So it would seem that we are on the right path, but we are not there yet.

The idea that heats of formation could be calculated by a method employing these additive bond energies and structural increments follows, as does molecular mechanics, from the fundamental assumption that organic molecules are built from repetitive structural fragments, and bond properties are transferable from one molecule to another. This scheme was highly developed over the years and is outlined in detail in a book by Benson.[3] Ordinary strainless compounds have their heats of formation well calculated by what has become widely referred to as *Benson's method*. Thus, we can conclude that the idea of transferability of energies of bonds and structural quantities is a useful one. It works much of the time to chemical accuracy. There are, however, some shortcomings to the method as so far described. A simple example would be cyclopentane. If one uses this method in the form so far outlined (with numbers from alkanes) to calculate the heat of formation of cyclopentane, one gets a number that is more negative than the experimental heat of formation by about 5 kcal/mol. The reason, of course, is that cyclopentane is strained by having some of the bonds partially eclipsed while also having the internal ring angles bent to smaller than the normal values. And, in general, strain will not be reflected in the Benson method as so far described, and accordingly heats of formation can be correctly calculated in this way only for strainless compounds. But many compounds, including most of the really interesting ones, are strained. So what to do? A simple possibility is to have a *structural increment* for cyclopentane rings, which is just added to the heat of formation along with the other terms, and the actual (more elaborate) Benson method does this. And this will work fine for compounds like methylcyclopentane, ethylcyclopentane, and the like. But suppose we fuse two cyclopentane rings together to give bicyclo[3.3.0]octane (Structure **1**):

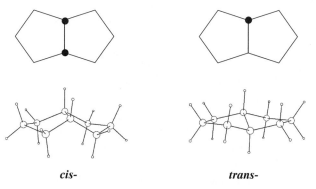

cis-   *trans-*

Structure 1. Bicyclo[3.3.0]octane.

First, note that this compound exists in cis and trans forms and that the latter is much more highly strained. (It may not be evident from the structures shown, but while the internal and external C–C–C bond angles at the bridgehead carbon are 106.0° and 115.4°, respectively, in the cis isomer, those in the trans are 102.3° and 125.8°.) If we use two cyclopentane increments, we can calculate approximately correctly the heat of formation for the cis isomer, where the total strain is approximately the same as that of two cyclopentanes. But the trans isomer is much more strained than that (about 6 kcal/mol more), due to very distorted bond angles, and the calculated heat of formation value would be very wrong.

The basic problem here is that if one can add a "proper" set of increments, one can reproduce a heat of formation, presumably for anything. But if the system gets sufficiently complicated, it was long thought that essentially another increment (parameter) would be needed for each different molecule. These increments would each have to be experimentally (or quantum mechanically) determined, and so the system would lose much of its predictive value. However, recent work indicates that it is not really that bad, and all hydrocarbons that have a known heat of formation have those heats calculated to an average accuracy of about 1 kcal/mol using a large parameter set for neighboring interactions.[3b]

At the time that Benson wrote his book, neither quantum mechanics or molecular mechanics was available in a practical sense for examining problems of this nature. But now one can ask the question: How can we use ab initio calculations to calculate heats of formation? With ab initio calculations we put together nuclei and electrons to make atoms, and then we put together atoms to make molecules. And we obtain energies at each step. Can we not obtain heats of formation in this way? In principle, of course, we can. In practice, we come up against the real problem mentioned earlier. The energies involved in putting electrons and nuclei together to make atoms are very large numbers. The energy of a single carbon atom (relative to the nucleus plus six isolated electrons) is −23,747.303 kcal/mol. Molecules, of course, have substantially larger (negative) energies. We will return to this problem later.

Molecular mechanics has an advantage over quantum mechanics in a problem like this because molecular mechanics assumes atoms themselves and the bonds between them as the starting point, and one can thereby bypass the problem of the very large energies that result when fundamental particles are put together to make atoms. Accordingly, the historical approach was to simply graft the molecular mechanics and the Benson method together in a straightforward way. If we return to our example of bicyclo[3.3.0]octane, what happens in determining the structure by molecular mechanics is that the energies for the different isomers are minimized. The resulting structures then become known, but we also know their stretching energies, bending energies, and the like on a bond by bond basis. We can accordingly add bond energies according to the Benson method, but also add in the "steric energies" (a measure of strain energies) summed over each individual bond, angle, and so forth as appropriate to that molecule. Will this give us correctly the heat of formation? The short answer is, yes it does. The equation needed then is expanded to that shown as Eq. (11.2):

$$H_f^\circ = \sum_i n\mathrm{BE}_i + \sum_i n\mathrm{ST}_i + E_S \quad (11.2)$$

where E_s is the *steric energy* of the molecule obtained from molecular mechanics.* The molecular mechanics method for calculating heats of formation takes us a big step beyond the Benson method and allows us to accurately calculate heats of formation for perhaps 95% of the hydrocarbons in which one might be interested. (It seems like there are always a few special cases.)

This method was used in early force fields in the 1970s. It worked pretty well, better than any previously existing method. There are, however, some better approximations that can be applied to this type of procedure. The Benson type of method, or the bond energy approach, assumes the additivity of bond energies over a sizable range. But "bond energies" are made up of component pieces, and, in principle at least, such a broad assumption may not be the best way to proceed. Let us look at this in a little more detail.

STATISTICAL MECHANICS

If one looks at a molecule such as heptane, for example, one can add all of the appropriate increments and calculate the heat of formation with acceptable accuracy by the method previously described. But there are a few things that are really not proper about that kind of calculation. Heptane in the gas phase at 25°C (where heats of formation are defined) is actually a complicated mixture (a Boltzmann distribution) of a great many conformations, most of which have different enthalpies and entropies. Additionally, each of these conformations is also a Boltzmann distribution over the possible translational, vibrational, and rotational states. The Benson method works adequately for many cases like this because these statistical mechanical terms can be lumped into the increments and averaged out, and they are not explicitly considered. By adjusting the values of the parameters in Eq. (11.1) or (11.2), much of the resulting error of neglecting the statistical mechanics can be canceled out, or at least minimized, in simple cases. But we would like for this scheme to work for more complex cases too. As the system becomes more complicated, errors tend to cancel out less well. So let us go back and approach this problem in a more proper way.

In MM4 we differentiate and calculate several different kinds of energies. These include the equilibrium energy (E_e) at the bottom of the potential well and also the zero-point energy in the lowest vibrational state. We also calculate the thermal excitation energies at 25°C (and can calculate them for any other desired temperature). We also explicitly calculate the entropy and hence properly differentiate enthalpy from free energy. All of these things are separately calculated with MM4, so that one may

---

*The steric energy of a molecule in molecular mechanics is the energy of the whole molecule as calculated by the force-field. It is consequently force field dependent. It contains the sum of each individual energy term for stretching each bond, bending each angle, and twisting about each torsion angle, plus the energies from each of the corresponding cross terms. It also contains the energies for each van der Waals interaction plus the total electrostatic energy from all of the charge and multipole interactions. For a given molecule, the differences in the steric energies of the conformations give directly the conformational energies. Steric energies are related to strain energies (later) and to heats of formation, but not in a simple way.

compare in a proper way by calculation whatever quantities are actually being measured experimentally. Note that the discussion here and following also applies to ab initio calculations. When one reads the literature, one finds ab initio structures are very frequently given, usually together with an energy and perhaps some properties. But such a structure is purely hypothetical. It is an approximation to a point on the potential energy surface. Real molecules do not exist as points on potential surfaces, they exist as collections of atoms that are vibrating somewhere above the potential surface (in the classical picture or, alternatively, have some probability of being found somewhere above the potential surface in the quantum mechanical picture). And this is always true for all quantum mechanical and classical mechanical structures, and it may or may not be important in any particular instance.

This difference between the hypothetical point on the potential surface and the real molecule can be well described by statistical mechanics. With respect to heats of formation, the ways for calculating them (by quantum mechanical methods or by molecular mechanical methods) require either the explicit inclusion of statistical mechanics (the details of this procedure have been spelled out in full[4] and will be outlined in the following) or else an implicit inclusion by lumping the statistical effects into the bond energies and hoping for the best, as in the Benson method.

Mechanical molecular models began with the ball-and-stick type, which gradually evolved into the molecular mechanics model that most chemists are familiar with today. But actually, the molecular mechanics model can be made very much more accurate than any kind of a physical mechanical model, and much more accurate than many chemists realize. Quite small molecules can have their structures very accurately calculated by quantum mechanics.[5] For larger molecules, molecular mechanics structures can in practice be more accurate than structures usually obtained currently by quantum mechanics.[6,7] And for molecular structures typically determined by X-ray crystallography, the molecular mechanics structures are also usually significantly more accurate. We can also calculate vibrational spectra by molecular mechanics that are at least competitive, and often more accurate, than those usually currently calculated by quantum mechanics. With large sets of test data on vibrational frequencies that we have compared with experiment, the approximate rms errors, respectively, are about $30\,cm^{-1}$ by MM4 and about $40\,cm^{-1}$ by the method of Baker and Pulay with proper scaling.[8] The corresponding rms errors obtained by experiment are quite variable. While these numbers can often be obtained experimentally to less than $1\,cm^{-1}$, when one simply looks at compilations where the same number is reported by different investigators, discrepancies of up to $10\,cm^{-1}$ are fairly common, especially in solution. (It should be remembered that $10\,cm^{-1}$ corresponds to only $0.028\,kcal/mol$.) All of this means that we can determine by molecular mechanics information that is sufficiently accurate that we can apply statistical thermodynamics in a useful way to our molecular mechanical model.

One of the things that we are most interested in obtaining when we do molecular mechanics calculations (or quantum mechanical calculations, or experimental measurements) on molecules is their conformational energies (and sometimes their thermodynamic functions). These will tell us in detail about the overall molecular structures and energies upon which many properties depend. And with a molecular mechanical model such as MM4, we can determine this information to the necessary degree of accuracy. For details as to how all of this is done, one can refer to standard texts on statistical mechanics,[2a,9] and especially to two studies that describe this particular subject in detail.[4a,10] We will give just an outline here. The program, and the computer, will actually do these calculations. The chemist is often only interested in the results. But it is sometimes helpful if the chemist understands, at least in an overview, just how these results are obtained.

The equation that we will need to solve to calculate the heat of formation using the above ideas in molecular mechanics is Eq. (11.3):

$$H_f^\circ = \sum_i n\text{BE}_i + \sum_i n\text{ST}_i + E_\text{S} + \text{POP} + \text{TOR} + \text{T/R} + E_\text{vib} \qquad (11.3)$$

It will be noted that the first three terms of Eq. (11.3) are the same as Eq. (11.2), with the additional last four terms from statistical mechanics added. Taking these last four terms in order, POP is the additional energy that comes from the Boltzmann distribution of conformations, TOR is needed to account for extra torsional energy that results when there are low barriers to rotation in the molecule, T/R is the energy from the translational and rotational degrees of freedom of the whole molecule plus R (the gas constant) to convert energy to enthalpy, and E_vib is the vibrational energy of the molecule. These will be discussed in turn.

Let us first consider the term that we have called POP, which is an abbreviation for the *population increment*. This is the amount by which the energy is increased because of the Boltzmann distribution of conformations. Most molecules in the gas phase at room temperature exist as mixtures of conformations. Normally, we are most interested in the lowest energy conformation, but in some cases we are also interested in the actual population of conformations present. In calculating the heats of formation we have to note that many simple compounds, say n-heptane, for example, in the gas phase at 25°C exists as a mixture of many conformations. The mixture will adjust its conformational composition so as to minimize its free energy. While the energy is a minimum for the most stable conformational species (the all-trans conformation in this case), if we promote some of the ground-state molecules to other (excited) conformations to yield a Boltzmann distribution, the free energy of the mixture is lowered because of the entropy of mixing (the $T\Delta S$ term is larger than the increase in the ΔH term). For many molecules, the resulting POP terms have significant values. For the n-heptane example, POP has a value of 0.92 kcal/mol, so the contribution of these POP terms to heats of formation is important.

Molecules can undergo three types of motion: translation, rotation, and vibration. At absolute zero in the gas phase, the molecules are all in their ground state, and their translational and rotational energies are zero. Their vibrational energy is the zero-point vibrational energy. At higher temperatures, molecules will be partially excited into higher translational and rotational states unless they are restricted by neighboring molecules, for example, in a crystal lattice. At still higher temperatures the crystal will melt, and the molecules will then undergo translational and rotational motions. These translational and rotational levels are very close together so their contribution to the energy can be calculated classically, as for a continuum, where there is ½RT/mol for each such degree of freedom. This is part of the T/R term (translation/rotation) in Eq. (11.3).

Since most larger molecules exist as mixtures of conformers, it is important to determine which conformers are of the lower energy. Since most work is carried out at or near room temperature, one may want to have the conformer of lowest energy (the global minimum), and also the energies of all of the other conformers up to about 3 kcal/mol above that, as these are the ones that will actually contribute significantly to the Boltzmann distribution. For an arbitrary molecule, we can calculate the relative energies (and importantly the free energies) for each conformer and for the compound without difficulty if we know all of the conformers involved. These can be found using a *conformational search* procedure as discussed in the Appendix. For now, just assume that we have all of the conformational structures that will be needed, and their energies and entropies.

The vibrational energy [the E_{vib} term in Eq. (11.3)] is a little harder to deal with because the vibrational levels are further apart, and the classical approximation may not be adequate. But this energy can be calculated knowing the frequencies of the normal vibrational modes of the molecule. From these frequencies (energies) one calculates the Boltzmann distribution over the vibrational states, and hence the total vibrational energy. Here is where the accuracy of the vibrational frequencies may become important. But as long as the errors are not systematic, they tend to cancel out when summed over the whole molecule, as when one is calculating thermodynamic properties. The result is that frequencies that have rms errors on the order of 30 cm^{-1} normally give thermodynamic quantities that are of chemical accuracy.

A major complication comes in at this point from something else, and it involves the torsional degrees of freedom. In order to calculate the energy, or other thermodynamic functions, with MM4 (or by statistical mechanics in general), we must know the vibrational frequencies of the molecule. We need to know these frequencies, first to get the zero-point energy, and second to calculate the energies of the higher vibrational levels that will be occupied by the Boltzmann distribution if the temperature is above 0 K. We can use the harmonic approximation to determine the higher vibrational levels for most frequencies. There is, however, sometimes a problem here with the harmonic approximation for the torsional vibrational levels. If the molecule can undergo torsional vibrations, as most organic molecules can, there is no problem when the potential well is deep, and the torsional frequency is high. Such a vibration is ordinary (approximately harmonic) from the thermodynamic point of view. There is also no problem if the potential barrier to the rotation is zero, whereupon the vibration becomes instead a free rotation. But there is a problem in the intermediate range, where we have barriers to rotation in the range of 0.5–5 kcal/mol, or so. The problem here is that while the cosine curve that describes the torsional motion is very close to harmonic right near the bottom, as the vibrational levels move up the curve, the cosine curve opens out increasingly more than a harmonic curve would. The resulting larger vibrational amplitudes lead to the vibrational levels becoming more closely packed, and thus more highly occupied, thereby increasing the energy content of the system more than calculated using the harmonic approximation.

What we would like to do here is straightforward in principle. We would like a simple, general way to deal with the vibrational levels for these hindered rotors, so that we can allow adequately for their Boltzmann distribution and obtain the torsional energy of the molecule. This problem was studied, and a solution to the general case was worked out by Pitzer and Gwinn in 1942.[11] As was typical before the availability of computers, the solution was presented as a series of numbers in tables. When we examine these, what we find is as follows. If the torsional barrier for the rotation is very high, then the vibration is an ordinary one, and a harmonic approximation is satisfactory. As the torsional barriers get lower, the energy levels become less harmonic, lower in energy, and more closely spaced. This in turn causes an increase in the population of molecules in these levels and, hence, an increase in the enthalpy of such molecules. The sizes of the changes involved are functions of the barrier height and of the foldness (twofold, threefold, etc.) of the barrier. Is there a simple way to deal with this problem? We decided to make the following approximation.[10] First, if the barrier has a height of 5 kcal or more, the tables presented by Pitzer and Gwinn[11] show that for thermodynamic purposes at room temperature, the harmonic approximation will suffice. If the barrier height is smaller than that, the levels become close enough together that some correction is desired. What we do for each barrier lower than 5 kcal/mol is to simply pick a single (overall) constant (empirically so as to get the best fit to the available data) and add that constant to the harmonic energy that is otherwise calculated. While this is a somewhat crude approximation, the constant is not very large, and the procedure appears to work adequately. The constant is called TOR (for torsional correction) and has a value of about 0.58 kcal/mol with MM4. To apply this to a real molecule, one has to know how many bonds in that molecule have hindered rotational barriers of less than 5 kcal/mol (excluding methyl groups as these have their own heat of formation parameter that can absorb the needed correction). This can normally be determined by inspection. For doubtful cases, one can use the driver routine (see Appendix) to obtain the actual barrier heights. This scheme works well for the most part, but it has its limitations. If there are very low vibrational frequencies in the molecule ($<40\,cm^{-1}$), it is advisable to specifically look at the details of the vibration and see that the general assumptions outlined are met. If not, then more arduous methods of statistical mechanics would have to be invoked, or one would have to accept larger error limits in the calculation.

All of the above calculations, while very tedious to carry out by hand, are quickly and accurately calculated by appropriate computer programs, such as MM4. Thus, when one determines a molecular structure, one can, with a little further expenditure of computer or user time, also determine these thermodynamic quantities for each conformation or molecule.

HEATS OF FORMATION OF ALKANES FROM MOLECULAR MECHANICS

In the MM4 calculation of heats of formation we proceed as in the usual molecular mechanics scheme, but we specifically include the statistical mechanical energy of the molecule[4,6,10] as in Eq. (11.3). Since the vibrational frequencies are already known from the MM4 calculation, this is straightforward. When the results were compared with those obtained by using the ordinary Benson method, it was found that overall the differences between the two methods are rather small for the most part. There were occasionally discrepancies between the two methods of up to about 2 kcal/mol, however, and the statistical mechanical method is better, as expected.

In Table 11.1 is given a list of compounds for which the heats of formation were calculated by fitting a total of 8 parameters for a representative set of 58 saturated hydrocarbons of which 52 were weighted in a least-squares fitting.[6] A weight of zero

TABLE 11.1. MM4 Heats of Formation (Gas, 25°), kcal/mol[a]

Wt.	H_f° Calc.	H_f° Exp.	Difference (Calc.–Exp.)	Compound
1	−17.89	−17.89	0.00	Methane
6	−19.75	−20.24	0.49	Ethane
9	−24.99	−24.82	−0.17	Propane
8	−29.97	−30.15	0.18	Butane
7	−35.03	−35.00	−0.03	Pentane
7	−40.12	−39.96	−0.16	Hexane
6	−45.16	−44.89	−0.27	Heptane
5	−50.21	−49.82	−0.39	Octane
4	−55.24	−54.75	−0.49	Nonane
9	−32.36	−32.15	−0.21	Isobutane
7	−36.69	−36.92	0.23	Isopentane
9	−40.67	−40.27	−0.40	Neopentane
7	−42.16	−42.49	0.33	2,3-Dimethylbutane
6	−49.01	−48.95	−0.06	2,2,3-Trimethylbutane
6	−49.70	−49.20	−0.50	2,2-Dimethylpentane
6	−47.86	−48.08	0.22	3,3-Dimethylpentane
6	−44.40	−45.25	0.85	3-Ethylpentane
6	−48.12	−48.21	0.09	2,4-Dimethylpentane
5	−52.85	−53.18	0.33	2,5-Dimethylhexane
5	−53.86	−53.92	0.06	2,2,3,3-Tetramethylbut
5	−56.75	−56.64	−0.11	2,2,3,3-Tetramethylpen
5	−57.59	−57.80	0.21	Di-*tert*-butylmethane
7	−55.33	−55.67	0.34	Tetraethylmethane
0	−54.06	−56.40	20.34	Tri-*t*-butylmethane
9	−18.59	−18.74	0.15	Cyclopentane
8	−29.59	−29.43	−0.16	Cyclohexane
7	−27.88	−28.22	0.34	Cycloheptane
7	−29.72	−29.73	0.01	Cyclooctane
6	−31.37	−31.73	0.36	Cyclononane
4	−36.74	−36.88	0.14	Cyclodecane
0	−53.49	−54.59	10.10	Cyclododecane
6	−33.02	−33.04	0.02	1,1-DiMethylcyclopen
2	−25.70	−25.27	−0.43	Methylcyclopentane
5	−30.35	−30.34	−0.01	Ethylcyclopentane

TABLE 11.1. *(Continued)*

Wt.	H_f° Calc.	H_f° Exp.	Difference (Calc.–Exp.)	Compound
9	−36.99	−36.99	0.00	Methylcyclohexane
6	−43.43	−43.26	−0.17	1,1-Dimethylcyclo-
6	−41.71	−41.13	−0.58	1-ax-2-eq-Dimethyl-
6	−43.34	−42.99	−0.35	1-eq-2-eq-Dimethyl-
2	−30.04	−30.50	0.46	Bicyclo[3.3.1]nonane
3	−22.72	−22.20	−0.52	cis-Bicyclo[3.3.0]oct
4	−15.31	−15.92	0.61	trans-Bicyclo[3.3.0]
6	−43.60	−43.54	−0.06	trans-Decalin
6	−40.91	−40.45	−0.46	cis-Decalin
5	−31.73	−31.45	−0.28	trans-Hydrindane
5	−31.01	−30.41	−0.60	cis-Hydrindane
1	−57.74	−58.12	0.38	tst-perHanthracen
0	−50.27	−52.73	20.46	tat-perHanthracen
5	−13.10	−13.12	0.02	Norbornane
5	−30.90	−30.62	−0.28	1,4-Dimethylnorborn-
2	−31.85	−31.76	−0.09	Adamantane
4	−66.55	−67.15	0.60	1,3,5,7-Tetramethylad-
2	−20.11	−20.54	0.43	Protoadamantane
3	−35.04	−34.61	−0.43	Congressane
7	−21.90	−22.58	0.68	Bicyclo[2.2.2]octane
3	−24.01	−24.46	0.45	Perhydroquinacene[e]
0	14.05	18.2[b]	4.15	Dodecahedrane
0	−58.34	(−59.22)[c,d]	0.88	2,2-Di-t-butylpropane
0	−28.14	—[c]	—	Tetra-t-butylmethane

[a]The standard deviation is 0.355 and the weighted standard deviation is 0.342, based on 52 equations.
[b]This value has been revised from that given earlier.[6]
[c]No experimental value is available.
[d]This is the MP4 value (later).
[e]Also called perhydrotriquinacene.

was given for a few compounds when the experimental value was questionable. (The last two compounds in Table 11.1 were subsequently added and were weighted zero to keep the statistics unchanged for comparisons with earlier work.) The weighted standard deviation (WSD) is 0.34 kcal/mol over that set, which is approximately the experimental error. There were discrepancies between the MM4 calculated and the experimental heats of formation of more than 1 kcal/mol for only 3 compounds out of the original set (56 compounds) (all of which are almost certainly due to experimental error, see later), and discrepancies of 0.7–1.0 kcal/mol for 2 more compounds. The results here are somewhat better than they were using the simple bond energy scheme without the statistical mechanics. If the statistical mechanics treatment is omitted (but the Boltzmann distribution is still included), the standard deviation increases to 0.36 kcal/mol. So the apparent overall improvement resulting from the statistics was small, but it is about 1 kcal/mol for a few compounds (and would be up to near

TABLE 11.2. MM4 Alkane Heat of Formation
Parameters[a] (kcal/mol)

C–C = −87.1067	Neo = −6.9273
C–H = −106.7763	R_6 = 4.9713
Me = 2.0108	R_5 = 4.4945
Iso = −3.3565	TOR = .5767

[a]The parameters given in part have different values from those originally published.[6] This is due to the fact that there was an error in the original work, where methane was accidentally attributed a POP term of −4.20 kcal/mol instead of zero. This led to no difference in the calculated heats of formation, but it did lead to the Me/Neo parameters having strange values. The parameter values given have that error removed and are to be regarded as the proper ones for MM4.

2 kcal/mol if the Boltzmann distributions were not included). The improvement in the WSD comes mainly from significant improvements with just a few compounds.

The least-squares fitting of the data for the compounds in Table 11.1 gave us the parameter values to be used in the MM4 heat of formation calculation. They are shown in Table 11.2. These parameter values are worth some discussion. Two parameters that are obviously required are those for the C–C and C–H bonds, which are known to have values of about −105 and −90 kcal/mol (bond energies).*

It is well known that branched-chain hydrocarbons tend in general to be more stable than less branched ones.[12] In determining the parameters here, we need three values, which we have chosen to be methyl, iso, and neo. Methane is considered a special case. The methylene value is not included, which is equivalent to choosing it as our zero point by default. The neo value is most negative (approximately −6.9 kcal/mol), followed by the iso (−3.3 kcal/mol), the methylene (zero), and the methyl value, which is positive (+2.0 kcal/mol).

It is also found that if there is a ring in the compound, with two hydrogens missing relative to the ordinary alkane, it raises the energy by about 4.9 kcal/mol. We have called this parameter R6, because it applies to six-membered rings (and also to all larger rings). The value for the five-membered ring, R_5, is similar to that for the six-membered ring (4.5 vs. 4.9 kcal/mol). It might be thought that it would be more positive than R_6, since cyclopentane is more strained than cyclohexane. But cyclopentane has its own bending and torsional parameters that already account for that strain.

We also need a value for the extra torsional energy that results when molecules have low rotational barriers (TOR). Putting that into the least-squares fitting gave us a value for TOR of 0.58 kcal/mol. We know that the value must be positive because the energy is raised as these extra vibrational levels are added. From the work by Pitzer

---

*There are different ways of defining bond energies, and the numbers obtained span large ranges, depending primarily on whether the definition is for the potential surface, or at 0 K, or at 25°C. The MM4 bond energies are for the potential surface and are also somewhat dependent on the values of the other parameters in Table 11.2.

and Gwinn,[11] a value of about +0.36 kcal/mol was estimated,[10] and the + 0.58 is the optimized MM4 value.

Three- and four-membered rings (cyclopropane and cyclobutane derivatives) were treated in heat of formation calculations by MM3. Each of them required its own structural parameter set and its own ring constants (R_3 and R_4, respectively). Four-membered rings have also been treated by MM4.[13] These small ring compounds present no particular difficulties and are well treated in the same way as are the alkanes.

The alkane heat of formation parameters have values more or less as we expect, but there are some peculiarities. Of course, they were obtained by a least-squares fitting of data points to experimental information, and this is the parameter set that gives the best fit. However, one wonders why the R_6 parameter is needed (it was not needed in early force fields when the statistical treatment was omitted; see later). And one might wonder why the TOR value is as large as it is, for example.

Tim Clark Story*

The heats of formation of adamantane and related compounds were of great interest in the 1970s. Compounds of this type show significant van der Waals interactions of carbon with carbon, whereas the effects involving hydrogen are relatively small. These compounds are related to diamond in more or less the same way that benzenoid compounds are related to graphite. The heat of formation of adamantane was determined experimentally several times, and the early numbers were not really in very good agreement with one another. The best value now (Table 11.3) appears to be that of Clark et al. (1975).[14] In that work a series of nine adamantane derivatives was prepared, and the heats of formation were determined by measuring their heats of combustion and heats of sublimation experimentally. These compounds are shown as Structures **1–9** in Table 11.3.

Clark et al.[14a] also noted in their work that force-field calculations would be a quick and easy way to obtain quantities such as the heats of formation of these compounds, if the force fields proved to be reliable. They accordingly used two of the available molecular mechanics force fields of that day to calculate these quantities[14b,c] and compared the results with experiment. They concluded that the force fields contained fairly large systematic errors, and while they gave useful information in terms of trends, they needed to be improved if they were going to be competitive in accuracy with experiments. Indeed, force fields were fairly primitive prior to 1975. These particular compounds are relatively free from stretching, bending, and torsional strain. They do, however, show significant van der Waals interactions. These are mainly between the carbons in the interiors of the molecules, which turn out to be mostly small repulsions. The sizes of these repulsions were not yet known very accurately in 1975.

The present author was particularly interested in this problem, and when the experimental heats of formation for compounds **1–9** were furnished to him by Dr. Clark before publication, he carried out calculations on the series. While it was found that the experimental heats of formation for the other compounds were well fit by the MM2 calculations,[15] compound **6** was found to be clearly out of line by about 2 kcal/mol.

---

*Used with the permission of Dr. Clark.

TABLE 11.3. Heats of Formation of Adamantane Derivatives (kcal/mol)

	1	2	3	4	5	6	7	8	9
$-\Delta H_f^\circ$ (C)	46.02 ± 0.09	56.72 ± 0.30	51.80 ± 0.29	86.54 ± 0.46	36.04 ± 0.41	55.53 ± 0.55	62.51 ± 0.17	62.24 ± 0.48	59.12 ± 0.78
ΔH_{sub}°	14.26	16.15	16.14	19.39	15.50	22.93	18.98	24.64	19.27
$-\Delta H_f^\circ$ (g)	31.76 ± 0.32	40.57 ± 0.34	35.66 ± 0.62	67.15 ± 0.50	20.54 ± 0.60	32.60 ± 0.58	43.53 ± 0.30	37.60 ± 0.58	39.85 ± 0.85

Source: From Reference 14a. Copyright 1975 by the American Chemical Society. Reprinted by permission of the American Chemical Society.

TABLE 11.4. Increments for Adding Methyl Groups to Adamantane, Diamantane, and Norbornane (kcal/mol)

		Calculations				
		Exptl[16]	EAS[14b]	MM2[15]	MM4	MP4[a]
(i)	Adamantane → 1-methyladamantane	−8.81	−9.32	−8.70	−8.65	
(ii)	Adamantane → 1,3,5,7-tetramethyladamantane/4	−8.85	−9.44	−8.73	−8.68	−9.15
(iii)	Diamantane → 4-methyldiamantane	−8.99	−9.45	−8.79	−8.70	
(iv)	Norbornane → 1-methylnorbornane		−9.56		−8.89	
(v)	Norbornane → 1,4-dimethylnorbornane/2	−9.12	−9.57		−8.90	−8.40

[a]Later.

The combustion work on these compounds by Clark et al.[14a] appeared to have been carried out carefully, and it was unclear to the present author what the trouble was with compound **6**.* Accordingly, he corresponded with Dr. Clark and indicated to him that the value published for compound **6** could not be correct.

Those who do calorimetry are extremely fussy about the purity of their compounds. For the adamantane type of compound, however, it can be extremely difficult to obtain a truly pure sample. Globular hydrocarbons tend to form solid solutions with one another and with many other compounds as well. These compounds are typically purified by chromatography, recrystallization, and sublimation (multiple times). But in spite of best efforts, problems can and occasionally do persist. The heat of combustion of **6** was then redetermined by Clark and co-workers and also duplicated by Good, on a more highly purified sample of diamantane, and they obtained closely similar results.[16] The two heats of combustion were 1942.06 ± 0.60 and 1942.58 ± 0.32 kcal/mol, which yielded a weighted value for the heat of formation of −34.54 ± 0.41 kcal/mol. The value shown for the heat of formation of **6** in Table 11.3 (32.60 ± 0.58) was then superceded by 34.54 ± 0.41 kcal/mol,[16] very close to that predicted by MM2 (34.30 kcal/mol).[15]

Clark and co-workers[16] devised another way to confirm the accuracies of certain of these heats of formation. If one looks at the structure of 1-methyladamantane and also at tetramethyladamantane (compounds **2** and **4** in Table 11.3), it is evident that the methyl groups in the latter compound are quite far apart and certainly must exert very small influences on one another. So if we take the difference in the heat of combustion of 1-methyladamantane minus the heat of formation of adamantane itself, we can obtain a *methyl increment* for an isolated adamantane bridgehead methyl group. In tetramethyladamantane, if the methyl increments are additive as one expects them to be, the heat of formation of the tetramethyl compound will be four times greater than the heat of formation of 1-methyladamantane, relative to adamantane itself. And indeed, this is found to be true to within experimental error (Table 11.4).

*Compound **6** is now called diamantane. It had the trivial name congressane in the older literature before the nomenclature for the adamantanes was settled on.

Now if we consider the heat of formation of 4-methyldiamantane (compound **7**), we can see that we have simply added a bridgehead adamantane-like methyl into a very similar environment in diamantane. Since the methyl group increments are additive in the adamantane case, we would expect that the number should carry over equally well to the 4-methyldiamantane example. And indeed this is found to be true (Table 11.4). Accordingly, we can now be fairly well convinced that these adamantane/diamantane numbers mentioned are all correct to within the stated experimental errors.

This *methyl technique* of Clark's is obviously general, and can be extended in principle to other groups of compounds and other substituents. In his work Clark discussed the use of this technique to differentiate several different values in the literature for the heats of formation of other methyladamantanes, and to show which are consistent and which are not, indicating likely errors. We will discuss only one of these cases.

The heats of formation were known at that time for both norbornane and 1,4-dimethylnorbornane.

Norbornane   1-Methylnorbornane   1,4-Dimethylnorbornane

From these data one can also calculate the heat of formation of 1-methylnorbornane, proceeding as previously. The environment for the methyl in this case is only slightly different from that in adamantane. As one would expect, the "methyl increment" for the methyl group in these compounds is very similar to that determined in the adamantane series.

These numbers are also shown in Table 11.4. The methyl increment number thus determined for the methylnorbornanes (−9.12 kcal/mol) is the same to within experimental error as the value for the adamantane methyl group (−8.81 to −8.99). And various other force field and MP4 calculations also are in agreement.

Clark et al.[14a] had done a masterful job in measuring the heats of combustion, sublimation, and formation for these compounds in the work described in his original paper, only by the expenditure of a great deal of time and effort. And, of course, much time and effort went into the syntheses of these compounds in addition. That the present author was able to conclude that one of the numbers was incorrect and also offer a value for that number from a day or so of computational work certainly impressed Dr. Clark. He accordingly ceased his work in thermochemistry and became a computational chemist. He was later the author of a well-known and widely used early book on computational chemistry.[17]

Thermodynamic Properties of Alkanes

The MM4 program calculates the geometry of a molecule (and hence its mass and moments of inertia) and its vibrational frequencies, and thus it is straightforward for it to also calculate the various thermodynamic properties at any desired temperature. For example, in Table 11.5 are given the entropies for a set of selected alkanes as calculated by MM4. These were each determined experimentally earlier, and the calorimetric values are included in the table for comparison. There are, of course, various problem cases, but the table shows that generally speaking, entropies can be calculated to within a few tenths of a kcal/mol K for a wide variety of hydrocarbons. The large uncertainties assigned in a few cases come from molecules that have very low vibrational frequencies. While there exist methods for dealing with these on a case-by-case basis, the default option used here is for the program to simply accept these large calculated uncertainties as shown.

TABLE 11.5. MM4 Entropies for Selected Alkanes (kcal/mol K at 298 K)[6]

Compound	S^a	S_{mix}	S_{tor}^b	$S°$	Exp.	Error
Methane	44.61	—	—	44.61	44.52	+0.09
Ethane	54.70	—	—	54.70	54.85	−0.15
Propane	64.59	—	—	64.59	64.51	0.08
Butane	72.55	1.27	0.40	74.22	74.12	0.10
Isobutane	70.46	—	—	70.46	70.42	0.04
Neopentane	73.50	—	—	73.50	73.23	0.27
Pentane	80.94	1.78	0.80	83.52	83.68	−0.16
Hexane	88.54	3.07	1.20	92.80	93.00	−0.20
Heptane	96.53	4.03	1.60	102.16	102.27	−0.11
2,3-Dimethylbutane	85.48	1.36	0.40	87.24	87.42	−0.18
2,2,3-Trimethylbutane	91.50	—	0.40	91.90	91.61	0.29
2,2,3,3-Tetramethylbut.	91.53 ± 3.25	—	—	91.53 ± 3.25	93.06	1.53 ± 3.25
Cyclopentane	68.20	—[c]	—[d]	70.09	70.00	+0.09
Cyclohexane	71.36	—	—	71.36	71.28	+.08
Cycloheptane	81.91	—	0.40	82.31	81.89	+.42
Methylcyclopentane	78.24 ± 2.71	1.16	0.40	79.80 ± 2.71	81.24	+1.48 ± 2.71
Methylcyclohexane	81.67	0.35	—	82.02	82.06	−0.04
Ethylcyclohexane	90.25	0.89	0.40	91.54	91.44	+0.10
1,1-Dimethylcyclohexane	87.01	—	—	87.01	87.24	−0.23
trans-Decalin	89.48	—	—	89.48	89.52	−0.04
cis-Decalin	90.03	—	—	90.03	90.28	−0.25

[a]The sum of the mole fraction weighted entropies.
[b]An entropy contribution of 0.40 kcal/mol K is added for each rotatable bond having a barrier height of less than 5.0 kcal/mol (excluding methyl groups).
[c]The ring is calculated to have C_s symmetry. If the C_2 conformation is studied, it in fact is found to have C_1 symmetry, but the C_2 form is higher in energy by only 0.001 kcal/mol. Therefore, the entropy of mixing the *dl* pair of C_1 forms (R ln 2 = 1.376 kcal/mol K) must be subtracted from the C_1 calculated value, and then the entropy is the same as that of the C_s form.
[d]The amount 1.891 (Pitzer's value for the pseudorotation) was added here.

It was expected, and found, that the entropies of hydrocarbons could be calculated to the accuracy shown in Table 11.5. This is a kind of independent test showing that all of the statistical mechanics employed in MM4 is correctly implemented and the approximations adequate, so that the results shown are obtained. Entropies are cumulative from 0 K up to the temperature of interest (in this case 25°C). Any sizable systematic errors would be revealed in these results.

HEATS OF FORMATION FROM QUANTUM MECHANICS: ALKANES

Heats of formation as we have so described them to this point have been measured experimentally, or calculated either by Benson's method or equivalent, or by a more modern extension of that, namely molecular mechanics. Let us return to the question of the calculation of heats of formation by quantum mechanics. As mentioned earlier, the basic problem here is that one wants these quantities to "chemical accuracy," say a fraction of a kcal/mol. As computers become ever more powerful (Moore's law), this approach using purely ab initio methods is gradually becoming feasible. But why can we not do here with quantum mechanics that which we routinely do with molecular mechanics, namely substitute a parameterization scheme that will permit us to carry out the quantum mechanical calculation of relatively small energy differences between compounds, so that the very large energies of putting atoms together in the first place will cancel out? This is certainly a feasible approach, and perhaps the most accurate practical method available at present for general extension in the near future. The idea was pursued in two somewhat different ways independently by Wiberg[18] and by Pople.[19] The method that is to be outlined here was introduced originally by Wiberg, and was used shortly thereafter by Ibrahim and Schleyer.[20] It has been subsequently extended, and more fully explored, and its scope and accuracy is now reasonably well understood.[4]

Basically, the application of this Wiberg–Schleyer method proceeds in a manner quite similar to the method used in molecular mechanics, previously described for the MM4 calculation of the heats of formation[4,6,10]. One needs to define a set of parameters that will be required in order to represent the molecules with appropriate generality and accuracy, and then one needs to calculate the energies of particular molecules using quantum mechanics, in place of molecular mechanics. The same eight parameters were used here as with MM4. If the quantum calculation could be done with sufficiently high accuracy, the parameters (other than the statistical ones and the numbers and kinds of bonds) would presumably tend to the value zero, and the heats of formation would be obtained with only minimal effort as outlined earlier.

The difficulty here, of course, is that we cannot usually at present carry out the calculations with sufficient accuracy so that these parameters will go to zero. Thus, we have to determine exactly what parameters are required, and to evaluate their numerical values for a specific level of quantum mechanical calculation (basis set/correlation). For sure one will want to use bond energies and structural features, presumably primary, secondary, and so forth, and the TOR parameter will be specifically required, as will Boltzmann distributions. We will here consider in detail the alkanes, and we can use our previous MM4 discussion as a starting point. We can first choose a basis set and level of correlation, and then carry out the quantum mechanical calculations by optimizing the geometries of the molecules as described earlier. Instead of using a steric energy in the heat of formation calculation, we will have a total quantum mechanical energy, which is utilized in the same way. Thus, we obtain Eq. (11.4):

$$\Delta H_f^\circ = \sum_j a_j n_j + E_{QM} + POP + TOR + T/R + E_{vib} \qquad (11.4)$$

where ΔH_f° is the heat of formation (gas, 25°C), E_{QM} is the quantum mechanical energy, a_j are the required structural parameters [identical to the bond and branching parameters BE and ST in Eq. (11.3)], and n_j is the number of times that each a_j appears in the molecule. This calculation was carried out initially using a set of energies obtained by 6-31G* Hartree–Fock calculations.[21] Then it was carried out again using B3LYP correlation added to the HF basis set, and the geometries were reoptimized to give a second set of energies, which were treated in the same way.

In more recent work we have extended those calculations to the same 58 compounds used as the standard for MM4 heats of formation (Table 11.1), so that we can make comparisons of the results obtained by these and other methods. Since the original quantum mechanical work was published,[21] computer capabilities have improved to where it has become possible to carry out MP2 and MP4 calculations on this same group of compounds. The optimized parameter set, together with the overall accuracy of the results, is shown for each of several methods in Table 11.6.[22a] There is

TABLE 11.6. Alkane Heat of Formation Parameters (kcal/mol) for Various Calculations[a]

Parameter	MM4[a]	HF[b]	MP2[b]	MP4[b]	B3LYP[b]	B3LYP/D[b]
C–C	−87.1067	11,885.826	11,928.521	11,934.276	11,960.060	11,958.022
C–H	−106.7763	6,301.245	6,325.032	6,331.240	6,351.951	6,344.581
Me	2.0108	2.540	−0.173	0.835	1.664	0.961
Iso	−3.3565	−3.537	0.704	−0.969	−2.767	−1.825
Neo	−6.9273	−8.028	2.594	−1.426	−6.366	−3.795
R_6	4.9713	8.127	1.409	2.349	4.171	2.693
R_5	4.4945	6.935	0.366	1.440	3.374	1.396
TOR	0.5767	1.002	0.121	0.254	0.559	0.265
WSD with stat.	0.34	0.92	0.61	0.42	0.55	0.35
WSD w/o stat.	0.45	0.82	0.65	0.43	0.47	0.44

[a]See footnote a to Table 11.2.
[b]The geometry of each molecule was optimized for each calculation (except as below) using Gaussian 2003[23] with the "tight" residual force criterion. The basis/correlation for columns 2 and 3 are as follows: RHF/6-31G*; MP2/6-311++G(2d,2p). Column 4 was a single-point calculation, MP4SDQ/aug-cc-PVTZ// MP2/6-311++G(2d,2p); column 5 was B3LYP/6-31G*; and the last column was the result of column 5 to which the MM4 dispersion energy was in each case multiplied by 0.5 and added.

no question but that the geometries obtained by MP2 calculations are in general more accurate for organic molecules than those obtained by comparable B3LYP calculations.[22b] But for energies the results are less straightforward. In some cases one method seems better and in different cases the other.

There is one significant systematic error in B3LYP calculations that needs to be noted. They do not properly account for dispersion energy, which is the van der Waals r^{-6} term in molecular mechanics. They consequently give gauche butane energies that are too high, and in general energies that are too high for molecules that are compact, wadded up, or ball-like relative to their stretched-out counterparts.

Table 11.6 gives the WSD for each method as calculated in two different ways. One way, to be discussed first, is to include the statistical mechanical calculations as previously described (with stat.). The other way is to omit these statistical calculations (without stat.*), as was routinely done in earlier (Wiberg, etc.) calculations. We will explain later why we want to do this.

A comparison of the parameter values shown in Table 11.6 is informative. The MM4 parameters were discussed earlier (Table 11.2). They are repeated here in the first column for convenience. What about the values of the other parameters. What can we learn from these numbers?

The first two lines in Table 11.6, the bond energies, are required to relate the calculations to the standard states of the elements as the zero of reference. The bond energies in molecular mechanics are just that because of the formalism. But the so-called bond energies in the quantum mechanical calculation also include the energies of the formation of the atoms from the nuclei and the electrons, and hence these numbers are very much larger.

The Neo/Iso/Me numbers are energies related to a secondary carbon as the zero point. First, consider the molecular mechanics numbers. In molecular mechanics we account for electron correlation energy (indirectly) in different ways. If it is between two atoms that are bound together, the effect is included in the bond energy. If it is between two atoms that are relatively distant (1,4 or greater, or intermolecularly), it is accounted for in the van der Waals attraction. But if the atoms in question are 1,3, that is, geminal or bound to a common center, the effects of *changing* this correlation energy are included in the angular bending terms, but the strainless correlation energy itself is not included. Therefore, if we wish to calculate the heat of formation of a molecule by molecular mechanics, we need to allow for the strainless correlation energy between the atoms bound 1,3 to one another. We know that the "branched-chain effect" in organic chemistry[12] yields Neo systems that are more stable than secondary by about 5 kcal/mol, with the Iso systems in between. These energy differences are largely from the correlation differences due to branching.** Those terms are missing in the energetics of an ordinary molecular mechanics calculation, and those terms are exactly what we are adding in here. The parameter value range or spread from Neo to Me is 8.9 kcal/

---

*The POP term is included here, however.
**There are also zero-point energy differences in the Neo/Me bonding arrangements. In the Benson or Wiberg–Schleyer scheme they are also lumped into the Neo/Me parameters. In the MM4 scheme, these zero-point energy differences are explicitly included in the E_{vib} and not in the Neo/Me parameters.

HEATS OF FORMATION FROM QUANTUM MECHANICS: ALKANES 277

mol in MM4, with the Neo being the stable end of the series. So it is clear why we have these terms in the heat of formation calculation in molecular mechanics. They are of the magnitude and sign required to account for the correlation energy differences between the various branched-chain substitution patterns.

First, note that with MM4, the WSD was 0.34 kcal/mol in the fit to the test set of 52 compounds. When the statistical mechanics treatment was omitted as in the Benson scheme (but the Boltzmann distribution was still included), the corresponding (reoptimized) parameter set gave a poorer result with a WSD of 0.45 kcal/mol.

Next, note that the WSD from the HF calculation (second column) without stat. is 0.82 kcal/mol, much poorer than the MM4 value. This larger error is mostly due to the long-range correlation error from the Hartree–Fock calculation. This correlation error can be dealt with in various ways, but let us first take an overview.

A Hartree–Fock calculation (second column) does not include any correlation, and so not surprisingly we need similar parameters with such a calculation in order to account for that part of the correlation energy. The Neo/Me spread is just a little larger in HF than in MM4 (10.6 vs. 8.9 kcal/mol). The 1,2 correlation is accounted for in the bond energy parameter in the HF heat calculation. So we have now included the 1,2 and 1,3 parts of the correlation in the HF heat calculation with parameters as we do in molecular mechanics. But this procedure neglects the longer range correlation completely, which, of course, limits the accuracy of the calculations at the HF level. This long-range correlation is accounted for in molecular mechanics with the van der Waals interactions.

The R_5 and R_6 terms are needed in the Hartree–Fock calculations for the same reason that they are needed in molecular mechanics (later). The numerical values of POP were simply carried over to the Hartree–Fock calculations (and subsequent calculations) from MM4, and the value of TOR was reoptimized. Note that TOR has quite a large value here (1.0 kcal/mol), and must be absorbing other errors.

A commonly used way to treat correlation is using perturbation theory (Møller–Plesset, Chapter 3). This approach is computationally demanding but can, in principle, be extended to any desired accuracy. We employed sequentially the HF, MP2, and MP4 levels of calculation (columns 2–4 in Table 11.6). These methods (with the statistical mechanical treatment included) gave results with the WSDs of 0.92, 0.61, and 0.42 kcal/mol, respectively. Thus, the better calculation gives a better result. The MP4 result is close to the MM4 result (0.34 kcal/mol) and close to the error limit imposed by the experimental data.

Note the spread of the values for the energy parameters Neo/Me previously discussed. The spread for HF, MP2, and MP4 was, respectively, −10.6, +2.6, and −2.3 kcal/mol. Thus, the need for these terms is large in the HF case where there is no correlation energy, but it is overdone at the MP2 level (opposite sign), and returns to a smaller but still significant value at MP4 (oscillatory convergence). The R_5 and R_6 terms show similar oscillatory behavior by being large at the HF level, smaller at MP2, and then slightly larger at MP4. Thus, all of these parameters appear to be converging, but perhaps not exactly to zero. The value for TOR is generally behaving as one would expect. It has values of 1.00 HF, 0.12 MP2, and 0.25 MP4, compared to the experimental value[10] of +0.36 kcal/mol. So this value also shows oscillatory convergence to near the expected value.

If we carry out an MP2 calculation on a molecule, it accounts for, to that level of approximation, the correlation energy of the whole molecule. As is often found, MP2 overaccounts for the correlation, and here the sign of the trend Neo/Me is reversed, but the actual magnitudes of the numerical terms are considerably smaller than the Hartree–Fock values. We might expect that at the Schrödinger limit these numbers would go to zero. But since they are adjustable parameters, they are optimizing to fit the data regardless of the physics of the situation. This means that they are not expected to go exactly to zero unless all of the other errors in the calculations (and experiments) also go to zero.

The need for the R_6 parameter is not completely understood.* This parameter is important in the MM4 calculation of the heats of formation. The R_5 number in molecular mechanics might be thought of coming from the fact that there is about 5 kcal/mol more strain in cyclopentane than is present in open-chain compounds or cyclohexane. But it is not that simple, as that part of the energy has been already accounted for, mainly by bending and torsion parameters. The carbon in a five-membered ring is given different parameters for bending, stretching, and the like relative to an alkane, so it is not surprising that some adjustment in the overall energy of the system is needed here. The value of R_5 is not very different from that of R_6 and appears to be converging toward it. [There are also R_4 and R_3 terms needed for systems containing small rings (not included in this study), and these have quite different values because of the large strain energies in such systems. These strain energies are also incorporated in the R_3 and R_4 numbers.] This R_5 number is also expected to disappear in an exact ab initio calculation or at least converge with the R_6 value. For further information on these R quantities see under Strain Energy and Ring Strain later in this chapter.

Another popular way to improve the results from Hartree–Fock calculations, which is particularly useful because of its computational speed, is *density functional theory* (Chapter 3). When the B3LYP procedure was applied to the present data (column 5 of Table 11.6), the weighted standard deviation in the results (without stat.) was reduced from 0.82 (HF) to 0.47 kcal/mol, indeed a major improvement.

Next, note that the statistical mechanical treatment required for the zero-point energy gives a certain level of results for each method (the with stat. line at the bottom of Table 11.6). If this procedure is not included (the without stat. line), the results are poorer as expected for MM4 and also for both the MP2 and MP4 results. But curiously for HF and B3LYP, the results are actually better when the statistical mechanical treatment is omitted (columns 2 and 5 of Table 11.6).

One might ask: Why can these heats of formation be fit more accurately with the Hartree–Fock or B3LYP methods if one leaves out the statistical mechanics addition of zero-point energy, for example? It appears that these results are not fortuitous

---

*It probably has to do largely with the difference in correlation energy between a ring and an open chain. An alkane has the formula C_nH_{2n+2}. If a ring is present (of any size), the formula of the cycloalkane is C_nH_{2n}. Thus, the cycloalkane has two less C–H bonds and one more C–C bond than the corresponding alkane. The 1,2 correlation changes would presumably be accounted for in the bond energies, as in the branching case, but again the 1,3 correlation changes would be omitted. If this parameter value is a result of correlation, it should go to zero at the Schrödinger limit.

because the discrepancies are too large. They are real and they seem to come about as follows.

Consider molecules divided into two general classes, highly congested ones and ordinary ones. Electron correlation will stabilize congested molecules more than it stabilizes the ordinary ones because the congested ones are, by definition, suffering from too many electrons in the same place at the same time. Correlation will consequently lower the energies for congested molecules more than it does for ordinary molecules.

On the other hand, congested molecules generally speaking have their zero-point energies raised by that congestion relative to those of ordinary molecules. The steric pressure of the different parts of the molecule on one another narrows their vibrational potential wells, causing the vibrational levels to be lifted up and spread out. Thus, if we start with the Hartree-Fock calculation and wish to improve it, adding electron correlation will lower the energy of the congested molecules more, but the vibrational part of the calculation will to some extent raise it back up again. So if we consider improving the HF calculation to the MP2 level, the stability gain here for a congested molecule from correlation is partly canceled out by the higher zero-point energy. But, if we simply add the statistical mechanical terms directly to a Hartree-Fock calculation on a congested molecule, and fail to include correlation, the systematic results are expected to be worse than they would have been if we had not added the statistical mechanical terms because we are increasing the energies of the congested molecules when we should be reducing them. And this is what is found.

The B3LYP calculations behave similarly to the HF calculations, apparently because an important part of the correlation, the dispersion energy, is not properly accounted for when the B3LYP method is used.[24] Thus, one wants to add the statistical mechanical terms only when one includes correlation (dispersion) more completely than in the B3LYP method. If one includes MP2 or MP4 correlation, the statistical terms improve the accuracy of the results. But if one does not include the correlation (Hartree-Fock) or includes it only at the B3LYP level, the results actually get worse when the statistical terms are included. Molecular mechanics does include the dispersion energies as previously discussed and, hence, behaves like an MP2 calculation and not like a Hartree-Fock calculation in this respect.

When a parameter set is optimized by the least-squares method, the result is obtained by allowing various errors to cancel one another insofar as possible. (This is inherent in the method and independent of the physics of the situation.) In the spirit of this method, then, while we know that omitting the vibrational calculations after a structure is calculated by the Hartree-Fock procedure is using one set of errors to cancel another set of errors, it seems the appropriate way to go because both sets of errors are largely systematic. (One might say that this is one reason why the Wiberg-Schleyer method works as well as it does with Hartree-Fock calculations.) In the present case of the calculation of heats of formation, the recommended procedure would accordingly be to include the vibrational calculation when the original quantum mechanical calculation is MP2 (or higher), but to omit it for calculations at the Hartree-Fock or B3LYP levels (as has traditionally been done in the past albeit only for reasons of computational simplicity!).

Actually, as computing power becomes increasingly available, it is doubtful that Hartree–Fock methods alone will be used in the future for the calculation of heats of formation. But B3LYP, and other density functional theory procedures, are still open to improvement. The problem of improving these functionals is under active investigation, so that newer methods can be expected to continue to be developed here. A simple improvement occurred to us, however, which begins with the well-tested B3LYP functional. Since there is a problem because of the inadequate inclusion of dispersion energies in this type of a calculation, why can't we calculate the dispersion energies separately and simply add them to the B3LYP results? Dispersion energies appear to be well calculated by MM4 for cases such as those at hand. These are simply the attractive parts of the van der Waals equation (the energy proportional to r^{-6} at longer distances). These longer range dispersion energies are well calculated by MM4 for hydrocarbons because they determine the heats of sublimation of crystals. And while this long-range part of the dispersion energy is clear enough, and well calculated, the short-range dispersion energies are even larger but rarely actually looked at. At rather short ranges (when atoms approach each other more closely than the sum of the van der Waals radii), these attractive dispersion forces become larger, but they are overwhelmed by the electron repulsions and buried from sight. It was mentioned in Chapter 8 for example in Fig. 8.2, that the high peaks, where there is serious repulsion on our contour maps, are expected to be too high in B3LYP calculations because of this lack of inclusion of dispersion. So the question to be asked is: Can we just add dispersion energies to B3LYP calculations and thereby improve their results noticeably?

We decided to try it. We simply added the dispersion energies* calculated by MM4 to the B3LYP energies from our test set of compounds listed in Table 11.1. It became evident that the full values of the MM4 dispersion energies were too large for use directly in the heats of formation, so they were scaled by a factor that was determined empirically to best be 0.50. We then redetermined a parameter set for this B3LYP + dispersion calculation (hereafter B3LYP/D) using our test set of 58 compounds given in Table 11.1. The parameters thus found are also listed in Table 11.6, column 6. First, note that the WSD (w/o stat.) is somewhat improved relative to the B3LYP calculations without the dispersion energy (0.44 vs. 0.47 kcal/mol). But, if we are now including the dispersion energy, then we should also include the statistics. And we see that the corresponding WSD goes from 0.55 to 0.35 when this is done. Thus, the B3LYP/D calculation gives a better overall result than the MP4 calculation! (Note that this statement applies to the case at hand, not necessarily in general.) And it is in fact an accuracy that is practically indistinguishable from that of MM4. We can also note that the Me/Neo range of numbers is reduced from the value for B3LYP (8.0 kcal/mol) down to 4.8 kcal/mol. Thus, much, almost half, of the extra energy lowering due to chain branching missing in the direct B3LYP calculations has been accounted for with the dispersion energy. The values for R_6 and R_5 have been noticeably reduced (consistent with these quantities being largely correlation corrections), and the value 0.26 for TOR is also better than previously.

---

*The numerical values of the MM4 dispersion energies are given later in Table 11.7 for reference.

Note that we now have three rather good calculational methods [excluding Hartree–Fock, MP2, and B3LYP (without dispersion)] that can be used to calculate heats of formation of alkanes (namely MM4, B3LYP/D, and MP4). With hydrocarbons and other well-studied classes of compounds, we feel that these three methods will for the most part give similar numbers, which will also usually agree well with the experimental numbers available. If all four of these numbers do agree, then the heat of formation of the molecule can be considered to be firmly established. But, if any one of them disagrees significantly, we have to wonder why. The three computational methods in their current form are partly, although not completely, independent of one another. If these three calculational methods disagree with one another, this is from a defect in one or another of the methods and probably not in the experiment. But, occasionally, it happens that the three computational methods all agree reasonably well and disagree with the experiment. This kind of situation is a strong indication that the experimental number is inaccurate, and, indeed, several repetitions of experiments in cases like that have in the past been resolved when a new experimental number did come into reasonable agreement with the calculated values.

It was mentioned at the beginning of this section that one method of dealing with heats of formation by ab initio calculations was that introduced by Wiberg, and that has been the method discussed to this point. Other methods of utilizing ab initio calculations exist and have been used. One is the use of isodesmic (and related) equations.[1,25,26] There was also a semiempirical program called PM3 that was published by Stewart[27] in 1989 that has been used for heats of formation calculations. The objectives of Stewart were rather different from those of Wiberg. The Wiberg approach is designed to deal with organic molecules. Stewart's approach was designed to deal with chemical compounds in general. The Wiberg method covers a much more limited class of substances, so that one would expect, and one finds, that it leads to higher accuracy.

The level of accuracy and area of applicability of Stewart's heat of formation calculations have, of course, improved very much with time, with the availability of more accurate semiempirical theory, more and better experimental data, and more powerful computers. The current version of his program is PM6.[28] It can be used for a wide variety of compounds covering 70 elements of the periodic table. When applied to 4492 diverse species, their heats of formation were calculated with an average unsigned error of 8.0 kcal/mol. When the calculations were limited to 9 common elements (H, C, N, O, F, P, S, Cl, and Br), the corresponding error was 4.4 kcal/mol, compared with 7.4 kcal/mol for HF 6-31G* and 5.2 kcal/mol for B3LYP/6-31G*.

Turning back to the ab initio calculation of heats of formation, John Pople and his co-workers introduced a different way of attacking this problem in 1989.[19] The method was referred to as the G-1 method, and it was mostly, but not completely, ab initio in nature. The basic idea was that one calculated energies of molecules using different basis sets in such a way that one could extrapolate the results to the Hartree–Fock limit. The electron correlation energies were then similarly calculated with some approximations to the Schrödinger limit, and then the molecular energies were converted into heats of formation. And, of course, one had to deal with the statistical mechanical problems. All of this is rather easy to say but rather hard to do. The current version (G-4[29]) is capable of giving an accuracy of about 1 kcal/mol over a large data set containing molecules up to the size of bromobenzene and has been applied to molecules containing atoms as far down the periodic table as bromine.

Strain Energy

Heats of formation are of interest to chemists for a wide variety of reasons, but probably the most common use has been for the study of strain. Molecules can, in general, be stretched, bent, twisted, and thus deformed by van der Waals or electrostatic forces in numerous ways and by varying amounts. Such deformations and interactions add strain energy to the molecule, and strain energy can have a profound effect on chemical behavior. Often, measurements of strain are carried out by studies of equilibria or in other ways that do not involve heats of formation. However, heats of formation are a fundamental way for examining and categorizing such phenomena. We will discuss here a few examples to illustrate some applications and usefulness of strain energies and some of the difficulties involved in determining them.

It is obvious that the strain energy in a molecule is an energy difference between the molecule and some (strainless) reference standard. That reference standard is arbitrary and various different standards have been used in the earlier literature. It is desirable to have a useful, clear, and precisely defined way of calculating strain energies. It should be simple, straightforward, and completely transparent. We introduced such a method for use with our 1971 force field[30] and used it in subsequent force fields (MM2, MM3, MM4). A detailed summary of the method is given by Burkert.[31]

The strainless standard for alkanes (MM4 and earlier) is defined by five parameters. These are the C–C and C–H bonds and three branching parameters, Me, Iso, and Neo, where the methylene group is a default quantity (with an energy value of zero). Methane is also given a strain energy of zero.

Our basic set of strainless compounds includes the anti conformations of the n-alkanes up to heptane, plus isobutane and neopentane. A set of parameters was then obtained by the least-squares method from the then available experimental data so as to give all of these compounds zero strain energies as closely as possible. To determine the strain energy for any given compound (alkane or cycloalkane) then, one uses the heat of formation (determined by any method chosen) and compares this to the energy of the (strainless) standard reference compound constructed by the summation of the parameters of an equivalent number of bonds and branching groups. This difference gives a quantity that we have referred to as the *inherent strain energy*. These are the numbers that should ordinarily be used when comparing *calculated* strain energies for different compounds.

However, real compounds frequently contain molecule specific and often large contributions to their energy, and hence to their strain energy, from their vibrational motions. These contributions are given by the terms POP and TOR that are used for thermodynamic calculations in molecular mechanics and in quantum mechanics. The relationship is given by Eq. (11.5).

$$E_S = E_{IS} + POP + TOR \qquad (11.5)$$

where E_S and E_{IS} are the *strain energy* and *inherent strain energy*, respectively. POP and TOR have their usual meanings. POP is the energy resulting from the Boltzmann distribution of conformations, and TOR is a vibrational correction for molecules that contain low torsional barriers. If one is referring to experimental measurements, these terms are real, often non-negligible, and they need to be taken into account. When they are, the strain energy thus obtained refers to the real molecule and has the meaning commonly associated with the term.

When one wants to compare molecular fragments, normally from theoretically calculated heats of formation or other thermodynamic quantities, one usually doesn't want these extraneous terms to get involved, and so one has to subtract them out in order to get the inherent strain energy. And in any discussion one should always make it clear whether one is discussing the strain energy or the inherent strain energy. The difference between these two energies is frequently nontrivial. In methylcyclohexane, for example, POP and TOR have the values 0.10 and 0.00 kcal/mol, respectively. In *n*-heptane, for comparison, the MM4 values for POP and TOR are, respectively, 0.92 and 2.28 kcal/mol. Thus, the difference in the strain present in these two compounds changes by 3.20 kcal/mol, depending on whether we are talking about strain or inherent strain. A collection of the values of the strain energies (and inherent strain energies) for our standard set of hydrocarbons is given in Table 11.7.[22]

TABLE 11.7. MM4 Heats of Formation, Strain Energies, and Dispersion Energies for Selected Alkanes[a,22a]

Compound	H_f°	E_{IS}^b	E_S^c	DE[d]
Methane	−17.89	0.00	0.00	0.00
Ethane	−19.75	−0.03	−0.60	−1.37
Propane	−24.99	0.00	0.00	−3.24
Butane	−29.97	0.02	0.86	−5.25
Pentane	−35.03	0.02	1.65	−7.29
Hexane	−40.12	0.01	2.40	−9.34
Heptane	−45.16	0.00	3.20	−11.40
Octane	−50.20	−0.01	4.00	−13.45
Nonane	−55.24	−0.03	4.81	−15.51
Isobutane	−32.36	0.00	0.00	−5.65
Isopentane	−36.58	0.99	1.62	−8.34
Neopentane	−40.67	0.00	0.00	−8.66
2,3-Dimethylbutane	−42.16	2.69	3.41	−12.01
2,2,3-Trimethylbutan	−49.01	4.87	4.87	−16.32
2,2-Dimethylpentane	−49.69	2.09	2.66	−14.26
3,3-Dimethylpentane	−47.85	4.50	4.50	−15.57
3-Ethylpentane	−44.40	3.67	5.49	−14.22
2,4-Dimethylpentane	−48.12	1.99	3.29	−13.88
2,5-Dimethylhexane	−52.85	1.88	4.40	−15.82
2,2,3,3-Tetramethylbutane	−53.86	8.33	8.33	−21.42
2,2,3,3-Tetramethylpentane	−56.74	11.29	11.29	−25.36

TABLE 11.7. (Continued)

Compound	$H_f^°$	E_{IS}^b	E_S^c	DE^d
Di-t-butylmethane	−57.58	9.29	10.45	−22.71
Tetraethylmethane	−55.33	8.72	8.72	−23.38
Tri-t-butylmethane	−54.05	42.86	42.86	−47.30
Cyclopentane	−18.59	5.58	6.15	−6.46
Cyclohexane	−29.59	0.53	0.53	−10.40
Cycloheptane	−27.88	7.50	8.07	−13.99
Cyclooctane	−29.72	12.08	12.08	−18.01
Cyclononane	−31.37	15.51	16.28	−21.94
Cyclodecane	−36.74	15.97	16.74	−25.64
Cyclododecane	−53.48	10.76	11.68	−30.49
1,1-Dimethylcyclopentane	−33.02	6.59	7.40	−12.66
Methylcyclopentane	−25.69	5.60	6.42	−8.79
Ethylcyclopentane	−30.35	6.15	7.61	−11.39
eq-Methylcyclohexane	−36.99	0.39	0.49	−13.03
1,1-Dimethylcyclohexane	−43.42	2.37	2.37	−17.35
1-ax-2-eq-Dimethylcyclohexane	−41.71	3.14	3.14	−17.29
1-eq-2-eq-Dimethylcyclohexane	−43.34	1.49	1.51	−16.50
Bicyclo[3.3.1]nonane	−30.04	7.65	8.24	−20.39
cis-Bicyclo[3.3.0]octane	−22.72	9.88	10.67	−13.24
trans-Bicyclo[3.3.0]octane	−15.29	18.10	18.10	−12.71
trans-Decalin	−43.60	0.52	0.52	−21.22
cis-Decalin	−40.91	3.21	3.21	−22.57
trans-Hydrindane	−31.73	7.03	7.03	−16.80
cis-Hydrindane	−31.01	7.70	7.75	−17.82
trans-syn-trans-Perhydroanthrancene	−57.73	0.40	0.40	−32.19
trans-anti-trans-Perhydroanthrancene	−50.26	7.87	7.87	−32.63
Norbornane	−13.10	14.45	14.45	−11.08
1,4-Dimethylnorbornane	−30.90	13.26	13.26	−16.18
Adamantane	−31.85	2.92	2.92	−22.69
1,3,5,7-Tetramethyladamantane	−66.55	1.46	1.46	−35.80
Protoadamantane	−20.11	15.13	15.13	−22.31
Congressane	−35.04	4.37	4.37	−35.19
Bicyclo[2.2.2]octane	−21.90	10.54	10.54	−16.37
Perhydroquinacene	−24.01	12.18	12.18	−18.11
Dodecahedrane	14.05	37.59	37.59	−34.41
Di-t-butylpropane	−58.34	25.37	25.37	−37.27
Tetra-t-butylmethane	−28.14	98.61	98.61	−75.50

[a] In kcal/mol.
[b] Inherent strain energies.
[c] Strain energies.
[d] Dispersion energies.

Ring Strain Energy

Molecular mechanics originally developed from a ball-and-stick model, which subsequently became weights and springs, and then further evolved into a complete classical mechanical model for molecules. The parallelism between molecular mechanics and quantum mechanics was always recognized. We know that Schrödinger theory works, and it was consequently necessary to take care to see that whatever was done in molecular mechanics was consistent with what was known from quantum mechanics. Innumerable ideas were investigated, many of which were never published because they were wrong or not useful. Molecular mechanics had to be robust, and of general predictive value, if it was to be useful. This meant that the boundaries of parameter transferability had to be firmly established. It has developed in the now popular presentations that five basic functions (and cross terms) are needed to describe molecules in general. These are functions that describe stretching, bending, and torsion in molecules and to those must be added van der Waals and electrostatic functions. Each of these functions has a zero point of reference. Energies are always relative. Van der Waals energies, for example, start with the zero point between two atoms at an infinite distance. When one calculates the energy of a molecule by molecular mechanics, it is relative to the zero points chosen for these functions. When one wants to calculate a heat of formation, the zero points are the standard states of the elements at 25°C. Hence, if the molecular mechanics force field is a good one, one can expect to obtain reasonable values for heats of formation from it by simply resetting the molecular mechanics zero points to those required for heats of formation.

Strain in a molecule is defined by an energy that is more positive than some standard because the system is located at a point that is above the defined energy zero point for that system. Thus, molecules can be bent, twisted, and the like with increases in strain energy. If there are rings in a molecule, the situation is somewhat different than for the open-chain case, and further attention has to be given to the problem of the zero points.

We will here confine ourselves to the ring systems defined by the cycloalkanes, but most of what is said is general and applies with appropriate modification to rings of any kinds of atoms with any kinds of hybridization.

Cycloalkane rings can be subdivided into several groups for practical purposes. We will consider small, common, medium, and large rings. The exact ring sizes that fit into each category depend to some extent upon what property one wishes to discuss, but generally 5- and 6-membered rings are regarded as common, medium rings are from 7 to about 12, with greater than 12 ring members constituting large rings. Cyclopropane and cyclobutane are considered small rings. The structures and attendant properties of the common, medium, and large rings have an extensive, convoluted, and quite fascinating history, but they have been well documented elsewhere[31,32] and will not be discussed in any detail here. We will just say that they show varying degrees of strain that are well understood and well described by molecular mechanics. For purposes of the calculation of heats of formation, we are here concerned specifically with the R_n parameters. These are required to set the zero points of the molecular mechanics calculation to the zero points required for heats of formation. The R_6 term results because of a change in bonding (the deletion of two CH bonds, and the creation of a CC bond, upon ring formation) as discussed earlier in this chapter. This is just a constant that applies to a ring of any size, 6-membered or larger. The R_5 term differs slightly from the R_6, for convenience and/or precision, as was discussed earlier. Here we wish to comment specifically on the R_3 and R_4 terms.

In the earliest molecular mechanics calculations regarding cyclopropane, cyclobutane, and other compounds containing small rings, it was originally thought that perhaps one could treat cyclopropane as simply an analogous propane system in which the C–C–C angle was bent down to give a 60° angle. Cyclobutane was similarly bent to about a 90° angle. It was quickly found that such large bendings gave very high energies, much too high to correctly calculate heats of formation. One had basically two choices. One could adjust downward either the bending constant for the large bendings found in these molecules or one could adjust the θ_0 value and thus bring the bending energies into agreement with experiment. Of these two choices, the bending constant was arbitrarily adjusted. It turned out (much later) that this was the wrong choice, although it could be made to fit a large amount of data on small ring compounds very well. What it did not fit were vibrational spectra. The MM2 force field, and most force fields of that time, did not fit vibrational spectra at all well anyway, so this was not recognized as a problem. Later with MM3, it became evident that one could well fit vibrational spectra in alkanes, and to fit the spectra of small ring compounds, a major reduction of the C–C–C θ_0 value was required. Now why should this be?

The rationalization involves something that is usually referred to as "orbital following." Imagine a carbon atom with tetrahedral geometry and sp³ hybridization. If one angle θ is pinched so that it has a small value and held that way by a constraint (the ring), the hybridization does not stay sp³, rather more p character is put into the bonds that are going to be within the ring, and the s character goes more into the bonds that are external to the ring. Thus, the ring C–C–C angle has a value for θ_0 of less than the open-chain (propane) value for cyclobutane, and a still smaller value for cyclopropane. The actual value chosen for θ_0 is somewhere in between the tetrahedral value and the 60° or 90° value that might have been expected. Thus, the orbitals bend or follow the distortion, but they do not bend all the way to the idealized values. Because of this orbital following, the energies of these ring systems are lowered relative to what they would otherwise have been in bent propanes, and the vibrational spectra are simultaneously fit.

A force field was thereby devised to fit the structures and vibrational spectra of these molecules. But, of course, if we change the zero of reference for our bending, then we have to expect that there will be required one or more parameters in the heat of formation calculations to allow for this change of reference. At a minimum, we expect that R_3^{33} and R_4^{13} parameters will be needed because the θ's here are substantially different from the usual values. These parameters are composed of two parts, one of which is from the same bonding change described for R_6. The second part involves the strain required to deform the geometry and reset θ_0 from tetrahedral to the small ring value.

Cyclopropane compounds are fairly rigid because they cannot change a bond angle without changing the bond length. It is much more difficult to stretch bonds than to bend angles in general, and the cyclopropane molecule has this bond-stretching type of rigidity imposing a similar rigidity on the bond angles. That complication does not occur in cyclobutane, and these are able to bend over a wider range in different molecules. There is no assurance that R_3, and especially R_4, will really be transferable constants. But it appears that they are, at least to within the accuracy that the problem has so far been investigated.

In quantum mechanical heat of formation calculations, the situation is less complicated in principle. R_3 and R_4 terms are expected to be desirable with relatively low-level calculations (in part because they help account for correlation as do R_5 and R_6), but they should go to zero at a sufficiently high level as correlation is better accounted for, and the orbital following energy will automatically appear in the E_{QM} term in Eq. (11.4).

Dodecahedrane

This is a truly unusual compound. It belongs to the point group I_h, which is a group that few organic chemists have even heard of.[34] It has numerous planes and threefold axes of symmetry. The MM4 structure gives the C–C bond length (r_g) as 1.5452 or 1.5451 Å for r_α. These are unusually similar in part because of the great rigidity of the molecule.

Since the dodecahedrane molecule has a geometry that overall is close to spherical, it seemed likely that it would pack into a crystal lattice with the individual molecules more or less completely disordered with respect to rotation about their centers of gravity. This proved not to be true. Rather it was originally concluded that the molecules are packed into the lattice with a maximally superimposed order between the point group and the space group.[35,36] (It may be remembered that fullerene does have this rotational disorder (Chapter 5).) It was subsequently shown, however, that the experimental structure is the mean of three equivalent arrangements of lower symmetry.[37]

This compound has been a "test case" for heat of formation studies for more than 30 years, and there are numerous studies on the subject. Early work is summarized in recent computational[26] and experimental[38] work. We will begin this discussion with the study by Clark and co-workers.[16]

Dodecahedrane is a 20-carbon globular molecule made up of only 5-membered rings, 12 of them in all, fused together.* As shown in Table 11.8, one may consider it starting with cyclopentane and building up to the larger structures shown, finally ending with dodecahedrane (**13**). At the time Clark et al. wrote their article, the heat of formation of this compound was not known experimentally. The synthesis proved to be quite laborious. A small amount of the compound was finally prepared in 1982,[39] but the synthetic chemists were not about to let anyone burn their treasured sample! Eventually, a sufficient amount of a derivative of the compound was prepared, and the heat of formation of that derivative was determined.[38]

The heats of formation of the remaining compounds given in Table 11.8 were already known experimentally at that time. The early predictions (1979) regarding the

---

*Actually, there are only 11 rings; see later.

TABLE 11.8. Heats of Formation (kcal/mol) of Dodecahedrane and Related Compounds[a]

	(C_s)	10 (C_s)	11 (C_3)	12 (C_s)	13 (I_h)
MM1	−18.09	−20.68	−19.74	14.61	45.28
EAS	−18.37	−22.61	−23.74	−8.64	−0.22
MM2	−18.27	−22.77	−22.08	−5.56	22.15
MM4	−18.59	−22.72	−24.01	−10.15	14.06
Experimental	−18.44	−22.30	−24.46	—	18.2
Strain[b]	5.58	9.88	12.18	28.59	37.59

[a] For clarity, only the carbon skeletons of the molecules are shown. Data from Clark et al.[16]
[b] Strain means MM4 inherent strain energies. These refer to the structures shown (the most stable conformation for each), without inclusion of the Boltzmann distribution energy for conformations (POP), or the energy lowering that comes from low potential barriers to torsion (TOR). These are strictly theoretical quantities (see previous section, Strain Energy).

heat of formation of **13** were highly divergent. They spanned a range from −0.22 to +22.15 kcal/mol, determined with better then-available force fields (Table 11.8).

There is some interesting additional information available now regarding the compounds in Table 11.8. Cyclopentane is well known to be puckered. The eclipsed (planar) form has a higher torsion energy, and by bending the C–C–C bond angles somewhat, this torsion energy can be reduced by more than the bending energy added, so the ring puckers. The barrier to planarity of cyclopentane is 4.23 kcal/mol.

Compound **10** (*cis*-bicyclo[3.3.0]octane) also has puckered rings, and it has a known heat of formation. (We feel that experimental heat of formation is a little too positive. The experimental value is −22.30 ± 0.50 kcal/mol. The recommended value is −22.86 kcal/mol, which is an average of MM4/B3LYP(D)/MP4 determinations.)

Compound **11**, perhydrotriquinacene (discussed earlier in Chapter 5), already had a known heat of formation in 1979. This compound is also puckered, as one would expect. If there is a cooperative puckering, the molecule can generate a C_3 rotational axis running through the tertiary carbon at the center of the molecule and the attached hydrogen. If one forces the molecule to have planar cyclopentane rings (C_{3v} symmetry), the energy increases by 12.56 kcal/mol (MM4), from the eclipsing of those rings (3.0 times the energy of eclipsing cyclopentane itself). Clark et al.[16] state: "Allinger states that perhydroquinacene* is a highly puckered system, i.e., the overall structure is C_3 not C_{3v}, with the implication that it has considerable strain energy, more than, say, norbornane." There is some misunderstanding here. The inherent strain in norbornane is 14.4 kcal/mol while that of perhydrotriquinacene is 12.2 kcal/mol (Table 11.7).

*Clarke[16] uses the name perhydroquinacene for the perhydro derivative of triquinacene.

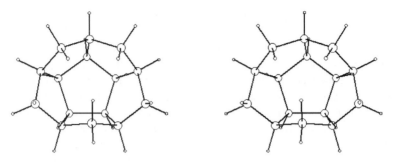

Figure 11.1. Peristylane (stereographic projection).

Cyclopentane puckers to lower its torsional energy. cis-Bicyclo[3.3.0]octane likewise puckers in both rings for the same reason. Similarly, perhydrotriquinacene can simultaneously pucker all three rings within the C_3 symmetry constraints, and this lowers the energy of the system a great deal. The ratios of the inherent strain energies (MM4) for these compounds cyclopentane – dodecahedrane are in order, 1.0:1.8:2.2 :5.1:6.7 kcal/mol. It might have been thought, just looking at models and counting the rings, that the strain energies for the whole series here would be in order 1:2:3:6:12 kcal/mol since in each case we appear to have a certain number of rings that appear to be free to pucker as they wish. But, it is obviously not that simple. The reason the strain energies go as they do is because, taking perhydrotriquinacene **11** as an example, if the strain energies of two puckered rings are counted, most of the energy of the third ring has already been counted as part of the puckering of the first two rings, and is not to be counted again. Hence, the total strain in this structure is only 2.2, rather than 3.0, cyclopentane units. (Perhydrotriquinacene also has a second conformation of C_1 symmetry, which is 1.76 kcal/mol higher in energy than the C_3 form, but it contributes little of interest.)

The conformation for compound **12** (peristylane) is not an immediately obvious one. The structure is shown as a stereographic projection (in Fig. 11.1). This structure has C_s symmetry, and three of the cyclopentane rings are puckered into half-chair conformations, while the other three are planar.* This was the only low-energy conformation for this molecule located by our conformational searching procedure (Appendix). One might have expected the inherent strain to be higher here than suggested by simple cyclopentane itself because three of the cyclopentane rings in pristylane are planar. This would appear to add another 12.7 kcal/mol to this molecule. It is, in fact, found that the strain energy is 28.59 kcal/mol, significantly *less* than 6 times the cyclopentane value.

---

*It is not easy to see which rings are planar in the peristylane projection in Figure 11.1, but it is like this. There is a planar five-membered ring at the bottom (plane of the paper) of the "cup" and we are looking down into that. The C_s plane is vertical, perpendicular to, and cutting through that five-membered ring at the "north" carbon, and the midpoint of the "south" C–C bond. The other two planar rings are those on the "northeast and northwest." The three rings around the southern part of the system are puckered, in alternating directions.

And dodecahedrane is calculated to be much more stable than one might have supposed, the strain energy being only about half (6.7/12) that which might have been expected, in spite of the fact that all of the rings in dodecahedrane are planar! (A planar cyclopentane has its inherent strain energy at 9.81 vs. 5.58 kcal/mol for puckered cyclopentane, for an increase of 76%. Thus, if you believe that dodecahedrane contains 12 planar cyclopentane rings (which turns out not to be true by the way, see Ring Counting in the Appendix), you might have reasonably expected a strain energy of $12 \times 9.81 = 117.7$ kcal/mol, vs. 37.6 kcal/mol observed. Obviously, we are missing something here.

When one looks across a row of calculations for these five compounds in Table 11.8 by a given method, for instance, the line labeled EAS,[14b] one sees that the magnitude of the heat of formation is slowly trending systematically more negative for cyclopentane and compounds **10** and **11**. But the value then turns and goes more positive for compound **12**, and still more positive for **13**. One might think that the strongly increasing values for compounds **12** and **13** are a result of excess strain, but the strain calculations indicate that this is not the case. The more positive heats of formation as we go across the table, especially for these last two compounds, is simply a result of the fact that the CH bond energy is much more negative than the CC bond energy (approximately 107/87 kcal/mol, Table 11.2). The number of C–H bonds decreases relative to the number of C–C bonds as we go to the right in Table 11.8, so the heats of formation of the compounds tend to become more positive, apart from any strain considerations.

Dodecahedrane (**13**) necessarily has planar five-membered rings. Thus, it has a great deal of torsional strain (75.1 kcal/mol), although minimal [but significant (3.8 kcal/mol)] bending strain. The total strain (37.59 kcal/mol) is, however, really small for a system containing 12 cyclopentane rings. The disparity comes largely from the van der Waals interactions. These total +2.94 kcal/mol in cyclopentane, but only +3.53 kcal/mol in dodecahedrane. Twelve times the cyclopentane van der Waals strain would be +61.3 kcal/mol. Thus, the actual dodecahedrane molecule is thus very much more stable than might have been expected. The low strain energy of dodecahedrane, and to a lesser extent that of peristylane, comes from van der Waals interactions. In cyclopentane there are sizable van der Waals repulsions between pairs of vicinal hydrogens, and no interactions at all between carbons. As we fuse together cyclopentane rings, there is a progression through the series of compounds **10–13** where the relative number of H/H repulsions are reduced and C/C attractions are added. That this occurs is expected, but that it is so dramatic probably not.

There was some discussion at the time as to whether or not it might be possible to trap a small molecule, atom, or fragment inside the dodecahedrane cage. There is not much room inside that cage, and any known stable particle that might be trapped inside would consequently exert a huge van der Waals repulsion on the cage. Such an entrapment attempt in this cage seems futile, based on what was then and is now known. However, with larger cages (much larger) such entrapment is not only possible but has been accomplished (see Section on C_{60} Fullerene, Chapter 5).

Looking at the numerical values from MM2 in Table 11.8, it is noted that the value for cyclopentane agrees well with the experiment, as does the *cis*-bicyclooctane number (where it is now believed that the experimental number is a little bit too positive; see above). But there is a notable discrepancy between the MM2 and experimental values for perhydrotriquinacene, compound **11**. We now believe that the MM2 value, which is −22.08 kcal/mol here, is more than 2 kcal/mol too positive. Why is this?

At the time that MM2 was developed, it was assumed that we could treat carbons in a five-membered ring with the same parameters we used to treat carbons in a six-membered ring or open chain. (It was well recognized at that time that special parameters were needed for carbons in three- and four-membered rings.) After further extensive work was done with MM2 on the structures and energies of five-membered rings, it was concluded that one simply could not fit the adjustable parameters required for five-membered rings to within chemical accuracy for a wide selection of compounds based on the assumption that the five-membered ring was just a distorted six-membered ring. That was a fair approximation, but not really good enough. Hence, the judgment was made that beginning with MM3, the five-membered rings really needed (and were given) their own parameter set. The problem of fitting parameters to structures is in general more easily solved than in fitting to energies, and especially to heats of formation.

In 1995 the experimental heat of combustion of a derivative of dodecahedrane was finally determined.[38] The experimental value quoted was deduced from the heat of combustion of the substituted compound, with corrections to allow for the effect of the removal of the substituents. The accuracy of these corrections is not completely certain.

Additional calculations have also been carried out subsequent to the experimental work. These various calculated heat of formation results (all of those calculated with sufficiently good methods that they cannot be dismissed out of hand) are only in rough agreement, ranging from 8.9 to 21.6 kcal/mol. The current value for this heat of formation thus appears to the present author to be 15.3 ± 6.4 kcal/mol.

HEATS OF FORMATION OF UNSATURATED HYDROCARBONS

The heats of formation of alkenes can be measured by heats of combustion, much as is done for determining those of alkanes, and very many good data have been obtained by this method. There are also many heats of formation available for alkenes obtained by measuring their heats of hydrogenation, when the heat of formation of the hydrogenation product was already known.[40,41] The heat of hydrogenation of a typical alkene is much smaller than its heat of combustion and can therefore be measured more accurately.

Normally, heats of formation in the gas phase are desirable as these are properties of the isolated molecule. For small molecules, such experimental heats of hydrogenation are relatively available.[40a] But for larger molecules it is normally more convenient to carry out the hydrogenation in solution, and the earlier available data are largely solution data.[40] To be widely useful, it is necessary to be able to convert these data to gas-phase values.

In calorimetry it is desirable to study rapid reactions because heat exchange with the surroundings otherwise becomes a larger source of error. Hydrogenation reactions in organic chemistry are often conducted in acetic acid solutions since the reactions are usually faster that way. The early work in determining heats of hydrogenation in solution was therefore mostly carried out using acetic acid as a solvent.[40] There is a substantial problem here in thermochemistry, however, because the π electrons of alkenes typically form hydrogen bonds with acetic acid. After the hydrogenation there is no such bond. Thus, what one is really measuring is not just the heat of hydrogenation but also the heat of desolvation. These latter numbers are typically of the order of 1 kcal/mol, but variable, and hence they introduce considerable uncertainty into the measured heats. In 1975 it was established that hydrogenations of alkenes in an alkane solvent occur sufficiently rapidly under proper conditions such that the procedure has subsequently been used in thermochemistry.[41a] Since the heat of solvation of an alkene and that of the alkane to which it is hydrogenated are very similar in an alkane solvent, this problem has been, if not overcome, at least minimized by such measurements. Almost all recent heats of hydrogenation have been measured in this way.[41b]

The MM4 calculation of the heats of formation of hydrocarbons containing double bonds is quite straightforward if the double bonds are isolated. The calculations for the structures are carried out first, and then the heats of formation are calculated. One needs heat of formation parameters for the double-bonded carbons and for their attached groups, with increments for primary, secondary, and so forth, and the calculations proceed along the same lines as those described earlier in this chapter for alkanes. The situation is less straightforward when the double bonds in the molecule are conjugated, and the addition of several additional parameters is required (Chapter 5). In this case the program also calculates the total π energy and uses it in the energy minimization to obtain the geometry. This minimized π energy is also one of the terms included in calculation of the heat of formation. (A total of 14 parameters are required for conjugated hydrocarbons. In addition, the statistical parameters from the alkanes are carried over, which include R_5, R_6, and TOR.) The MM4 heats of formation were calculated for a total of 111 compounds, and the parameter set was optimized by a weighted fit to the experimental values for 42 alkenes and 69 conjugated hydrocarbons.[7b] A representative (truncated) list of some of these molecules is given in Table 11.9. It contains several simple molecules such as ethylene, propene, and the like, and then

TABLE 11.9. MM4 Calculated and Experimental Heats of Formation of Alkenes and Conjugated Hydrocarbons (kcal/mol)[7b]

Wt.	Compound	ΔH_f° Calc.	ΔH_f° Expt.	Expt. Error	Calc.–Expt.
8	Ethylene	12.52	12.56	0.08	−0.04
8	Propene	4.99	4.88	0.16	0.11
8	*gauche*-1-Butene	0.01	0.02	0.19	−0.01
6	*cis*-2-Butene	−1.63	−1.67	—	0.04
6	*trans*-2-Butene	−2.58	−2.67	—	0.09
8	Isobutene	−4.01	−4.04	0.22	0.03
6	*cis*-2-Pentene	−6.65	−6.60	0.17	−0.05
6	*trans*-2-Pentene	−7.58	−7.60	0.21	0.02

HEATS OF FORMATION OF UNSATURATED HYDROCARBONS 293

TABLE 11.9. (Continued)

Wt.	Compound	ΔH_f° Calc.	ΔH_f° Expt.	Expt. Error	Calc.–Expt.
8	2-Methyl-2-butene	−10.38	−10.12	0.24	−0.26
6	Cyclopentene	8.70	8.56	—	0.14
6	Methylenecyclopentane	2.72	2.78	0.26	−0.06
6	3-Methylcyclopentene	2.04	1.76	0.16	0.28
6	Methylcyclohexene	−10.31	−10.36	0.19	0.05
6	Methylenecyclohexane	−8.03	−8.22	0.33	0.19
6	Cyclohexene	−1.07	−1.20	0.17	0.13
6	Cycloheptene	−2.37	−2.20	0.26	−0.17
4	Norbornene	21.65	21.40	0.28	0.25
3	trans-Cyclooctene	5.85	4.68	0.71	1.17
3	cis-Cyclooctene	−5.78	−6.20	—	0.42
5	1,4-Cyclohexadiene	25.73	26.30	—	−0.57
5	1,4-Pentadiene	24.58	25.26	0.31	−0.68
10	trans-Butadiene	25.77	26.01	0.19	−0.24
8	trans-Isoprene	17.98	18.06	0.24	−0.08
4	2,3-Dimethylbutadiene	11.38	10.79	0.26	0.59
6	trans-Pentadiene	18.20	18.12	0.16	0.08
6	cis-Pentadiene	18.52	19.13	0.24	−0.61
6	1,3-Cyclopentadiene	32.52	32.12	0.28	0.40
6	1,3-Cyclohexadiene	25.25	25.38	0.22	−0.13
6	1,3-Cycloheptadiene	23.13	22.55	0.27	0.58
1	1,3,5-Cycloheptatriene	44.55	43.90	0.36	0.65
1	1,3,5,7-Cyclooctatetraene	69.83	70.30	0.31	−0.47
10	Benzene	19.88	19.81	0.14	0.07
8	Toluene	12.30	12.06	0.14	0.24
8	Ethylenzene	7.22	7.15	0.22	0.07
8	o-Xylene	4.78	4.56	0.26	0.22
8	m-Xylene	4.13	4.14	0.18	−0.01
6	t-Butylbenzene	−5.73	−5.40	0.29	−0.33
3	Hexamethylbenzene	−21.06	−20.75	0.62	−0.31
2	Indene	38.16	39.03	0.61	−0.92
6	Styrene	35.80	35.31	0.29	0.49
1	Azulene	73.86	73.53	0.82	0.33
8	Naphthalene	35.40	35.85	0.30	−0.45
2	trans-Stilbene	57.56	56.44	0.30	1.12
2	cis-Stilbene	59.29	60.31	0.47	−1.02
2	Anthracene	54.07	55.03	0.48	−0.96
2	Phenanthracene	50.51	49.88	0.58	0.63
1	4,5-Dimethylphenanthrene	44.87	46.26	1.44	−1.39
1	9,10-Dimethylphenanthrene	37.51	39.93	2.02	−2.42
1	2,2-Paracyclophane	59.87	58.48	0.50	1.33
1	2,2-Metaparacyclophane	52.24	52.21	0.40	0.03
0	Tetracene	74.23	72.30	1.25	1.93
0	[18]Annulene	123.42	124.0	5.5	−0.58

more complicated compounds such as norbornene and *trans*-cyclooctene. The conjugated molecules run the gamet, beginning with butadiene, benzene, and the like and then benzenoid systems such as napthalene, phenanthrene, the paracyclophanes, annulenes, and so on.

Over the whole set of 111 compounds, the unweighted rms errors calculated by MM3 and MM4 gave values of 0.62 and 0.68 kcal/mol, respectively. The weighted averages give more realistic values of 0.53 and 0.47 kcal/mol, respectively, where the weightings were generally inversely proportional both to the stated experimental errors and to the molecular size. These numbers are regarded as being in good agreement with experiment. A number of values (19) were included in the full list, but weighted zero for various reasons. Our previous experience has indicated that when the heats of formation of these "outliers" are redetermined experimentally, the new experimental value is almost always in better agreement with the MM3/MM4 value than was the older experimental value. The calculated results here are expected to be limited by the accuracy imposed by the Hartree–Fock calculations employed in MM4 for the conjugated systems. The weighted average of 0.47 kcal/mol over 111 compounds suggests that the overall results here are largely limited by the accuracy of the experimental results, however.

MM3/MM4 are intended to deal with aromatic rings that are bent out of planarity, and they appear to do so to within the limitations of the experimental data available. These calculations have been extended to quite large molecules (C_{60} and above), and the MM4 values for heats of formation are probably the best currently available for compounds of this size. We will here briefly discuss two specific cases of special interest, [18]annulene and fullerene.

[18]Annulene, Aromaticity

The word "aromatic" means benzene-like to an organic chemist. [18]Annulene, although it fits Hückel's rule, is definitely not "benzene-like" in that it is reactive like an ordinary polyene, and similarly unstable toward air, and the like. The heat of formation is calculated by MM4 to well within the experimental error of 5.5 kcal/mol,* and the structure contains alternating bond lengths (with some reservations regarding the X-ray structure as previously discussed in Chapters 5 and 10). Accordingly, we believe that the molecule should be considered to be nonaromatic, the NMR evidence notwithstanding. Dewar calculated that [18]annulene was aromatic, but negligibly so.[43] The MM4 calculations indicate it to be antiaromatic by 0.84 kcal/mol.[7b] Both of these numbers are zero to within the accuracy of any kinds of measurements or calculations that have been made. Thus, we conclude that the magnetically induced ring current shown by [18]annulene, while very real and quite large, is just that. It is a magnetically induced ring current, and it does not appear to be related to what we would call the aromaticity of the molecule.

---

*An attempt was made to determine the heat of formation of [18]annulene by combustion in the usual way, but it was later concluded that the sample oxidized significantly during the preliminary calorimetric work, and that the value obtained was in error. The heat was subsequently determined[42] in a much more convoluted way, which led to the value and error as stated.

Fullerene

This compound (discussed in some detail previously in Chapter 5) presents a formidable challenge as far as its heat of formation is concerned.[44] The determination of the latter by any method has to overcome a number of serious difficulties. First, there is the size of the molecule to consider. When the heat of formation is determined by combustion, other things being equal, the accuracy of the result is inversely proportional to the molecular weight. Fullerene itself contains 60 carbons, considerably more than molecules for which we have accurate experimental heats of combustion.

Because of the high molecular weight of fullerene, the physical properties of the compound do not make the isolation and purification of it an easy task. It is high melting and rather insoluble. There are two special properties in addition that tend to make handling of the compound difficult. The first is the simple fact that polycyclic aromatic hydrocarbons tend to react with the oxygen in air, among other things, and accordingly must be handled with some care. The severely nonplanar benzene rings will further contribute to the significant expected reactivity here. Of equal concern is the fact that materials in solution are frequently purified by filtering them through charcoal (largely graphite). The purification results because the graphite adsorbs a wide variety of materials on the surface of the aromatic rings. The same thing will tend to happen here, further complicating the purification of the compound.

Heats of sublimation are usually difficult to measure for high-melting compounds because of their low vapor pressures. One can calculate heats of sublimation with rather good accuracy for aromatic hydrocarbons in general, but fullerene has a volume/surface ratio that is markedly different from that found with essentially all other aromatic hydrocarbons (except other fullerenes), so that the reliability of this calculation is less certain here than usual.

Taken together, all of the above present a formidable series of obstacles for the calorimetric determination of the heat of formation of fullerene.

What about the calculation of the heat of formation? The simple size of the molecule takes it out of the realm where we can expect to carry out a reliable quantum mechanical calculation. How reliable do we expect that a molecular mechanics calculation would be here? MM4 calculations have been carried out on a wide variety of aromatic systems, most of which are planar, and all of which are considerably smaller than fullerenes.[7b] Similar calculations on nonplanar unsaturated systems are based on limited experimental measurements involving much smaller compounds. Thus, considering both the size of the molecule, and the nonplanarity, and the considerable extrapolations required from known model compounds, a sizable error is expected here.

In 2001, there were available seven reports[45] that gave values for the heat of formation of C_{60}. They gave an average value of 599.7 kcal/mol with a standard deviation of 13.1 kcal/mol. The average reported experimental uncertainty was 3.25 ± 0.62 kcal/mol. While the experimental uncertainty is reasonable for a compound of this molecular weight, the standard deviation in the average value attests to the experimental sample difficulties discussed earlier.

Let us look a little further at the experimental information here. For reasons outlined above, and perhaps others as well, it can be seen that there is considerable difficulty in obtaining a "pure" sample of the material. If the material being burned is less than 100% pure, then apart from the thermochemical measurements themselves, there will be variations in the results that depend upon the nature and amount of the impurity. We might, for example, look at the study by Steele and co-workers.[46] They went to great lengths to assure that they had as pure a sample as they were able to obtain, and they compared the heat of combustion of the fullerene with that of graphite determined with the same equipment and procedure as a standard. Their result was given as 634.8 ± 6.0 kcal/mol. This may be compared with the MM4 value of 645.4 kcal/mol, and also with the average value and standard deviation from the seven reports mentioned above of 599.7 ± 13.1 kcal/mol.

The stated experimental error in the result obtained by Steele (±6.0 kcal/mol) can be focused on as a typical result by an experienced group, working with well-calibrated equipment, for example. The 6.0 kcal/mol value is quite high compared to values that one usually sees reported for heats of formation of hydrocarbons in general, but the molecule is about six times larger than most of those and so is in the range of expectations. But the actual H_f° value found (634.8 kcal/mol) differs quite seriously from the average value found by seven other groups (599.7 kcal/mol). Most of the difference here must be due to the differences in the samples burned. These heat of formation values are really large and indicate that the molecule has quite a high strain energy (E_{IS} is 422.2 kcal/mol from MM4). Likely impurities in the molecule from adsorption, and from air oxidation, for the most part would tend to lower the measured heat of combustion, so the average value of 599.7 value is probably too low because of the difficulties involved with the sample handling.

HEATS OF FORMATION OF FUNCTIONALIZED MOLECULES

The extension of heat of formation calculations by molecular mechanics to functionalized alkanes is in many cases straightforward. One simply needs to determine a few bond energies and chain branching parameters for each different functional group, and also parameters that connect the functional group to the alkane fragment. Thus, the MM4 method has been extended to include alcohols/ethers,[47] amines,[48] sulfides,[49] aldehydes/ketones,[50] carboxylic acids/esters,[51] and amides.[52] The method could be easily extended to any other functional group, as long as sufficient experimental and/or quantum mechanical heat of formation data are available for that group to define the required parameters.

Ab initio calculations have also been applied in some detail to the calculation of heats of formation for a few functionalized molecules, including alcohols/ethers[4] and amines.[53] The calculations were first carried out using both Hartree–Fock and B3LYP methods, and the latter generally gave somewhat better results as expected, and as indicated for hydrocarbons in Table 11.6. MP4 and/or B3LYP(D) calculations are expected to give still better results, similar to those discussed earlier in this chapter for alkanes. It would seem that extension of the heat of formation calculations to functionalized aromatic compounds by molecular and quantum mechanical methods would also be pretty straightforward, although this has not yet been done.

Polyfunctional molecules, on the other hand, are another matter. The ordinary Benson scheme is not generally of chemical accuracy when applied to polyfunctional molecules. The problem with these molecules is that two functional groups in the same molecule are likely to interact *through bonds* and *through space* in such a way as to change the enthalpy of the molecule. If that interaction enthalpy is not taken into account, the accuracy of the calculation will be diminished. If the functional groups are distant from one another in terms of connectivity, the through-space interaction between them is taken into account by molecular mechanics as van der Waals and electrostatic interactions, and the scheme should work well. But, if the functional groups are close together by connectivity, they will interact through bonds via inductive and resonance effects. They will become in effect, a single extended functional group (a superfunctional group), in a sense, and in that case the interaction between them needs to be included, and it is probably not experimentally well defined. The basic problem here is easy to understand. There are many different kinds of functional groups, perhaps 20 or so common ones. Each pair of these would form one or more extended functional groups, and each triplet still more extended functional groups, and so on. And many of these extended groups would have stereochemical components to their interactions. To extend the method to this very large possible combination of functional groups, one would have to have a considerable amount of information on each of these combinations. The outcome is obvious. The experimentally based method cannot, at least at present, be extended in a practical way much beyond the case of isolated functional groups. So here is an area where computationally based methods will potentially be very useful.

REFERENCES

1. (a) W. J. Hehre, L. Radom, P. v. R. Schleyer, and J. A. Pople, *Ab Initio Molecular Orbital Theory*, Wiley Interscience, New York, 1986. (b) L. A. Curtiss, P. C. Redfern, and D. J. Frurip, *Theoretical Methods for Computing Enthalpies of Formation of Gaseous Compounds, Reviews in Computational Chemistry*, Vol. 15, K. B. Lipkowitz and D. B. Boyd, Eds., Wiley, New York, 2000, p. 147.
2. (a) K. B. Wiberg, *Physical Organic Chemistry*, Wiley, New York, 1964. (b) Y. Fan, Heats of Formation, in *Encyclopedia of Computational Chemistry*, Vol. 2, P. v. R. Schleyer, N. L. Allinger, T. Clark, J. Gasteiger, P. A. Kollman, H. F. Schaefer III, and P. R. Schreiner, Eds., Wiley, Chichester, UK, 1998, p. 1217.
3. (a) S. W. Benson, *Thermochemical Kinetics*, Wiley, New York, 1976. (b) A recent and powerful extension of the Benson method is the SPARC program, T. S. Whiteside and L. A. Carreira, *J. Theor. Comput. Chem.*, **3**, 451 (2004).
4. (a) N. L. Allinger, L. R. Schmitz, I. Motoc, C. Bender, and J. Labanowski, *J. Am. Chem. Soc.*, **114**, 2880 (1992). See especially the Supplementary Material. (b) U. Burkert and N. L. Allinger, *Molecular Mechanics*, American Chemical Society, Washington, D.C., 1982.
5. (a) F. Pawlowski, P. Jorgensen, J. Olsen, F. Hegelund, T. Helgaker, J. Gauss, K. L. Bak, and J. F. Stanton, *J. Chem. Phys.*, **116**, 6482 (2002). (b) W. D. Allen, A. L. L. East, and A. G. Csaszar, in *Structures and Conformations of Non-Rigid Molecules*, J. Laane, M. Kakkouri, B. van der Veken, and H. Oberhammer, Eds., Kluwer, Dordrecht, The Netherlands, 1993, p. 343.

6. N. L. Allinger, K. Chen, and J.-H. Lii, *J. Comput. Chem.*, **17**, 642 (1996).
7. (a) N. Nevins, K. Chen, and N. L. Allinger, *J. Comput. Chem.*, **17**, 669 (1996). (b) N. Nevins, J.-H. Lii, and N. L. Allinger, *J. Comput. Chem.*, **17**, 695 (1996). (c) N. Nevins and N. L. Allinger, *J. Comput. Chem.*, **17**, 730 (1996). (d) N. L. Allinger, K. Chen, J. A. Katzenellenbogen, S. R. Wilson, and G. M. Anstead, *J. Comput. Chem.*, **17**, 747 (1996).
8. J. Baker and P. Pulay, *J. Comput. Chem.*, **19**, 1187 (1998).
9. (a) G. Herzberg, *Molecular Spectra and Molecular Structure II. Infrared and Raman Spectra of Polyatomic Molecules*, Van Nostrand, New York, 1945. (b) C. Garrod, *Statistical Mechanics and Thermodynamics*, Oxford University Press, New York, 1995. (c) R. K. Pathria, *Statistical Mechanics*, Butterworth-Heinemann, Boston, 1996. (d) B. Widom, *Statistical Mechanics: A Concise Introduction for Chemists*, Cambridge University Press, New York, 2002.
10. D. H. Wertz and N. L. Allinger, *Tetrahedron*, **35**, 3 (1979).
11. K. S. Pitzer and W. D. Gwinn, *J. Chem. Phys.*, **10**, 428 (1942).
12. Discussed in textbooks on organic chemistry. It should be mentioned that although the fact that branched-chain hydrocarbons are more stable than straight chains has been recognized for almost 100 years, the reason why this is so is still under discussion. See (a) M. D. Wodrich, C. S. Wannere, Y. Mo., P. D. Jarowski, K. N. Houk, and P. v. R. Schleyer, *Chem. Eur. J.*, **13**, 7731 (2007); (b) S. Gronert, *Chem. Eur. J.*, **15**, 5372 (2009).
13. K.-H. Chen and N. L. Allinger, *J. Mol. Struct. (Theochem)*, **581**, 215 (2002).
14. (a) T. Clark, T. M. Knox, H. Mackle, M. A. McKervey, and J. J. Rooney, *J. Am. Chem. Soc.*, **97**, 3835 (1975). (b) E. M. Engler, J. D. Andose, and P. v. R. Schleyer, *J. Am. Chem. Soc.*, **95**, 8005 (1973). (c) N. L. Allinger, M. T. Tribble, M. A. Miller, and D. H. Wertz, *J. Am. Chem. Soc.*, **93**, 1637 (1971).
15. N. L. Allinger, *J. Am. Chem. Soc.*, **99**, 8127 (1977).
16. T. Clark, T. M. Knox, M. A. McKervey, H. Mackle, and J. J. Rooney, *J. Am. Chem. Soc.*, **101**, 2404 (1979).
17. T. Clark, *Handbook of Computational Chemistry*, Wiley-Interscience, New York, 1985.
18. K. B. Wiberg, *J. Comput. Chem.*, **5**, 197 (1984); K. B. Wiberg, *J. Org. Chem.*, **50**, 5285 (1985).
19. J. A. Pople, M. Head-Gordon, D. J. Fox, K. Raghavachari, and L. A. Curtiss, *J. Chem. Phys.*, **90**, 5622 (1989).
20. M. R. Ibrahim and P. v. R. Schleyer, *J. Comput. Chem.*, **6**, 157 (1985).
21. J. Labanowski, L. R. Schmitz, K.-H. Chen, and N. L. Allinger, *J. Comput. Chem.*, **19**, 1421 (1998).
22. (a) J.-H. Lii and N. L. Allinger, *J. Mex. Chem. Soc.*, **53**, 95 (2009). (b) A. St-Amant, *Density Functional Methods in Biomolecular Modeling, Reviews in Computational Chemistry*, Vol. 7, K. B. Lipkowitz and D. B. Boyd, Eds., Wiley, 1996, p. 217.
23. M. J. Frisch, G. W. Trucks, H. B. Schlegel, G. E. Scuseria, M. A. Robb, J. R. Cheeseman, J. A. Montgomery, Jr., T. Vreven, K. N. Kudin, J. C. Burant, J. M. Millam, S. S. Iyengar, J. Tomasi, V. Barone, B. Mennucci, M. Cossi, G. Scalmani, N. Rega, G. A. Petersson, H. Nakatsuji, M. Hada, M. Ehara, K. Toyota, R. Fukuda, J. Hasegawa, M. Ishida, T. Nakajima, Y. Honda, O. Kitao, H. Nakai, M. Klene, X. Li, J. E. Knox, H. P. Hratchian, J. B. Cross, C. Adamo, J. Jaramillo, R. Gomperts, R. E. Stratmann, O. Yazyev, A. J. Austin, R. Cammi, C. Pomelli, J. W. Ochterski, P. Y. Ayala, K. Morokuma, G. A. Voth, P. Salvador, J. J. Dannenberg, V. G. Zakrzewski, S. Dapprich, A. D. Daniels, M. C. Strain, O. Farkas, D. K.

Malick, A. D. Rabuck, K. Raghavachari, J. B. Foresman, J. V. Ortiz, Q. Cui, A. G. Baboul, S. Clifford, J. Cioslowski, B. B. Stefanov, G. Liu, A. Liashenko, P. Piskorz, I. Komaromi, R. L. Martin, D. J. Fox, T. Keith, M. A. Al-Laham, C. Y. Peng, A. Nanayakkara, M. Challacombe, P. M. W. Gill, B. Johnson, W. Chen, M. W. Wong, C. Gonzalez, and J. A. Pople, Gaussian (03), Revision C.02, Gaussian, Inc., Wallingford CT, 2004.

24. (a) S. Grimme, *J. Chem. Phys.*, **124**, 034108 (2006). (b) T. Schwabe and S. Grimme, *Phys. Chem. Chem. Phys.*, **9**, 3397 (2007).
25. S. E. Wheeler, K. N. Houk, P. v. R. Schleyer, and W. D. Allen, *J. Am. Chem. Soc.*, **131**, 2547 (2009).
26. R. L. Disch and J. M. Schulman, *J. Phys. Chem.*, **100**, 3504 (1996).
27. J. J. P. Stewart, *J. Comput. Chem.*, **10**, 209, 221 (1989).
28. J. J. P. Stewart, *Mol. Model*, **13**, 1173 (2007).
29. L. A. Curtiss, P. C. Redfern, and K. Raghavachari, *J. Chem. Phys.*, **126**, 084108 (2007).
30. N. L. Allinger, M. T. Tribble, M. A. Miller, and D. H. Wertz, *J. Am. Chem. Soc.*, **93**, 1637 (1971).
31. U. Burkert and N. L. Allinger, *Molecular Mechanics*, American Chemical Society, Washington D.C., 1982, p. 184.
32. E. L. Eliel, N. L. Allinger, S. J. Angyal, and G. A. Morrison, *Conformational Analysis*, Wiley-Interscience, New York, 1965, p. 189ff.
33. P. Aped and N. L. Allinger, *J. Am. Chem. Soc.*, **114**, 1 (1992).
34. F. A. Cotton, *Chemical Applications of Group Theory*, 2nd ed., Wiley, New York, 1971.
35. J. C. Gallucci, C. W. Doecke, and L. A. Paquette, *J. Am. Chem. Soc.*, **108**, 1343 (1986).
36. N. L. Allinger, H. J. Geise, W. Pyckhout, L. A. Paquette, and J. C. Gallucci, *J. Am. Chem. Soc.*, **111**, 1106 (1989).
37. J. L. M. Dillen, *S. Afr. J. Chem.*, **44**, 62 (1991).
38. (a) H.-D. Beckhaus, C. Rüchardt, D. R. Lagerwall, L. A. Paquette, F. Wahl, and H. Prinzbach, *J. Am. Chem. Soc.*, **117**, 8885 (1995). (b) H.-D. Beckhaus, C. Rüchardt, D. R. Lagerwall, L. A. Paquette, F. Wahl, and H. Prinzbach, *J. Am. Chem. Soc.*, **116**, 11775 (1994).
39. R. J. Ternansky, D. W. Balogh, and L. A. Paquette, *J. Am. Chem. Soc.*, **104**, 4503 (1982).
40. (a) J. L. Jensen, *Prog. Phys. Org. Chem.*, **12**, 189 (1976). (b) R. B. Turner and W. R. Meador, *J. Am. Chem. Soc.*, **79**, 4133 (1957).
41. (a) D. W. Rogers, P. M. Papadimetriou, and N. A. Siddiqui, *Mikrochem. Acta*, 389 (1975). (b) D. W. Rogers, *Heats of Hydrogenation*, World Scientific, Hackensack, NJ, 2006.
42. J. F. M. Oth, J.-C. Bünzli, and Y. de Julien de Zelicourt, *Helv. Chim. Acta*, **57**, 2276 (2004).
43. M. J. S. Dewar, *The Molecular Orbital Theory of Organic Chemistry*, McGraw-Hill, New York, 1969, p. 176.
44. Eight references on the experimental thermochemistry of fullerenes and 24 references on the theoretical thermochemistry of fullerenes are cited by (a) M. E. Minas da Piedade, Ed., *Energetics of Stable Molecules and Reactive Intermediates,* NATO Science Series, Series C: *Mathematical and Physical Sciences*, Vol. 535, Kluwer Academic, Dordrecht, The Netherlands, 1999, p. 48. Also see (b) J. Cioslowski, *Electronic Structure Calculations on Fullerenes and Their Derivatives*, Oxford University Press, Oxford, 1995, p. 154.
45. S. W. Slayden and J. F. Liebman, The Energetics of Aromatic Hydrocarbons: An Experimental Thermochemical Perspective, in *Chemical Reviews*, Vol. 101, P. v. R. Schleyer, Ed., American Chemical Society, Washington, D.C., 2001, p. 1541.

46. W. V. Steele, R. D. Chirico, N. K. Smith, W. E. Billups, P. R. Elmore, and A. E. Wheeler, *J. Phys. Chem.*, **96**, 4731 (1992).
47. N. L. Allinger, K.-H. Chen, J.-H. Lii, and K. Durkin, *J. Comput. Chem.*, **24**, 1447 (2003).
48. K.-H. Chen, J.-H. Lii, Y. Fan, and N. L. Allinger, *J. Comput. Chem.*, **28**, 2391 (2007).
49. N. L. Allinger and Y. Fan, *J. Comput. Chem.*, **18**, 1827 (1997).
50. C. H. Langley, J.-H. Lii, and N. L. Allinger, *J. Comput. Chem.*, **22**, 1476 (2001).
51. J.-H. Lii, *J. Phys. Chem.*, **106**, 8667 (2002).
52. C. H. Langley and N. L. Allinger, *J. Phys. Chem.*, **106**, 5638 (2002).
53. L. R. Schmitz, K.-H. Chen, J. Labanowski, and N. L. Allinger, *J. Phys. Org. Chem.*, **14**, 90 (2001).

CONCLUDING REMARKS

In the early days of molecular mechanics, say 1960–1975, there were constant objections from the naysayers: Force constants are not transferable. There were those who simply refused to accept that molecular mechanics would be a workable procedure. (Unfortunately, many of them were reviewers for grant proposals sent by this author to various granting agencies, so this was then a serious matter.) Well, literally, what they said was correct. *Force constants* are not transferable, but *force parameters* are. And they are much more transferable than we had thought at the time. For strainless molecules, a harmonic and quadratic force field can give a reasonably good approximation to a correct molecular structure. For strained molecules, and those that are awash with the steric and electronic effects that are so common in organic chemistry, we need more. We need what Hagler described as a *class 2 force field*. We need more than the harmonic term, and we need more than diagonal elements in the force constant matrix. And often, things that we take as constants are not really constants, although that may be an adequate approximation over a short range. But really, when one looks over a longer range, they turn out to be functions, though usually simple ones. Chemists generally accept now that molecular mechanics "works." It is only a question of whether or not it gives an accuracy that is adequate for the problem at hand. But of equal importance is the fact that the valence bond pictures so popular among physical organic chemists since the 1930s can now be tied to a computational method that yields quantitative values for experimental observables. Certainly, not all problems have been solved. There are still enough "problem compounds" to keep chemists busy for a long while. But the general outline of the structures of organic molecules, and how that outline can be represented with molecular mechanics and valence bond theory, is now pretty clear. There are probably still lurking in the shadows more "effects," which will presumably come to light as more data are collected and examined. Heats of formation are intimately connected to molecular structures, and their accurate calculation is also on its way to being tidied up. As with the structures themselves, being able to calculate accurate heats of formation is important, and being able to understand those quantities is of similar importance.

Molecular Structure: Understanding Steric and Electronic Effects from Molecular Mechanics,
By Norman L. Allinger
Copyright © 2010 John Wiley & Sons, Inc.

APPENDIX

INTRODUCTION

There are a number of miscellaneous items that are of interest and importance with regards to molecular structure that are mentioned but not really discussed elsewhere in this book. They are more or less well known to those working in this field. Since not all of them are well known to everyone, however, we have accumulated some of those topics here and will briefly describe them, giving some leading references for the convenience of the reader.

Jargon

This abbreviated form of terminology is essential for efficient communication in science at the practical level, but it can also lead to a formidable communication barrier. We have made every effort here not to introduce any jargon, but it could not be avoided in a few cases.

In several places, particularly where we talk about *chemical effects*, it is essential to describe just what a particular effect does in terms of the observable results. We do this by defining two versions of a program. Thus MM4 means the program in its current version [MM4(08)]. We also need to refer to the "original" version MM4O, for the case where the effect is not included. Historically, this is the way it actually happened. It was found that MM4O gave an unsatisfactory answer, and so the effect was added in. (One can always remove the effect from MM4 by simply reading in zeros for the parameters concerned.)

In quantum mechanics it is always important to define a calculation in terms of the basis set used. The "standard" basis set used for a calculation has always been dependent on the time period in which the calculation was carried out. The earliest calculations were mostly carried out at the STO-3G level. Later calculations were usually at the 4-31G level, followed still later by MP2/6-31G*. This last is the lowest level that has been commonly used in recent years, and to which we often refer. The better level that is now usually feasible in almost all cases adds diffuse functions to the basis set, and our standard basis for a long time has been 6-31G++(2d,2p). We refer locally to

---

Molecular Structure: Understanding Steric and Electronic Effects from Molecular Mechanics, By Norman L. Allinger
Copyright © 2010 John Wiley & Sons, Inc.

this basis as "big" and represented by the symbol B. Thus, the meaning of MP2/B should be clear and is frequently used throughout this book.

This B basis set (like most) leads to some systematic truncation errors in bond lengths for organic molecules. The actual bond lengths in molecules (r_e), to the extent that they are currently known, can be obtained from an MP2/B calculation by adding relatively small correction terms. The expression MP2/BC refers to calculated geometries wherein the bond lengths have been corrected to the r_e values. These are the only corrections that have been made to warrant the C ending terms. The changes in other geometric (angular) quantities from basis set truncation are quite small, perhaps not very constant, and have consequently been ignored. No change in the energy is made in this correction process. Only the bond lengths are changed.

The corrections to be added to the B bond lengths to convert them to BC bond lengths are as follows (Å): C–H 0.0056; N–H 0.0079; C–N –0.0032; C–O –0.0077, C–F –0.0063; C–Cl –0.0146; C–Br –0.0124; C=O –0.0116. These r_e values were not very accurately known in 1990 when this MM4 work was begun, but this is the way they have been fit. If one is to seriously read this book, it will be helpful to remember these three pieces of jargon, B, BC, and MM4O.

BASIS SET SUPERPOSITION ERROR[1,2]

This is a problem that comes up when one is using relatively small (ordinary) basis sets in a quantum mechanical calculation to describe the interaction between two molecules or between two parts of the same molecule. As two molecules approach one another, their orbitals increasingly overlap. The familiar van der Waals curve results with respect to the energy of the system. At longer distances the charges feel one another, and the total energy of the system goes down from correlation effects. As they come closer together, orbitals begin to overlap more significantly, and the transfer of electron density from a filled orbital in one molecule into an empty orbital in the other molecule starts to occur. As the molecules come still closer, there are too many electrons to fit comfortably into the same space, and the energy rises rapidly due to their mutual repulsion. If very large basis sets are being used to describe the system, one can obtain an accurate appraisal of the situation, which can be summarized as attraction at longer distances, which is superseded by repulsion, and finally increasingly strong repulsion at short distances.

At relatively short distances where orbital overlap between the two molecules is significant, a special problem arises if one is using small (ordinary) basis sets. Because a molecule is being described with a too small basis set, it will tend to make excessive use of the other molecules' empty orbitals as places into which to delocalize electrons. Of course, it should do this to some extent, but it does it to too great an extent when its own orbital description is incomplete.

Here we must ask two questions with respect to any case that is under examination: Is this effect sufficiently significant that we need to worry about it? And, a second question (if the answer to the first question is affirmative) is: How should we correct for this?

For an approximate overview, the stabilization energy due to basis set superposition error in the case of relatively small basis sets (6-31G*) will likely be of the order of 1 kcal/mol for small molecules containing three or four first-row atoms and increasingly larger for larger molecules. For larger basis sets, the problem tends to be reduced in importance, but it really doesn't go away for basis sets commonly used at present.

The theory of basis set superposition error in detail is relatively complex, and there are a variety of ways by which it may, in principle, be eliminated or at least minimized.[1] The most commonly used method for larger molecules is referred to as the *counterpoise correction method*, which was proposed long ago by Boys and Bernardi.[2] Their idea is that one takes the low-lying empty orbitals from one molecule and places them at the energy minimum position with respect to the second molecule. This is done in both directions and yields a measure of the energy lowering that occurs in the system under study with the particular combination of basis set and correlation methods treatment in use. While there are many problems in principle, and an expected overcorrection in general, in practice the method is widely used and works pretty well. Further discussion and examples are given in Chapter 9, under Amine–Carbonyl Interactions.

CARBOHYDRATE CONFORMATIONAL NOMENCLATURE[3]

This is a complicated subject that is usually covered in more or less detail in textbooks on organic chemistry. For purposes of this book it is only necessary to know which names go with which structures. Just why it works like that can be found in textbooks if one is interested.[3] A few points not usually found in textbooks will be mentioned here.

With only rare exceptions, hexapyranose molecules that could exist in boat conformations do not. They are found in chair conformations. Additionally, they normally are found in what is called the $^4C_1$ conformation, as opposed to the $^1C_4$ conformation. As an example, α-D-glucose is shown (Structure **1**):

α-D-Glucose ($^4C_1$)

1

The major determinant of the conformational stability is the –CH$_2$OH group (carbon 6 and its attached hydroxyl). This group is much "larger" than any of the hydroxyls or other substituents that are normally found attached to the rings in sugars. It, therefore, preferentially assumes an equatorial position, here, and in most carbohydrate molecules. The molecule can, in principle, invert to give a second chair form, where all of the groups that are equatorial in Structure **1** would become axial and all of the axial groups would become equatorial (Structure **2**). Such a structure would require the –CH$_2$OH group to assume an axial position. While possible in exceptional cases, such arrangements are quite uncommon. The latter type of structure is called $^1C_4$ in the usual nomenclature.[3] Reeves[4] used a nomenclature in which the ordinary $^4C_1$ structure was called C1, and $^1C_4$ was called 1C. Both of these types of nomenclature suffer from the disadvantage that if one changes from the D series to the L series, the symbols are interchanged.

α-D-Glucose ($^1C_4$)

2

This information is provided for the convenience of the reader who wishes to pursue the subject further but will not be utilized in this book.

CONFORMATIONAL SEARCH ROUTINE[5–7]

Most organic molecules, and especially larger ones, do not exist as just a single structure, but rather they exist as mixtures of different conformers. The number of possible conformations usually increases rapidly with molecular size. For simple cases it is relatively easy for the chemist to decide which conformations are likely to be rather stable, and to limit one's studies to such structures. But as the molecules become larger and/or more complicated, this becomes more difficult to do. Accordingly, different methods have been devised for carrying out *conformational searches*, which are designed to find all of the possible conformers that exist and their relative energies. There are three important methods that we will discuss here. In each case, we want to search the potential energy surface of the molecule in such a way as to be reasonably sure that we have found all of the possible conformations of interest. In principle, it would seem that the most simple method would just be to choose a conformation, and using a molecular dynamics routine, let the molecule undergo vibrational motions (at high temperatures to speed the process), and examine all of the low-energy points that are found. This would be done by taking the structures corresponding to those points and optimizing their geometries in the usual way. This could be done by molecular mechanics, or it could be done, at least in principle, by quantum mechanics. This molecular dynamics search method works but it is extremely slow.

Of the two methods that are actually most often used in practice, one is a systematic search and the other is a stochastic search, although others, and numerous variants, exist.[5–7]

The *systematic search* is very straightforward in principle. In the following discussion, we will consider molecules that contain only sp³ hybrid bonding (saturated molecules). One can easily see how to extend this to other cases.

For the arrangements about a typical bond in a saturated molecule, there are three energy minima corresponding to the anti and the two gauche conformers in butane. So one can begin from some starting structure at one point in the molecule, say, the all-anti structure, calculating the minimum energy structure and then using a driver routine (next section) to rotate the end C–C bond systematically by 60° increments, and at each increment recalculate the energy minimum, and save each of the (three) minimum energy structures. If the molecule under examination was pentane, for example, one could then reset the first bond to the original minimum, and repeat the process on the second bond, while keeping the first bond at the first energy minimum. Then one moves the first bond to the second minimum and later to the third minimum, repeating the locating of the three energy minima of the second bond for each minimum of the first bond. For a still larger molecule one does the same thing for the third bond and so on. Since there are three possible energy minima for each nonterminal bond, the total number of conformers that are expected will be 3^n, where n is the number of nonterminal bonds. We have only one nonterminal bond in butane, and we have $3^1 = 3$ possible conformations. If we have two nonterminal bonds, as in pentane, we will have 9, some of which are enantiomers, for example. But even at the $n = 2$ case of pentane, where we expect 9 conformations, there is already a problem that develops. Namely, what one would have naively expected to be a particular minimum (the gauche–gauche′ conformation) actually splits into two conformations, where the torsion angles differ in opposite directions from the ideal 60° value. It will be worth a little digression at this point to discuss this particular case. This is a simple example of something that is very common in conformational analysis, where a single conformer is expected, but in fact it splits into two different conformers.

If one looks at a Dreiding model (or equivalent) of pentane in the gauche–gauche′ conformation (where the signs of the two torsion angles are opposite), in an idealized conformation where the torsion angles are +60° and −60°, we will obtain a molecule with a plane of symmetry, the carbon skeleton of which is shown labeled C_s in Figure A.1. Actually, the torsion angles are not exactly ±60° (because the bond angles are not exactly tetrahedral, but they are reasonably close, being ±76°.

But the two methyl groups here are much too close together, and there is a large van der Waals repulsion between them. While the molecule could deform in different

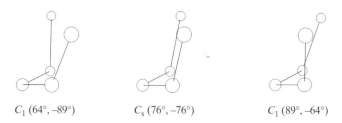

Figure A.1. Pentane conformations.

ways to move those methyl groups apart, the deformation that requires the least energy is torsional in nature. The methyl groups, therefore, twist away from each other, and the torsion angles become +89, −64°, or in the other direction and they become +64, −89°. These two structures are also shown in Figure A.1. They each have C_1 symmetry, and they constitute a *dl* pair of conformers. The C_s structure is the transition state separating the two conformers. Of course, the molecule minimizes its energy by deforming in all of the available degrees of freedom. This means that the central bond angle opens a little bit, and the torsion angles involving the methyl groups change quite a bit as they twist the offending hydrogens away from one another. But in general what is occurring here is that the structure that might have been expected to be an energy minimum (C_s) splits into a pair of conformers. This means that each of these C_1 conformers now corresponds to an energy minimum, while the C_s structure is an energy maximum (a saddle point) separating the C_1 structures. And the situation gets worse as the molecule gets larger, as there will be increasing opportunities for various repulsions to lead to a profusion that we might call "conformational splittings."

All of this means that one cannot move the torsion angles in 60° increments in a conformational search if one wants to be sure that the search is complete. One has to increment more finely than that, and something like 10° increments are typically used. The energy of the structure apart from the angle being driven is minimized for each of these increments. While this remains straightforward as long as there are only a few bonds to consider, it is evident that as the molecule becomes larger, the amount of computation required becomes larger, and the relationship between the number of bonds and the required computing time is exponential.

The *stochastic search*[6,7] is the most widely used method for trying to avoid this sort of exponential problem, and it works in the following way. One has a starting structure for the molecule in question. One then moves (or "kicks") each atom in the molecule individually in a random direction and by a random amount (up to some limit, usually 2–3 Å). One then minimizes the energy of the resulting structure. Depending on the kick sizes, the molecule will frequently fall back into the potential well from which it was generated, but will sometimes end up in an adjacent potential well. If the process is repeated a great many times, one at first finds more and more conformers, but after a while one finds only the same conformers that have already been found before, and no new ones are added. (The search routine usually is designed to compare the new conformer to those previously found by comparing their energies. If the energies are the same to within the minimization error, it is often expedient to also check that the moments of inertia are the same.) One can become more confident (but never absolutely certain), with increasing search time, that all of the significant conformers have been found.

Sometimes it can be shown for a particular case, or perhaps a similar related case, that the probability of having successfully completed a conformational search is extremely high. A simple elegant example was described by Saunders.[7] He examined cycloundecane, the 11-membered carbocylic ring, for the minimum energy conformer (which has C_1 symmetry). There should in fact have been found in his stochastic search 11 structures of the same conformer that differ only by the numbering system. That is, any particular position in the ring (none of which are equivalent) can be occupied by atom number 1 in that structure. So a complete search would have 11 copies of every conformation, differing only by their numbering systems. And, in fact, there need to be 22 copies if *dl* pairs are individually counted because the numbering system can go in either direction. Accordingly, the search was run and the collection of conformers obtained was monitored. After 48 hours, he concluded that the set was complete as all 22 conformers of the lowest energy structure had been found. This was done in 1986, so a 2009 PC would perhaps require a few minutes for such a calculation (Table 3.1). If the search were sufficiently complete so as to find all different 22 copies of the global minimum energy conformer, and 21 of the next lowest energy conformer (as it did), it is exceedingly unlikely that any other low-energy conformer could still be there and not have had a single copy found.

For molecules that contain more than a few carbon atoms, the stochastic search may become fairly time consuming, so that molecular mechanics rather than quantum mechanics is ordinarily used here for the energy minimization. While it is true that one can never be completely certain that all conformers have been found for a particular structure, just because no more are found over some time interval, the probability of any significant conformer being found goes asymptotically to zero with increasing time. There are significant and insignificant conformers. Significant ones are within 2 or 3 kcal in energy from the ground state. Conformers that do not fall within this limit will have a negligible effect on the conformational populations, and on molecular properties, and can usually be neglected anyway. A conformer also must lie in a potential well that is deep enough to accommodate at least one vibrational level. (If it does not meet the second criterion, it is a conformation, but it is not a conformer.) It exists only in the classical mechanical approximation, not in the real world.

If one is dealing with a structure that is far from "ordinary," then it may be preferable to use a *quantum mechanical stochastic search*. For example, in the *secondary-*butyl cation, the structure is far from ordinary in terms of bond lengths, angles, and torsion angles.[8] But the potential surface can be searched very well, and it was established that all of the significant minima had been found. In this case it was desirable to use a quantum mechanical geometry optimization procedure because the minimum energy structures are so different from the structures of ordinary molecules.[9] At the outset at least, one would want to have an idea what to look for in a systematic search. Further, one has to worry about using a large enough basis set so that one is satisfied that one has all of the "significant" minima. It may happen in such a search that shallow minima may be found with one basis set but disappear with a larger basis or vice versa. Such minima are almost never deep enough to contain a vibrational level. They have often been argued about in the literature but are usually of negligible importance in the real world.

DRIVER ROUTINE[10,11]

Ordinarily, the only points on a potential surface that are of use are a select group of the so-called *stationary points*. These are points where all of the first derivatives of the energy with respect to the coordinates are equal to zero. They include all of the energy minima, that is, individual points where energies are associated with particular conformers, and they also include all of the saddle points or transition states that lead from one minimum to another. Other stationary points on the potential surface (two simultaneous saddle points, hilltops, etc., which are identified by multiple imaginary vibrational frequencies) are not normally of interest. By definition, a stable conformer not only has all of the first derivatives with respect to the energy equal to zero, but it also has all of the second derivatives positive. This means that all of the vibrational frequencies are real numbers. A transition state differs from a ground state only in that exactly one of the second derivatives is negative. The vibrational frequency corresponding to that derivative is an imaginary number.

One frequently wonders about how the molecule moves from one energy minimum to another one. How do we find that out with molecular mechanics or quantum mechanics? There are different possibilities. To take a simple example, consider going from the ground-state staggered conformation of ethane to the eclipsed molecule, and then back to another ground state in which one methyl group has rotated 120° with respect to the other. In this case it is easy enough to sketch geometries that approximate the ground and transition states. And if one uses a proper energy minimization scheme, one can arrive at the structures and energies of each of those states easily enough from approximate initial geometries. But in more complicated cases, how does one proceed? Consider the inversion of cyclohexane, for example. One can look at a Dreiding model and decide that in the transition state four of the carbon atoms must be in a plane, or nearly so, and the molecule can move in either of two directions from there to get to the different (boat or chair) minima. If one can guess the transition state well enough, one can start with that approximate transition state and optimize the structure to find the actual transition state. That the transition state has actually been found can be checked by noting that exactly one of the vibrational frequencies is imaginary. But, as the molecules get more complicated, this becomes an increasingly difficult approach.

Another alternative[10,11] is to use what is called *the driver procedure*. What one does in this case is to take a molecule, say cyclohexane, where carbons 1 and 4 have a torsion angle of about 60° from one another, and "drive" that torsion angle in the molecule in angular increments toward a torsion angle of 0°, and then keep driving it on the other side of the potential barrier until one gets to another 60° angle. How does one do this? With a handheld mechanical model, one simply grabs the model and twists it. That is, one applies a large torsional force, which drives the torsion angle from one value to another. With the computer model, one does substantially the same thing. One only needs to put a very large artificial negative torsional dip on the potential surface and move that dip along in such a way as to force (drive) the torsion angle in increments from where it is (+60°) through 0° to −60°. One calculates the energy at points along the way, where one minimizes the structure completely at each point as one moves along. When finished, one goes back and looks at the points along the pathway, which now consist of energy versus torsion angle data. At each torsion angle, one subtracts the energy of the artificial potential at that point. Then what is left is the correct torsion potential for the molecule as it moves along the path from +60° to 0° and on to −60°

with changing the torsion angle. All of the rest of the structure has been minimized at each point. And this gives one of the possible potential paths that the molecule may follow from +60° to −60°. Then to find accurate structures and energies at the stationary points on both ends and in the middle of this trajectory (near +60°, 0°, −60°), one then does an ordinary geometry optimization at each of those points. It might also be noted that MM4 makes use of the Newton–Raphson method[12] for accurate energy minimizations. Once one is close to a stationary point, this method converges very rapidly to an accurate result. When one optimizes a structure with this method, the final structure obtained is the one that corresponds to the stationary point nearest to the starting structure. Hence, one may obtain either a ground state or a transition state with more or less equal ease, as long as one is careful to start from a point near where one wants to end up. Some other methods in common use tend to preferentially give ground states, and with these methods, transition states are found only with more or less difficulty.[13]

While the driver procedure is straightforward and widely used, there are complications in it that may cause problems to the unwary. Let us consider again the ethane case. The ethane barrier, as we usually think of it, has the molecule in an eclipsed (D_{3d}) conformation. That is, each hydrogen of one methyl is eclipsing a hydrogen of the other methyl. But, if we use the driver routine to generate a torsion potential, what we obtain is something different from that. The methyl groups of the ethane do not retain three-fold symmetry as one of them moves along in response to the driver force the way that a handheld model would. The reason is that one of the hydrogens is fixed in torsion angle, and the other two are allowed to energy-minimize to wherever they turn out to be. And they will tend to fall downhill from what would have been obtained if we had done a rigid rotation. A rigid rotation would correspond to going by one possible pathway along the potential surface to the transition state. The driver routine gives the molecule a slightly different pathway. But, if we optimize the structure at both ends of the path, where the first derivatives of the energy with respect to the coordinates are all zero, we will get both proper structures and proper energies. The structures and energies that we get directly from the driver routine in between the stationary points are normally distorted somewhat from what would have resulted from a rigid rotation. The directly obtained points (after subtraction of the artificial driver potential) do correspond to possible points along a path on the potential surface where the driver routine happened to position the system, but they usually are not exactly the numbers that we want or might have expected. To locate the transition state acurately, we have to release the artificial potential and energy minimize the molecule in the usual way.

When one reads books or articles wherein there is a discussion about torsion, think of butane as a simple example, one see's plots of energy vs. torsion angle. The diagram is usually quite symmetrical, and looks like the cosine type curve that one expects (see Figure 4.4). The actual curve generated by the driver is different. It will be unsymmetrical, and more or less distorted from that "textbook" type of diagram. To generate the symmetrical curve from the driver curve, it is only necessary to optimize the structures at all of the stationary points, and then to connect those points on the energy diagram using a Fourier series. The "textbook" and driver curves represent just two (of infinitely many) pathways that interconnect the stationary points.

MOLECULAR MECHANICS PROGRAMS

There are major differences at the practical applications level between quantum mechanics and molecular mechanics. A general listing of books published on the topics of computational chemistry is given by Lipkowitz and Boyd.[14a] A general reference to molecular mechanics force fields is available in the *Encyclopedia of Computational Chemistry*.[14b] Also see several other entries in the *Encyclopedia* under Force Fields. The quantum mechanical calculations have the disadvantage that they are orders of magnitude more time consuming than the molecular mechanics calculations. But the latter have the disadvantage that one must have a great deal of information before one can carry out the calculation. Specifically, many "constants" have to be known. These include the stretching and bending and torsion constants for the various bonding arrangements in the molecule, plus the van der Waals and electrostatic interactions. These quantities are to some extent force field dependent, and they have to have been determined from simple molecules for a given force field before one can start. So for the molecular mechanics method to be practically useful, all of that information had to be assembled for the common structural units that are met with in organic molecules. This was first done in a very general way in the 1970s and resulted in the MM2 (Molecular Mechanics–2) program.[15] It may be of interest that when the MM2 program first became available there were perhaps a few hundred computational chemists worldwide. Most chemists did not have access to computers in those days. Computers were located in computer centers at that time, and they were used mainly by computer specialists and inaccessible to the average chemist. But in the first 9 months after MM2 was available, the Quantum Chemistry Program Exchange (QCPE) distributed more than 1500 copies of that computer program. Hence, the recognition of the usefulness and the wide acceptance of molecular mechanics was dramatic and occurred quite suddenly at this time. This was really the point at which computational chemistry went public, as it were. There are innumerable molecular mechanics computer programs that have been described in the literature or are commercially available or both.*

Much of the discussion in this book revolves about a set of molecular mechanics programs published by this author's research group. We have mentioned MM2 (1977). The subsequent programs were MM3 (1989)[16] and MM4 (1996).[17] Any molecular

---

*Initially, computer programs such as MM2 were given away for free or, as with QCPE, for only a handling charge. And the source code was distributed because computational chemists usually want to modify things and experiment with the program. But very shortly thereafter, modified MM2 programs began to be sold commercially, by other than the original authors, and often at quite high prices. The versions of these programs that were sold commercially were usually in binary code only, which made it essentially impossible for others to copy and resell them, but also made it impossible for the user to actually understand exactly what it was that the program did and how it did it. There were at the time many "improved" versions of MM2 available from different sources. In some cases there were actual improvements in the program because the new authors made special efforts to better fit something that they regarded as particularly important. However, program changes such as those usually caused serious errors in other places, which were often overlooked by those who made the "improvements." Thus, the results produced by such "improved" programs needed to be looked at with circumspection.

mechanics program is written to accomplish certain objectives, frequently to deal with structures and properties of a specific type of compound. It is always possible to improve the results of a program for a particular group of compounds, as newer and more accurate experimental and computational data are becoming available in a continuing stream. It is also possible to improve a program in more fundamental ways as our understanding of how nature works, and how we can describe those workings, improves. But the nature of chemistry, and science in general, has always been such that it must be possible for an experienced person to repeat the "experiment." In order to repeat a computation, one normally must have available the appropriate computer program. Hence, it is not desirable that a continuing stream of small improvements be added to a program, as the program will be constantly changing, and reproducing the former calculations will be inconvenient and probably impossible. After we published the MM2 program, we tried to never make changes in it (at least almost never; we may have fixed a typo or something similar here and there). The later versions of the program give the same results as the earlier versions for those things that could be run on the earlier version. We did expand the program, however, to cover new classes of compounds. But the intent was anything that could be calculated with a given published version of the program could always thereafter be calculated unchanged.

The MM2 program was obviously going to be useful for studies of all different sorts of organic molecules. But when the program was first released, relatively few such studies had yet been made. Parameterization of the force field to handle these myriad compounds was therefore not immediately available. Many users indicated a desire to have approximate values for these parameters, so that they could carry out approximate calculations, later to be susceptible to refinement. Accordingly, we adopted the policy of putting "temporary" parameters into the program and flagging them. Thus, the part of the "experiment" that was reproducible was the part for which "final" parameters only were used. Once these final parameters were chosen, they were not subsequently changed.

While the MM2 program was optimized to fit a broad selection of diverse data, after several years it became evident that it gave some significant errors. Because our understanding of molecular mechanics had improved over the years, we could then do what we regarded as a much better job of writing a molecular mechanics program. Hence, all of these accumulated changes and information were incorporated into a new program, called MM3 (1989), and again, it was intended that continuing subsequent small changes would not be made. It turned out that the MM3 program was rather near to, but not quite up to, what we now regard as a kind of "plateau" in the calculations, where certain important groups of changes relative to MM2 had been made, but not quite all of the changes that were really desired. Accordingly, there was later published the MM4 program (1996).

At this writing (2007–2009) we can do calculations on alkanes (and thus on practically everything else) with slightly better accuracy than that given by MM4 because of the continuing availability of more accurate experimental and computational data on molecules (previously discussed under Alkanes Summary in Chapter 4). But the improvements that could be made are not judged as sufficient to warrant the effort required to prepare a new molecular mechanics force field.

NUCLEAR EXPLOSION PREVENTER

When two atoms come very close together, the use of an exponential function in the repulsive part of the van der Waals function (as when a Buckingham potential is used) causes the energy to tend toward a large but finite value. The r^{-6} term, on the other hand, goes to an infinite negative value. Hence the function works only up to a point, as discussed earlier (Dr. Miller's Nuclear Explosion in Chapter 4). During ordinary calculations of van der Waals energies, the part of the van der Waals curve used is just fine, but when two hydrogens approach each other to a distance of 0.99 Å, the energy goes through a maximum (of about 35 kcal/mol) and then plunges to $-\infty$ (a nuclear reaction/explosion). In practice, no problem occurs using the MM4 program under normal circumstances. However, when one constructs an initial geometry using standard bond lengths, angles, and the like in congested or deformed molecules, one may accidently place two atoms within the critical distance. The way used to avoid this nuclear fusion in MM4 is as follows.

The van der Waals interaction between two nonbonded atoms i and j is described by Eq. (A.1):

$$\begin{aligned} E_{vdw}^{i-j} &= \varepsilon\left(1.84 \times 10^5 \exp(-12.0/P) - 2.25 P^6\right) & \text{if } P \leq 3.02 \\ &= \varepsilon 192.870 P^2 & \text{if } P > 3.02 \end{aligned} \quad (A.1)$$

where $P = r_0/r$.

The second equation comes into use at very short distances to prevent two atoms from getting too close and mathematically fusing.

QUANTUM CHEMISTRY PROGRAM EXCHANGE

The QCPE was originally formed, as its name implies, to distribute to potential users computer programs that had to do with quantum mechanical calculations. The scope of the distribution later expanded to include the entire area that subsequently became known as *computational chemistry*. The driving force behind the formation of this organization was a theoretician at Indiana University, in Bloomington, Indiana, named Harrison Shull. The QCPE formally came into existence in April of 1962. A detailed article about the organization was published in 1999 by Boyd and Lipkowitz.[18]

Prior to 1962, theoretical chemists were the ones who did the most computationally intensive type of calculations in chemistry, and most of those calculations were in the area of quantum mechanics. In and prior to the 1930s, tables of logarithms were an important aid to computations. These were more or less replaced by desk calculators, over the years through the 1940s, and then subsequently by computers beginning in the late 1950s (see Chapter 3). Through the 1960s, people doing computations generally made wide use of tables that had been assembled by other workers in the field. Overlap integrals and two-centered Coloumb integrals, for example, were utilized in this way. But a big change occurred as computers became faster and more available, and it became feasible to simply calculate these things as you needed them for a given calculation. All of the large tables that were so essential for the early computer calculations were put aside, and we entered into a new era. At this point practical computations required very much less time for looking up numbers and instead required more time for writing programs. Shull, and several others, recognized around 1960 that many people were spending a considerable amount of effort to write programs that others had already written elsewhere, and it became important to establish some kind of a distribution system so as to avoid repetitive reinvention of the wheel. The QCPE was the result. This was an organization, formally founded in 1962, that one could join that published a newsletter and a catalog of available programs. By 1965, the QCPE membership had grown to 425 individuals. The organization by then had a library of 71 programs, and some 500 copies had been distributed.

In those days, programs were "written" on punched cards, and they were distributed as source code. The IBM and CDC computers were the most popular in universities, but there were numerous other kinds of computers that were widely used as well. In general, one could not take a program written for any particular machine and run it on a different machine. It required some time and effort, sometimes quite a bit of time and effort, to get a program to run on a different machine. The problem was that each computer center wrote its own operating system (set of general instructions that told the machine how to process the input data). Today, there are only a few operating systems in wide use, Windows being the most common. So programs are now easily transferable from one machine to another. One of the things that QCPE had to spend a lot of time doing was facilitating not only the flow of programs but also the flow of information required to get them to work.

In 1967, Richard Counts became the administrator of QCPE. He was to retain this position for 33 years and really became Mr. QCPE. Peggy (Dr. Margaret) Edwards was formally his secretary, but also his right hand (she had a Ph.D. degree in English). Dr. Stanley Hagstrom, a faculty member and theoretician from Indiana University at Bloomington, was also available for consultation on technical problems, and several others also contributed. This author had extensive contact with these people and well remembers how they really put it all together and made it work. In the late 1970s the QCPE began a series of summer workshops, which normally lasted for about a week. Many were held in Bloomington, but a few others were held in diverse places such as Oxford University, England, and in Geneva, Switzerland. This author had the privilege of participating in most of those workshops, and they were highly successful. They really presented computational chemistry for the masses. Most of the attendees were Ph.D. chemists who were some years, in some cases very many years, away from the university and with widely diverse backgrounds and specialties. They all had an interest in learning about the subject. Some of them were, or became, computational chemists. Up to that time computational chemistry was really not taught as such in universities. There were no departments of computer science either. (The Computer Science Department at the University of Georgia, for example, was formed in 1984.) People who were interested in these subjects had to pretty much go out on their own and dig up information where they could. This usually resulted in a group of would-be users that spent a great deal of time in the computer center learning about computers from the staff there, and from each other, and subsequently teaching what they had learned to younger members of the group. Then they also needed to learn about applications in their own fields, such as computational chemistry. The QCPE workshops immensely simplified the latter process. By 1980 the QCPE was distributing about 1000 program copies per year. And that number continued to increase for many years thereafter, reaching a maximum of about 2500 programs per year.

In the late 1970s a company called Molecular Design Limited (MDL) was formed in California and began the distribution of commercial software for use in chemistry. This was the beginning of the commercial software business in chemistry. For the distribution of the MM2 program, the decision was made that QCPE would provide this program only to nonprofit organizations, and MDL would provide commercial copies. It was determined within a few years that molecular mechanics had a big future in the chemical industry, and especially in the pharmaceutical industry. As computers continued to be developed over the next two decades, so did chemical software develop to take advantage of the increasingly available computing power. But commercial companies are just that. They earn their place in the world by selling a product, in this case computational chemistry programs. Over time such programs became increasingly powerful but also increasingly proprietary. What these programs did and how they did it became largely a matter of commercial trade secret rather than academic science. The importance of QCPE therefore diminished through the 1990s, and it effectively went out of business about 1999.

RING COUNTING

The dodecahedron is one of the *Archimedean solids* known since ancient times. There is no question but that the figure is made up of 12 identical five-membered faces. This has led chemists to assume that the *dodecahedrane* molecule contains 12 cyclopentane rings as its carbon skeleton. This latter statement turns out not to be true. This curiosity is a result of the definition of a *ring* in organic chemistry.

An alkane, by definition, has the formula $C_nH_{(2n+2)}$. Cycloalkanes and polycycloalkanes, by definition, contained fewer hydrogen atoms and instead contain rings. The number of rings in a (poly)cycloalkane is determined as follows:

Take the formula for the ring compound of interest, and the formula of the alkane with the same number of carbons, and subtract the number of hydrogens in the former from the number of hydrogens in the latter. Divide the resulting number by 2. The number thus obtained is the number of rings in the (poly)cycloalkane. As an example, consider dicyclohexylmethane:

$C_{13}H_{24}$ $C_{13}H_{24}$ $C_{13}H_{28}$

The original compound has the formula $C_{13}H_{24}$. The corresponding alkane would have the formula $C_{13}H_{28}$. This is easily determined by breaking a C–C bond in each ring and replacing the broken bond with a hydrogen on each end. This yields an open-chain compound, $C_{13}H_{28}$. Therefore, the ring compound has four fewer hydrogens, divided by 2, equals two rings.

Now let us turn to dodecahedrane. This molecule has the formula $C_{20}H_{20}$. A saturated compound with 20 carbons would have 42 hydrogens. Hence, 42 minus 20 equals 22, divided by 2 equals 11 rings.

So this is a matter of definition. Dodecahedrane contains 11 rings, not 12, according to the definition. Now how can this be? If we look at the dodecahedrane molecule as an actual physical model and count the faces, there are 12 five-membered rings. But, by definition, there are only 11.

Perhaps it is easiest to illustrate this with a simple example. Suppose we start with ethylcyclohexane. The ethyl group is attached to carbon 1. Suppose we loop it over toward the adjacent carbon (carbon 2). Then remove two hydrogens, one from the methyl group, and one from ring carbon 2 on the ring, to give the diradical shown. Then join those two radicals. The compound that we obtain is bicyclo[4.2.0]octane:

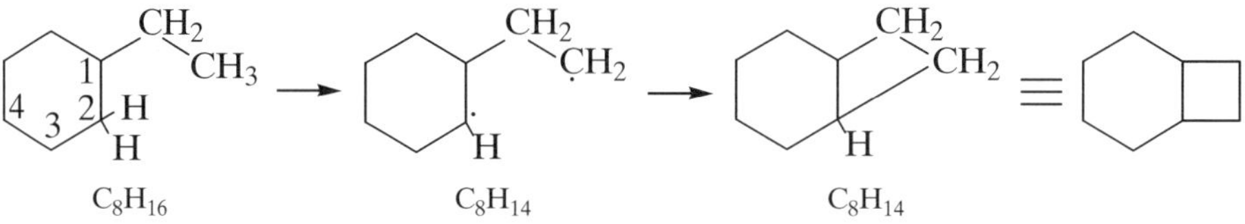

The formula for ethylcyclohexane is C_8H_{16}. Hence, it already contains one ring. When we remove the two hydrogens to make the diradical (prior to cyclization), the formula has two hydrogens less, that is C_8H_{14}. After we cyclized it, we then have two rings, by definition. This is indicated by the prefix "bicyclo-." Nothing unusual here.

Now consider going back to the ethylcyclohexane again and removing two hydrogens. However, this time remove one hydrogen from the methyl group and the other hydrogen from ring carbon 4, rather than from ring carbon 2, which gives a diradical. Close the ring, and we obtain bicyclo[2.2.2]octane:

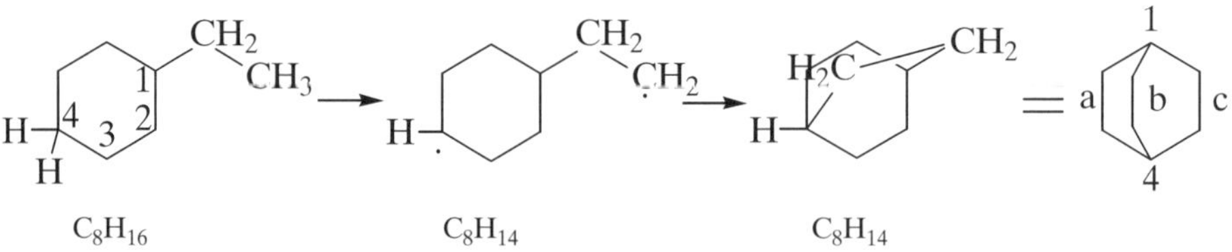

Again, the name of the compound indicates that it is a bicyclo-substance and hence contains two rings. The formula is again the same for its previously prepared isomer, C_8H_{14}. Hence, the compound contains two rings. But if you look at it, you might be inclined to say it contains three rings. There are three different ways that you can begin with carbon 1, and count around a ring, and come back again and join to carbon 1. (They are 1a4b1, 1a4c1, and 1b4c1 in the right-hand structure above.) There are only, by definition, two rings. We sort of got one for free because of the way we did the ring closure.

One can presumably carry out a similar exercise with dodecahedrane and come to the conclusion that there are only 11 rings. We got the other one for free (this author has not actually carried out this procedure).

Thus, we have two different methods for determining how many rings there are in a molecule. We can apply the formal definition from organic chemistry and conclude from the formula how many there are. But we can also simply look at the molecule and count how many there are. Which method is the "correct" method? We can put the methods to test by calculating heats of formation of some bicyclic molecules (or polycyclic), with and without this kind of problem. Then we can see which way works. It turns out that if you want to correctly calculate heats of formation, you must stick with the formal definition. In several cases that we have tried, where there are believable experimental data with which we can compare the result, the method of counting rings (like 12 in dodecahedrane or 3 in bicyclo[2.2.2]octane) fails. The difference in the calculated energies from these two methods is usually of the order of 5 kcal/mol, depending on the exact calculation used (the value of the R5 or R6 term is described together with Table 11.2), and, of course, also on the experimental data. This is a pretty large number from an experimental viewpoint, and hence it's easy to see. Norbornane, for example, consists of two five-membered rings from this point of view. There is a six-membered ring present also, but that one you get for free and don't count. (If there are small and six-membered or larger rings to choose from, one wants to choose the number of small rings and take a larger ring for free. How this would work out if one had enough data has not, to this author's knowledge, been put to a serious test. But bicyclo[2.2.2]octane, perhydrotriquinacene, and the two norbornanes discussed previously have their heats of formation well calculated by the method indicated and would show errors of ca. 5 kcal/mol if the alternative ring counting method were to be used.

STEREOGRAPHIC PROJECTIONS

In Chapter 1 some of the early history of molecular models was discussed. Ball-and-stick models and their more recent counterpart, Dreiding models, are still widely used by organic chemists because it is often convenient to have an actual model in hand, rather than just to try to visualize something in one's mind. There are serious limitations with this sort of model, however. The mechanical model worked fine for systems that contain rather few atoms. However, when one worries about very large molecules such as proteins, mechanical models present overwhelming practical problems. With the aid of modern computers, we can generate models that can be projected onto a computer screen rather quickly and easily. However, if there are many atoms present in that 2-dimensional figure, the projection can be quite difficult to interpret. This problem can sometimes be solved by simply rotating the model to a better perspective. But still, there are limits beyond which a two-dimensional projection of a three-dimensional structure becomes hard to interpret. Of course, we can also easily generate stereographic projections with the aid of computers. These are two-dimensional projections that are observed pairwise, one with each eye separately, in such a way as to appear three dimensional. And, of course, these can be rotated and oriented in any desired perspective. The author has always told his students that often "a picture is worth a thousand Cartesian coordinates." Sometimes a simple three dimensional picture can save a lot of wear and tear.

The idea behind a stereographic projection is simple. The reason we see things in three dimensions is because we have two eyes, and they look at the same object at the same time from two slightly different perspectives. Our mind then constructs the three-dimensional structure from these two images.

The ideas of stereographic projection were well understood at least back into the nineteenth century, when photographs were obtained from slightly different perspectives, and mounted in a viewer in such a way that the observer could see an actual (apparent) three-dimensional projection by looking into this viewer. Nowadays, one typically uses what look like reading glasses but that can be polarized in perpendicular planes, or they can be colored different colors for each eye, so there are thus various ways of seeing these structures. But for the average chemist it is convenient to simply look at these projections and see them in three dimensions. One does not need any kind of a projector or other device to do this.

So how does one look at two projections at the same time separately with each eye and see things in three dimensions? This is like asking, "How do you ride a bicycle?" The answer is that the average person has the ability to ride a bicycle, but it is a skill that has to be learned. If you can look at the structures shown in Figures 2.5 and 2.6 and see them in three dimensions, fine. You need read no further here. But, if you cannot, this is a skill that you should take the time to learn. It can be said at the outset that this skill is nowhere near as difficult as riding a bicycle, but nonetheless, it is something that must be learned. It is a skill that should be taught in grade school, but that may be for the next generation. For now, you can learn to see these figures in three dimensions (if you can't already) by understanding the following, and then practicing it.

The human eyes are typically about 2.5 inches apart, and they can be focused from something like 8 inches away from the eyes, out to infinity. One does this subconsciously by simply looking at something. The eyes automatically focus on the thing being looked at and move into the proper positions.

To see the two stereographic projections combined into a three-dimensional figure, what one must do is look at the left figure with the left eye, the right figure with the right eye, and not look at either figure with the wrong eye. If one simply holds the book page showing the figures at a normal viewing distance, and focuses the eyes at infinity, the left eye sees primarily the left figure, and the right eye sees primarily the right figure, and a three-dimensional image appears, and at the sides will be two (now) weakly visible figures that can be ignored. That's all there is to it.

Various devices exist that help (like training wheels on a bicycle) to learn to do this. One can use a viewer that simply blocks the unwanted view from each eye. For example, place a piece of cardboard in between the two figures and the eyes perpendicular to the plane of the figures so that each eye can see only the desired figure. This produces a single three-dimensional view. Then one can remove the piece of cardboard without refocusing the eyes and still see the subject in three dimensions. Or, one can look at the page on which the two projections are, and put one's finger behind and to the top of the page where it is visible. Then keep both eyes focused on the finger and move it gradually backward away from the page, thus allowing the eyes to focus more toward infinity. If one keeps one's eyes on the page figures at the same time, they will merge into a three-dimensional figure between the two two-dimensional projections.

That's all there is to it. The rest cannot be taught, it has to be learned by practice. And once you learn it, it's like riding a bicycle—you will know how to do it.

REFERENCES

1. N. R. Kestner and J. E. Combariza, *Basis Set Superposition Errors: Theory and Practice, Reviews in Computational Chemistry*, Vol. 13, K. B. Lipkowitz and D. B. Boyd, Eds., Wiley, New York, 1999, p. 99.
2. S. F. Boys and F. Bernardi, *Mol. Phys.*, **19**, 553 (1970).
3. Discussed in textbooks on carbohydrates, for example, J. F. Stoddart, *Stereochemistry of Carbohydrates*, Wiley-Interscience, New York, 1971. Also see the official carbohydrate nomenclature website: http://www.chem.qmul.ac.uk/iupac/2carb/.
4. R. E. Reeves, *J. Am. Chem. Soc.*, **72**, 1499 (1950).
5. A. R. Leach, *Survey of Methods for Searching the Conformational Space of Small and Medium-Sized Molecules, Reviews in Computational Chemistry*, Vol. 2, K. B. Lipkowitz and D. B. Boyd, Eds., Wiley, New York, 1991, p. 1.
6. M. Saunders, K. N. Houk, Y.-D. Wu, W. C. Still, M. Lipton, G. Chang, and W. C. Guida, *J. Am. Chem. Soc.*, **112**, 1419 (1990).
7. (a) M. Saunders, *J. Am. Chem. Soc.*, **109**, 3150 (1987). (b) J. Chandrasekhar, M. Saunders, and W. L. Jorgensen, *J. Comput. Chem.*, **22**, 1646 (2001).
8. V. Vrcek, O. Kronja, and M. Saunders, *J. Chem. Theory Comput.*, **3**, 1223 (2007).
9. (a) M. Saunders, *J. Comput. Chem.*, **25**, 621 (2004). (b) P. P. Bera, K. W. Sattelmeyer, M. Saunders, H. F. Schaefer, and P. v. R. Schleyer, *J. Phys. Chem.*, **A110**, 4287 (2006).
10. K. B. Wiberg and R. H. Boyd, *J. Am. Chem. Soc.*, **94**, 8426 (1972).
11. U. Burkert and N. L. Allinger, *J. Comput. Chem.*, **3**, 40 (1982).
12. U. Burkert and N. L. Allinger, *Molecular Mechanics*, American Chemical Society, Washington, D.C., 1982, p. 65.
13. (a) G. Fogarasi and P. Pulay, in *Structures and Conformations of Non-Rigid Molecules*, J. Laane, M. Dakkouri, B. van der Veken, and H. Oberhammer, Eds., Kluwer, Dordrecht, The Netherlands, 1993, p. 377. (b) T. Schlick, *Optimization Methods in Computational Chemistry, Reviews in Computational Chemistry*, Vol. 3, K. B. Lipkowitz and D. B. Boyd, Eds., Wiley, New York, 1992, p. 1.
14. (a) K. B. Lipkowitz and D. B. Boyd, in *Reviews in Computational Chemistry*, Vol. 17, Wiley, New York, 2001, p. 255. (b) N. L. Allinger, J. R. Maple, and T. A. Halgren, in *Encyclopedia of Computational Chemistry*, Vol. 2, P. v. R. Schleyer, N. L. Allinger, T. Clark, J. Gasteiger,

P. A. Kollman, H. F. Schaefer, III, and P. R. Schreiner, Eds., Wiley, Chichester, UK, 1998, p. 1013.
15. N. L. Allinger, *J. Am. Chem. Soc.*, **99**, 8127 (1977).
16. N. L. Allinger, Y. H. Yuh, and J.-H. Lii, *J. Am. Chem. Soc.*, **111**, 8551 (1989).
17. N. L. Allinger, K. Chen, and J.-H. Lii, *J. Comput. Chem.*, **17**, 642 (1996).
18. D. B. Boyd and K. B. Lipkowitz, in *Reviews in Computational Chemistry*, Vol. 15, K. B. Lipkowitz and D. B. Boyd, Eds., Wiley-VCH, New York, 1999, Preface, p. v.

INDEX

Ab initio calculations, on particular molecules, *see* the particular molecule
Ab initio methods 37
Accuracy, chemical, *see* Chemical accuracy
Acetaldehyde, hyperconjugation in 157
Acetone dimer, molecular mechanics calculation on 239
Acetone-trimethylamine interaction 233, 234
Acetonitrile 24
Acids
  general 212
  Lewis 212
Adamantanes, heats of formation of 269, 270
Adenine 125
Aldehydes, hydrates of 191
Aldehydes, hydration of 191
n-Alkanes, crystal conformation of 241
Alkanes
  heats of formation of 257 ff
    by molecular mechanics 265 ff
    by quantum mechanics 274 ff
    table of 266 ff
  MM4 entropies of 273
  thermodynamic properties of 273
Alkenes
  heats of combustion of 291
  heats of formation of 291
    table of 292, 293
  heats of hydrogenation of 291, 292
Alkyl bromides, reactivity in S_N2 reactions 135
Alpha-helix, stereographic projection of 16
AM1 method 37

Amdahl (computers) 30
Amine-carbonyl interactions 228
Amine-ketone reaction, reaction coordinate diagram of 230, 231
Ammonia, hydrogen bond to water 213
Ammonia-formaldehyde interaction 228, 232 ff
Andrews, D. H. 45
Anharmonicity 21
[18]Annulene
  aromaticity of 252, 294
  bond lengths in 250, 251
  crystal packing of 249
  Dewar resonance energy of 252
  heat of formation of 294
  proton NMR of 249 ff
  structure of 116 ff, 249 ff
    quantum mechanical calculations of 250
  symmetry of 249 ff
Annulenes, bond lengths in 116
Anomer energy differences, for glucose diastereomers 200 ff
Anomeric effect, 167 ff, 189 ff
  comparison with Bohlmann effect 168
  definition of 192
  in 2-methoxytetrahydropyran 176
  in carbohydrates 167, 192
  in dimethoxymethane 169
  in organic chemistry 88, 135
  solvent dependence of 192
Anthracene 106
Antiaromatic compounds 116
Apple (computers) 30

Molecular Structure: Understanding Steric and Electronic Effects from Molecular Mechanics, By Norman L. Allinger
Copyright © 2010 John Wiley & Sons, Inc.

Aromatic compounds, 104, 112 ff, 294 ff
  bond lengths in 105
  electric ring current in 105
  kekule forms of 106, 107
  thermochemistry of 105
  ultraviolet spectra of 123

B3LYP calculations
  systematic correlation error in 276 ff
  with dispersion energy 275, 280 ff
B3LYP functional 42
B3LYP vs. MP2 results 42
Badger, R. M. 145
Badger's equation 145
Baker, J. W. 151, 161
Baker-Nathan effect 151
Barton, D. H. R. 167
Bases
  general 212
  Lewis 212
Basis set superposition error 303 ff
Basis sets 302, *see also* Orbitals
  STO-3G 38, 39, 302
  4-31G 302
  MP2/6-31G* 302
  6-31G++(2d,2p) 302, 303
  B 303
  MP2/B 303
  BC 303
Bender C. F. 37
Benson, S. W. 259
Benson's method, for calculation of heats of formation 258, 259
Benzene, 93 ff
  bond lengths in 100, 101
  calculated to be nonplanar 254
  C-H bond, dipole of 244
  crystal structure of 243 ff
  crystal structure, stereographic projection of 245
  Hückel calculations for 112
  kekule forms for 113
Benzene crystal, heat of sublimation of 245
Benzene dimers 243
Benzyl carbocations, hyperconjugation in 153
Bicyclo[2.2.2]octane, rings in 316, 317
Bicyclo[3.3.0]octane, heat of formation of 259, 260

cis-Bicyclo[3.3.0]octane, heat of formation of 267, 288
Bicyclo[4.2.0]octane, rings in 316
Biphenyl, crystal structure of 245, 246
Bletchley Park 29
Boggs, J. E. 23
Bohlmann, F. 163
Bohlmann bands, in infrared spectra 163
Bohlmann effect, 163 ff
  comparison with anomeric effect 168
  definition of 163 ff
  in ethanol 166
  in ethylamine 164
  in fluoroethane 166
  in methyl ethyl sulfide 167
  in organic chemistry 88, 135
Bohr, N. 33, 34
Bohr model, of hydrogen atom 34
Bohr-Sommerfeld model 34
Boltzmann distribution, *see also* POP
Boltzmann distribution, from conformational search 263, 264
Bond angles, and electronegativity effect 138
Bond energies, definitions of 268
Bond length—bond order relationship, *see* Bond order—bond length relationship
Bond length corrections, *see also* Bond lengths, definitions of
  B to BC 302
  B to r_e 302
  X-ray 14
Bond lengths
  accuracy of 10
  and electronegativity effect 138
  C-C, *see* C-C Bond length
  C-H, *see* C-H Bond length
  definitions of 21
  definition of r_a 25
  definitions of r_α, r_e, r_g, r_s, r_z 23 ff
  definitions of r_e, r_g 21
  in annulenes 116
  of typical 24
  versus vibrational frequencies 144, 145
Bond order(s)
  and hyperconjugation 150, 176, 207
  and kekule forms 106 ff
  definition of 94
Bond order—bond length relationship 98, 99, 105, 108, 158, 159

Bond order—torsion barrier relationship 99, 100
Bond stretching, from hyperconjugation 155
Boyd, D. B. 43
Branched-chain effect, in heats of formation 258, 268, 276
Bregman, J. 255
α-Bromo, see 2-Bromo
2-Bromo-4-*t*-butylcyclohexanone, stereoisomers of 179
2-Bromocyclohexanone, conformers of 177 ff
  electrostatics in 178
  energetic analysis of 185
  molecular mechanics model of 182
  structures and vibrational frequencies 183
Buckingham, A. D. 68, 313
Buckingham potential 68, 313
Bürgi, H. B. 230
Burkert, U. 282
Burroughs (computers) 30
Butadiene 93 ff, 147
trans-Butadiene 100, 101
n-Butane
  bond lengths in 24
  conformations of 54
  rotational barrier in 40
  torsional angles of 54
  torsional energy of 55
2-Butanone 24
1-Butene, hyperconjugation in 154, 156
2-Butene 24
sec.-Butyl cation, conformational search for 308
2-Butyne 24

C_{60}-Fullerene, see Fullerene
Calculators, desk 313
Cambridge Structural Database 12, 15
Carbohydrates, nomenclature of 304
Carbohydrates, stereochemical effects in 189
Carbon atom, energy of 260
Carbonyl effect 157
Carbonyl vibrations, infrared 60
C-C Bond length
  average value 87
  definitions, various 24
  electronegativity effect on 138 ff
  in [18]annulene 249 ff

  in 2-bromocyclohexanone conformers 183
  in 1,2-dichloroethane 9
  in 1,2-difluoroethane 142
  in dodecahedrane 287
  in ethyl fluoride 87, 138 ff
  in ethylamine conformers 165 ff
  in hexafluoroethane 142
  in various molecules 43, 155 ff
  stretching constant 60
  stretching parameter 60, 61
  typical 21
C-Cl Bond length 9
C-D Bond, vibrational frequency of 22
C-D Bond length 22
CDC (computers) 314
Cellobiose analog, potential surface of 202, 203, 208, 209
Cellobiose analog, structure of 201
Cellulose 189
Central processing unit (CPU) (computer) 41
C-H Bond length 8, 9, 21, 144
C-H Bond, in benzene, dipole of 244 ff
C-H Bond, properties of 144
C-H Bond, vibrational frequency of 22
C-H Hydrogen bond 102
Chemical accuracy
  definition of 64, 181
  of bond lengths 52
  of calculated heats of formation 257, 258
  of macro molecule structures 126
  of microwave geometries 10, 40
  of quantum mechanical calculations 2, 10, 257, 258
  of vibrational energies 264
Chemical effects, see Effects, in organic chemistry
α-Chloro, see 2-Chloro
2-Chlorocyclohexanone, conformers of 180
Chloromethane 24, 25
Chlorophyll a 128
Cioslowski, J. 40
Circular dichroism 124
Clark, T. 269
Classical mechanics, see Mechanics, classical
CNDO method 37, 96
Colossus (computer) 29
Coloumb integrals, tables of 313
Compaq (computers) 30

Complete neglect of differential overlap, *see* CNDO method
Computational chemistry 33 ff
Computational methods, 33 ff
　Ab initio 37 ff
　Density functional theory 41 ff
　Molecular mechanics 42
　Semi-empirical 34 ff
Computational models 20
Computer(s)
　Amdahl 30
　Apple 30
　Burroughs 30
　CDC 30, 314
　Compaq 30
　Cray 30
　Dell 30
　Digital Equipment 30
　Gateway 30
　history of 28 ff
　IBM 30 ff, 314
　mainframe 30
　punched cards 314
　mechanical 29
　VAX 31
Computer centers 30
Computer memory 31 ff
Computer operating system 314
Computer programs, *see also* molecular mechanics programs
　CRSTL 242
　Gaussian 37, 96
　MM4 51, 68 ff, 312
Computer Science Departments 314
Conformational analysis 46
Conformational search, 305
　for Boltzmann distribution 264
　Molecular dynamics 305 ff
　Systematic 305 ff
　Stochastic 305 ff
Conformational splitting 306
Conformations, definition of 4
Conformers, definition of 55
Congressane 271
Conjugate gradient method 47
Conjugated heterocycles, structures of 124
Conjugated hydrocarbons, electronic spectra of 122
Conjugated molecules 34, 92 ff

Conjugation
　isovalent 147
　sacrificial 147
Control Data Corporation (computers) 30
Coordinate-covalent bonds 213
Coordinates
　Cartesian 3 ff, 318
　external 3
　internal 3
Coppens, P. 17
Corannulene 108, 109
Corey, E. J. 177
Corey resonance 180
Coronene 106, 108
Corrections, bond length, *see* Bond length corrections
Correlation energy. *See also* Dispersion energy; Møller-Plesset; Coupled cluster; MP2; MP4; DFT methods
　branched-chain alkane energy patterns 277. *See also* Branched-chain effect
　in annulenes 249 ff
　in congested molecules 279, 280
　in 1,2-difluoroethane 193, 194
　in heats of formation 201 ff, 276 ff
Correlation, electron, definition of 39
Coulomb integrals, *see* Integrals, two-electron
Coulson, C. A. 215
Counts, R. W. 314
Coupled cluster method 39
CPU, *see* Central Processing Unit (computer)
Cray (computers) 30
Crick, F. H. C. 213
Cross terms, in **F** matrix 63 ff
CRSTL (computer program) 242
Crystal
　lattice forces in 16, 242
　packing in 240
Crystals, molecular mechanics applications to 242
Crystal structure, unit cell in 240
Crystal structures
　calculations of 241
　accuracy of 13 ff
Cubane, stereographic projection of 17
Cycloalkane rings, 285
　Small 285, 291
　Common 285, 291

Medium 285
Large 285
Cyclobutadiene
  Hückel calculations for 114
  kekule forms of 113
Cyclobutane 39, 285 ff, 291
1,3-Cyclohexadiene 101
Cyclohexane
  conformations of 2
  transition state for inversion 309
Cyclooctatetraene
  geometry of 115, 116
  Hückel calculations for 114
  kekule forms of 113
Cyclopentane, 289
  heat of formation of 259, 290
Cyclopropane 286, 287
Cytosine 125
Cyvin, S. J. 24, 25

Dehydrobutane
  definition of 57
  force constant matrix for 57, 63
Dell (computers) 30
Delta-two effect, 197 ff
  in D-mannose 198
Density functional theory (DFT) 41, see also B3LYP
  in heat of formation calculations 275 ff
Deoxyribonucleic acid (DNA) 125
Desk calculators 313
Dewar, M. J. S. 37, 38
Dewar resonance energies, 115
  in annulenes 116
DFT, see Density functional theory and B3LYP
D,L-3,4-di(1-adamantyl)-2,2,5,5-tetramethylhexane 32
Diamantane 271
2,2'-Dibromo-4,4'-dicarboxybiphenyl 45
1,2-Dichloroethane, radial distribution function of 8
1,2-Difluoroethane, gauche effect in 193, 194
Digital Equipment (computers) 30
Dimethoxymethane, 168 ff, 204 ff
  anomeric effect in 169
  conformational energies of 169
  conformers of 170, 171, 206
  F matrix of 172
  resonance forms of 168
  methyl rotational barriers in 206
Dipole moments, of α-bromocyclohexanones 178 ff
Dipole moments, of alkanes 89
Dipole moments, of alkenes 89
Dirac, P. 34, 40
Direct methods (X-ray) 12
Dispersion energies, see also Correlation energies
  in water dimer 216 ff
  lack of in B3LYP 276 ff
  MM4 279
  MM4, table of 283, 284
  short-range 204, 280
Distance geometry 19
Ditrityl ether
  crystal structure of 247
  geometry of 247
  stereographic projection of 248
DNA 124, 125
Dodecahedrane, 287 ff
  endohedral complexes of 290
  heat of formation of 287 ff
  ring counting in 316
  strain energy of 290
Double-zeta orbitals 36
Dreiding, A. S. 44
Dreiding models 44, 47, 309
Driver routine, 308
  complications using 310
  for potential surface of cyclohexane 309
Duell, C. H. 30
Dunitz, J. D. 230
Dyes, colors of 124

Edwards, M. 314
Effects, in organic chemistry, 134 ff, 302
  Anomeric 167 ff
  Baker-Nathan 151 ff
  Bohlmann 163 ff
  Delta-two 197 ff
  Electronegativity 137 ff
  External-anomeric torsional 204 ff
  Gauche 193 ff
  α-Haloketone 177 ff
  Hyperconjugation 147 ff
  Negative hyperconjugation 162 ff
Electrodynamics 1

Electron correlation, *see* Correlation, electron; Correlation energy
Electron density 34, 41, 94
Electron diffraction 7 ff
Electron diffraction structures, accuracy of 13
Electron scattering 8
Electronegativity bond length corrections, table of 140
Electronegativity bond shortening 138 ff
Electronegativity constants 140
Electronegativity effect, 135, 137 ff
  in organic chemistry 88, 135
  on bond angles 138, 143
  on bond lengths 138
  induction of 141
Electronic effects, stereochemical consequences of 137
Electronic integrals, *see* Integrals, two-electron
Electronic spectra, $\pi \to \pi^*$ transitions 124
Electronic spectra, $n \to \pi^*$ transitions 124
Electronic spectra 122 ff. *See also* Ultraviolet spectra
Electronic spectrum, of C_{60}-fullerene 123
Electrostatic energy, variation with dielectric constant 178, 179
Electrostatics, 1
  in anomeric effect 173 ff
  in benzene crystal 243 ff
  in benzene dimers 243
  in crystals 241
  in α-haloketone effect 178 ff
  in molecular mechanics 88 ff, 182 ff
  in nucleic acid bases 126 ff
Encyclopedia of Computational Chemistry 28
Energy, derivatives of 52
Energy gradient 52
Energy minimization, 56
  Newton-Raphson method for 310
Energy minimum, global 56
Energy, multiple minimum problem 239
Enigma code 29
Entropies, MM4, for alkanes 273
Equilibrium bond length (r_e) 21, 23
Equilibrium energy (E_e), definition 261
Ermer, O. 249
Ethanol, Bohlmann effect in 166

Ethylamine, conformers of, 164 ff
  Bohlmann effect in 164
  vibrational frequencies of 165
Ethylcyclohexane 316
Ethylcyclopentane, heat of formation of 259
Ethylene, 93 ff
  structure of 100, 101
Ethylene glycol, 193, 223 ff
  conformations of, 193, 224
    energies of 225
  gauche effect in 193
  geometry of 226
E_{vib} (statistical mechanics) 263, 264
Experiments, *in silico* 245
External anomeric torsional effect 204
  energetics of 205 ff

F matrix
  cross terms in 63 ff
  in electronegativity 139
  in hyperconjugation 158
  of a molecule 3, 5, 6
  of dehydrobutane 57–65
  of dimethoxymethane 172
  stereoelectronic effects in 148
Fingerprint region (infrared) 60
Fischer, E. 192
Fischer projections 190
Fluoroethane, 140, 142
  Bohlmann effect in 166
Fluoroethanes, 140
  bond lengths in 142
Focal-point analysis 40
Force 52
Force constant matrix, *see* **F** matrix
Force constant, definition 52
Force constants vs. force parameters 60 ff
Force field parameters, *see* Molecular mechanics parameters
Force field
  class 2 301
  definition of 44
  MM4 51
Formaldehyde—ammonia interaction 228, 232 ff
Fourier series, for a torsion potential 310
Fullerene, 109
  electronic spectrum of 123
  endohedral complexes of 111
  heat of formation of 295 ff

Fullerene-helium complex 111
Functional groups, definition of 135

G-1 method 281
G-4 method 281
Gates, B. 32
Gateway (computers) 30
Gauche effect, 193
  in 1,2-difluoroethane 193
  in ethylene glycol 193, 224
Gaussian orbitals 37, 38, 96
Gaussian program 37, 96
Geometry optimization 38, 44
Global energy minimum 56
Globin 128
Glucose, 189
  mutarotation of 191
  structure of 190
Glucose diastereomers, anomeric energy
    differences of 200 ff
α-D-Glucose ($^1C_4$), stability of 305
α-D-Glucose ($^4C_1$), stability of 304
α-D-Glucose, conformational formulation of
    192
α-D-Glucose, nomenclature of 304
D-Glucose, energetics of mutarotation of 198
D-Glucose, Fischer projections of 190
GPU, *see* Graphical processing unit
    (computer)
Graphical processing unit (GPU) 41
Graphene 33, 109
Graphite 109
Guanine 125
Gwinn, W. D. 265

Hagler, A. T. 301
Hagstrom, S. A. 314
α-Halo ketone effect 177 ff
α-Halo. *See also* 2-Bromo; 2-Chloro
Hamiltonian operator 34, 35
Harmonic approximation 51
Hartree-Fock method, definition of 35
Hartree-Fock method, for conjugated systems
    95, 96
Hartree-Fock method, limitations of 39
Hauptman, H. 12
Heat(s) of formation 257 ff
  Benson's method 258
  bond energies in 258
  branched-chain effect in 258, 268, 276

methyl technique for 272
MM4, table of 283, 284
molecular mechanics calculation of 260
Møller-Plesset calculations in 277
Heat of formation calculations
  by molecular mechanics (MM4)
    of alkanes 265 ff
    of functionalized molecules, 296
    of alkenes, 291 ff
    of congested molecules 279
    of conjugated hydrocarbons, table of
      292 ff
    of polyfunctional molecules 297
    of unsaturated hydrocarbons 291
  by quantum mechanical methods
    Hartree-Fock 274, 275
    B3LYP 275, 276
    B3LYP plus dispersion 275, 276
    MP2 275
    MP4 275
    of alkanes 296
    of functionalized molecules 296
  correlation energy in 278
  density functional theory in 278
  inclusion of anharmonic vibrations in
    264
  test for accuracy of 281
  transferability of parameters in 259
  using statistical mechanics 261 ff
  Wiberg-Schleyer method 274
Heat of formation parameters
  analysis of 275 ff
  MM4 table of 275
Heat(s) of formation of
  adamantanes 269, 270
  dodecahedrane 287
  dodecahedrane and related compounds,
    table of 288
  fullerene 295, 296
  norbornanes 272
Heats of sublimation, of crystals 248
Helium atom, calculations on 37
Helium-fullerene complex 111
α-Helix, in proteins 213
Heme 128
Hemiacetal formation 191
Hemiacetals 190
Hemoglobin 128
Hendrickson, J. B. 46
Heptane, statistical mechanics of 261

Heterocycles, in variable electronegativity self-consistent field calculations 124
Hewlett Packard (computers) 30
Hexafluoroethane 142, 143
Hexamethylbenzene
  crystal structure of 245, 246
  heat of sublimation of 245
Hexopyranoses 189
High-angle data (X-ray) 17
Hofmann, A. W. 2
Hohenberg-Kohn theorem 41
Homoaromaticity 118
Homoconjugation 119
Hooke's law 51, 53
Hückel calculations 112
  for benzene 112
  for cyclobutadiene 113, 114
  for cyclooctatetraene 113, 114
Hückel method 95
Hückel, E. 34
Hückel's $(4n + 2)$ rule, for aromaticity 114
Hydrogen atom, calculations on 34
  Bohr model of 34
  Schrödinger model of 34
Hydrogen bond(s), 212
  between water and ammonia 212
  C-H 102
  definition of 214
  energies of 212 ff
  in biological processes 213
  in infrared spectra 214
  in liquid water 73, 213
  in ultraviolet spectra 214
  MM4 model of 218 ff
  of water dimer 215
  quantum mechanical description of 215
Hydrogen fluoride dimer, 221
  geometry and energy of 221
Hydrogen molecule, calculations on 34
Hydrophobic effect 73
Hydroxyl groups
  geminal 193
  vicinal 193
Hyperconjugation, 147
  and bond order 150
  bond stretching from 155
  in 1-butene 154, 156
  in 2,2-paracyclophane 155
  in acetaldehyde 157

  in benzyl carbocations 153
  in F matrix 158
  in molecular mechanics 158
  in propene 148 ff, 156, 157
  in toluene 152, 155
  in various compounds 155, 156
  negative 163, 168

IBM 650 (computer) 31, 32
IBM computers 314
in silico, Experiments 245
Inductive effects 138, 141
Infrared spectra, *see also* Vibrational spectra
  calculation of 3
  range of 59
Ingold, C. K. 46
Integrals, two-electron 34 ff, 96
Interactions, amine-carbonyl 228, 232 ff
International Business Machines (IBM), *see* under IBM
Isotopic substitution method (microwave) 11
Isovalent conjugation 147

Jargon 302

Karle, J. 12
Kasha, M. 214
Kekule forms
  and bond orders 106 ff
  of aromatic compounds 107
  of benzene 113
  of cyclobutadiene 113
  of cyclooctatetraene 113
Kelvin, Lord (W. Thomson) 28
Kraitchman's equations (microwave) 11
Kuchitsu, K. 24, 25

Latimer, W. M. 213, 214
Lattice forces, of a crystal 16, 242
Lennard-Jones, J. 215
Lennard-Jones potential 67
Leonard, N. J. 229
Lewis bond(s), 212 ff
  examples of 227 ff, 234 ff
  in 11-methyl-11-azabicyclo[5.3.1]
    undecane-4-one 229
  in 1-methyl-1-aza-cyclooctan-5-one 228
  in molecular mechanics 235, 236
  structures showing 235

INDEX

Libration (X-ray) 15
Liljefors, T. 43
Lipkowitz, K. B. 43
Liquid crystals 19
Liquid-phase structure, calculations on 240
Liquids, bulk properties of 240
Logarithms, tables of 29, 313
Low-angle data (X-ray) 17

Mainframes (computers) 30
D-Mannose, delta-two effect in 198
D-Mannose, energetics of mutarotation 198
in-[3^{4,10}][7]Metacyclophane 102
Matrix, **F**, *see* **F** matrix
Matrix, force constant, *see* **F** matrix
Matrix elements
   diagonal 57
   off-diagonal 63
McKean, D. C. 145
McKean's equation 145
Mechanics, classical 1, 3
Mechanics, molecular, *see* Molecular
     mechanics
Memory, computer 31 ff
Methanol dimers, geometries and energies of 223
Methoxyethoxymethane, rotational barriers in 207
2-Methoxytetrahydropyran
   anomeric effect in 176
   conformers of, 173 ff
     electrostatics in 174
     energies of 174
     solvation of 173, 175
Methyl ethyl sulfide, Bohlmann effect in 167
Methyl technique, for heats of formation 271, 272
Methylamine, Bohlmann effect in 168 ff
11-Methyl-11-azabicyclo[5.3.1]
     undecan-4-one
   Lewis bond in 229
   stereographic projection of 230
1-Methyl-1-aza-cyclooctan-5-one, Lewis bond in 228
2-Methylbutadiene, rotational barrier in 100
Methylcyclopentane, heat of formation of 259
Methylene radical, structure of 37
MicroVAX II (computer) 31, 32

Microwave spectroscopy 7, 10
Microwave structures, accuracy of 10 ff
Miller, M. A. 313
MINDO method 96
MINDO/3 method 37
MM2 calculations, on particular molecules,
     see the particular molecule
MM2 program. *See also* MM3 program;
     MM4 program
   description of 68 ff, 311
   parameters 312
MM3 calculations, on particular molecules,
     see the particular molecule
MM3 program, description of 68 ff. 312,
     See also MM4 program
MM4 calculations 84 ff
   for acetylenes 86
   for alkanes 84
   for alkenes 86, 88 ff
   for allenes 86
   for conjugated hydrocarbons 92 ff
   for fluorides 87
MM4 calculations, on particular molecules,
     see the particular molecule
MM4 entropies, for alkanes 273
MM4 force field, overview of 51 ff
MM4 parameters
   for alkane heats of formation 268
   van der Waals 71 ff
MM4 program
   description of 68 ff, 312
   availability of 27
MM4O program, definition of 302
MNDO method 37
Models
   ball-and-stick 2, 20, 262, 317
   computational, *see* Molecular mechanics
   Dreiding 44, 317
   hard sphere 20
   mechanical 317
     chemical accuracy 2
   molecular 2
   of cyclohexane, 309
     computer 309
     mechanical 309
   soft sphere 20
   space filling 3, 20
Modified neglect of differential overlap, *see*
     MNDO 37

Molecular Design Limited (MDL) 315
Molecular mechanics 51 ff. *See also* MM2; MM3; MM4
Molecular mechanics model, overview of xvi ff, 262
Molecular mechanics parameters, transferability of xviii, 60, 69, 75, 126, 127, 201, 210, 301
Molecular mechanics programs 311, *see also* specific program names
Molecular mechanics
  as a model 43
  books about 5
  definition 1
  description of 5, 42 ff
  in crystallography 16
  in crystals 242
  of alkanes 51 ff
  of alkenes 88 ff
  of conjugated hydrocarbons 92 ff
  of functionalized molecules 89 ff
  of hydrogen bonds 218 ff
Molecular models, *see* Models
Molecular orbital theory, definition of 106
Molecular orbital theory, semi-empirical 34, 96 ff
Molecular structure, *see also* the particular molecule, structure of
  calculation versus experiment 243
  differences in, resulting from methods used 242
Molecule, definition of 1
Molecule, structure of, definition 3, 20, 56
Molecules, congested
  heats of formation of 279
  structures of 75 ff
  zero-point energies of 279
Molecules, models of, *see* Models
Molecules, polymorphic 241
Molecules, structures of in crystals 240
Molecules, thermodynamic properties of 3
Møller-Plesset calculations, 39, 99
  in heats of formation 277
Moments of inertia, of molecules 10, 40
Monte Carlo methods 240
Moore, G. E. 33
Moore's law 33, 40, 274
Morokuma, K. 215
Morse curve 21, 24

MP2 perturbation calculation 99. *See also* Møller-Plesset calculations; Correlation energy
MP2 vs. B3LYP results 42
MP2, in molecular mechanics (MM3) pi system calculations 127 ff
Mulliken, R. S. 148
Mulliken bond order 94, 149
Mulliken charge 94, 149, 150
Multi-centered integrals, *see* Integrals, two-electron
Multiple minimum problem (energy) 239
Mutarotation energetics
  of D-glucose 198
  of D-mannose 198
Mutarotation, definition of 192

Naphthalene, 106
  crystal structure of 245
Neopentyl bromide, reactivity in S_N2 reactions 135
Neutron cross-section (crystallography) 19
Neutron diffraction 7, 17 ff
Newman projection 54, 55
Newton, M. D. 214
Newton-Raphson method, of energy minimization 47, 310
NMR spectra 19
No-bond resonance 150
Norbornane, 76 ff, 288
  geometry of 77
  rings in 317
Norbornanes, heats of formation of 272
Normal modes, of vibration 59
Nuclear explosion preventer 313
Nuclear magnetic resonance spectra 19
n-Octane, quantum mechanical energy of 257

Operating system, computer 314
Optical rotation, measurement of 192
Optical rotatory dispersion 124
Orbitals
  Gaussian 37, 38, 96
  Slater 36
    double-zeta 36
    triple-zeta 36
ORTEP, computer program 16

Oscillator strengths, in ultraviolet spectra 123
Overlap integrals, tables of 313

[2,2]Paracyclophane, hyperconjugation in 155
Parameters, heat of formation, see Heats of formation
Parameters, molecular mechanics, see Molecular mechanics
Pariser-Parr-Pople method 36, 38, 96 ff
Pauling, L. 213
Pauling atomic radii 72
1,4-Pentadiene 118
Pentane, conformational splitting in 306
Perhydroquinacene, see Perhydrotriquinacene
Perhydrotriquinacene
 conformational analysis of 288
 heat of formation of 120, 288
 rings in 317
 strain energy of 289
Peristylane
 stereographic projection of 289
Phase problem, (X-ray) 12 ff
Phenanthrene
 addition reactions in 106
 geometry of 106
Pi systems, in quantum mechanical calculations 34 ff
Pi systems. See also Aromatic compounds; Conjugated molecules
Pimentel, G. C. 213
Pitzer, K. S. 265
β-Pleated sheet, in proteins 213
PM3 program 37, 281
PM6 program 37, 281
POE, see Polyoxyethylene
Polymorphic compounds 241, 248, 249
Polyoxyethylene (POE) 195 ff
POP (statistical mechanics)
 definition 263
 numerical values of 263, 268
 in heat of formation calculations 274
 in strain calculations 282 ff
Pople, J. A. 37, 39, 215, 274, 281
Popular Mechanics magazine 30, 32
Porphin 127 ff
Porphyrins 127
6/12 Potential 67

Potential energy surface, 51
 of a molecule 4
 of a water molecule 53, 54
 of cellobiose analog 202, 203, 208, 209
PPP method, see Pariser-Parr-Pople method
Propanol 24
Propene
 hyperconjugation in 148 ff, 156 ff
 rotational barrier in 100
Proteins, folding in 213

QCPE, see Quantum Chemistry Program Exchange
Quantum Chemistry Program Exchange, 311 ff
 history of 313
 workshops 314, 315

r_α, see Bond lengths
r_a, see Bond lengths
Radial distribution function, 8, 10
 of 1,2-dichloroethane 8
Raman spectra 3, see also Vibrational spectra
r_e, see Bond lengths
Reaction coordinate diagram, for amine-ketone reaction 230, 231
Relativistic effects 41
Remington-Rand (computers) 30
Resonance
 no bond 150
 energies 115
r_g, see Bond lengths
Ribonucleic acids, (RNA) 125
Rigid body motion (X-ray) 14 ff
Ring counting 315
Ring strain energy 285
RNA 124, 125
Roothaan, C. C. J. 35
Roothaan method 35
Rotational constants 10
Rotational oscillations, in crystal lattice 14, 15
r_s, see Bond lengths
r_z, see Bond lengths

Sachse, H. 2
Sacrificial conjugation 147
Saddlepoints 5

Saunders, M. 307
SCF method, *see* Self-consistent field method
Schaefer, H. F. 37
Schleyer, P. v. R. 274
Schrödinger, E. 216
Schrödinger equation 34
Schrödinger model, of hydrogen atom 34
Self-consistent field method 35 ff, 95, 96, 114
Semi-empirical molecular orbital theory 34, 96 ff
Semipolar double bonds 213
Shrinkage, (electron diffraction) 9
Shull, H. 313, 314
Sigma-pi separation 94, 96, 127
Single point calculations 38
6/12 Potential 67
Slater-type orbital 36
Sperry-Rand (computers) 30
Starch 189
Stationary points, from driver routine 310
Stationary points, of a molecule 308
Statistical mechanics, in heats of formation 261
Stereoelectronic effects 137, 148. *See also* Effects, in organic chemistry
Stereographic projections, 16
   how to view them 317
   benzene crystal structure 244, 245
   cubane 17
   ditrityl ether 248
   alpha-helix 16
   11-methyl-11-azabicyclo[5.3.1]undecan-4-one 230
   peristylane 289
   tris(bismethano)benzene 122
   2,6,15-trithia[$3^{4,10}$][7]metacyclophane 102
Steric effects, and molecular distortions 135 ff
Steric energy (molecular mechanics) 261
Stewart, J. J. P. 281
STO-3G, *see* Basis sets
Stochastic conformational search 305, 307 ff
   quantum mechanical 308
Strain energies, MM4, table of 283, 284
Strain energy 259, 261, 282 ff
Strain energy, inherent 282 ff
Stretch-bend effect 65
Stretching constant 60

Stretching parameter 60
Structural energy increments, for heats of formation 258
Structure, of a molecule, *see* the particular molecule, structure of

T/R (statistical mechanics) 263, 264
Tetracyanoethylene, structure of 18
Tetracycanoethylene oxide, structure of 18
Tetracyclododecane, 76 ff
   structure of 77
   vibrations of 81
Thermal ellipsoids (X-ray) 14 ff
Thermal excitation energy, in heat of formation calculations 261
Thermal motion (X-ray) 14, 15
Thermodynamic properties, of alkanes 273
Thermodynamic properties, of molecules 3. *See also* Entropies; Heats of formation
Toluene, hyperconjugation in 152, 155
TOR (statistical mechanics) 265, 268, 269
   definition 263
   numerical value of 268
Torsion angle driver, *see* Driver routine
Torsion potentials
   "textbook" type of 310
   using a Fourier series 310
Torsional angle 54
Torsional energies, of alkanes 55
Torsional vibrations, anharmonic 264, 265
Transition states 5
1,1,2-Trihydroxyethane 199
Trimethylamine 212
Trimethylamine–acetone interaction 233, 234
Trimethylboron 212
Triple-zeta orbitals 36
Triquinacene, 118
   hydrogenation products of 119
2,6,15-Trithia[$3^{4,10}$][7]metacyclophane 103
Trityl cation, conjugation in 155
Tunneling 54

Ultraviolet spectra, *see also* Electronic spectra
   hydrogen bonds in 214
   of aromatic systems 123
   of conjugated hydrocarbons 122
   of vibronic transitions 123

oscillator strengths in 123
Woodward's rules for 123
Unit cell (crystallography), 240
anticipation of 241
Uracil 125
Urey-Bradley force field 66

Vacuum tubes (computer) 31, 33
Valence bond method 147 ff, 301, *see also* Hyperconjugation
as a basis for molecular mechanics xviii, 148 ff, 162 ff, 301
definition of 106 ff
van der Waals energy 67 ff
van der Waals force, description of 3
van der Waals force(s), 66 ff
between non-identical atoms 74
for atoms 74
for carbon-hydrogen 102 ff
for first-row elements 71
for graphite 74
for helium 70
for hydrogen 70, 73
for neon, 71, 72, 74
illustration of 67
van der Waals radii, 66 ff
definitions of 72
van der Waals radius, for hydrogen 73
van der Waals repulsions, large 75 ff, 80 ff, 102 ff, 279 ff
Van't Hoff, J. H. 2
Variable electronegativity self-consistent field method, 98, 123
for conjugated heterocycles 124
Variational principle 41
VAX (computer) 31
VESCF method, *see* Variable electronegativity self-consistent field method
Vibrational amplitudes 8, 21

Vibrational bands, intensities of 60
Vibrational energies, chemical accuracy of 264
Vibrational frequencies, 3
of compressed hydrogens 80
versus bond length 144, 145
normal modes of 59
Vibrational spectra, theory of 58 ff
Vibrational spectra 3, 58. *See also* Infrared spectra; Raman spectra
Vibrations, anharmonic 264
Vibronic transitions, in ultraviolet spectra 123

Water dimer
geometry and energy of 215–223
hydrogen bond of 215 ff
Water molecule
hydrogen bond to ammonia 213
structure of 53
translational entropy of 191
vibrations of 53, 61
Watson, J. D. 213
Watson, T. 30
Wertz, D. H. 274
Westheimer, F. H. 45
Westinghouse (computers) 30
Wiberg, K. B. 47, 274
Williams, D. E. 244
Windows (computers) 314
Woodward's rules, for ultraviolet spectra 123

X-ray crystallography 11 ff
X-ray diffraction 7
X-rays, resolution of 253

Zero-point energy, definition 261
Zero-point energies, of congested molecules 279
Zero-point vibrations 58